Topology in Condensed Matter

An Introduction

Topology in Condensed Matter

An Introduction

Miguel Araújo
University of Évora, Portugal

Pedro Sacramento
University of Lisbon, Portugal

World Scientific

NEW JERSEY · LONDON · SINGAPORE · BEIJING · SHANGHAI · HONG KONG · TAIPEI · CHENNAI · TOKYO

Published by

World Scientific Publishing Co. Pte. Ltd.

5 Toh Tuck Link, Singapore 596224

USA office: 27 Warren Street, Suite 401-402, Hackensack, NJ 07601

UK office: 57 Shelton Street, Covent Garden, London WC2H 9HE

British Library Cataloguing-in-Publication Data
A catalogue record for this book is available from the British Library.

TOPOLOGY IN CONDENSED MATTER
An Introduction

ISBN 978-981-123-721-8 (hardcover)
ISBN 978-981-123-722-5 (ebook for institutions)
ISBN 978-981-123-723-2 (ebook for individuals)

For any available supplementary material, please visit
https://www.worldscientific.com/worldscibooks/10.1142/12286#t=suppl

Desk Editor: Nur Syarfeena Binte Mohd Fauzi

Preface to the Portuguese Edition

The introduction of the first theoretical model of a topological superconductor by Kitaev in 2001, and the invention of the topological insulator by
Kane and Mele in 2005 have led to a renewed interest in condensed matter
topological systems. This followed earlier studies of magnetic moments in
low dimensional systems, such as in the context of the Kosterlitz-Thouless
transition and spin chains, and also in the context of the quantum Hall
effect. Indeed, we have witnessed a great proliferation of scientific papers
in the last 10 years on the subject, which has been one of the main research
areas in the theoretical group of the Center for Physics and Engineering of
Advanced Materials at Instituto Superior Técnico (IST) in Lisbon.

Over the years, we increasingly felt the need to provide graduate stu-
dents at IST with a text allowing them to quickly learn recently developed
concepts that are still very disperse in the literature.

Chapters 1 and 2 present the basic concepts relevant to the classification
of topological properties. While the first chapter may be omitted on a
first reading, we consider chapter 2 indispensible for the understanding of
the remainder of the book. The remaining chapters do not need to be
read sequentially, so the reader may directly choose to read those which
most interest him or her. The first five chapters contain exercises that will
help consolidate the understanding of the concepts and techniques. The
remaining chapters contain applications of the basic concepts.

Chapters 3, 4, and 5 are devoted to the three classes of fermionic topological systems this book is mostly concerned with, namely, the topological
insulators, superconductors and semimetals.

Spin and photonic systems are discussed in chapters 6 and 7, respectively. Under certain conditions, these systems also have topological properties. In chapter 8 we discuss the application of quantum information
methods as an alternative way to understand the properties of topological
systems. Finally, the robustness of out-of-equilibrium systems' topological properties is studied in chapter 9. The book includes two appendices
discussing some complementary aspects to the presented subjects.

Interacting systems have purposedly been left out of this introductory
text, owing both to their complexity and to the fact that they are a still
developing subject.

This text serves as a pedagogical introduction to the theoretical concepts on the subject, allowing the advanced student or researcher to acquire
the basic knowledge necessary to access the specialized literature. It does

not attempt to provide a general review of this extensive field. Instead, we refer the interested reader to the review articles which include recent experimental and theoretical developments.

The bibliography at the end of each chapter does not aim to be, and could not be, comprehensive. It includes those papers which most influenced the authors in the writing process, or which were found to best serve as complementary reading.

There are several books and review articles which cover topological systems and can be used as complementary reading to this introductory text. We suggest, for instance,

- D.J. Thouless, Topological quantum numbers in nonrelativistic physics, World Scientific (1998).
- B.A. Bernevig, T.L. Hughes, Topological insulators and topological superconductors, Princeton University Press (2013).
- S.-Q. Shen, Topological Insulators (Springer, Berlin, 2012).
- J. K. Asbóth, L. Oroszlány, and A. P. Pályi, A Short Course on Topological Insulators (Springer, 2016).
- M. Z. Hasan, and C. L. Kane, Rev. Mod. Phys. 82, 3045 (2010)
- X.-L. Qi, and S.-C. Zhang, Rev. Mod. Phys. 83, 1057 (2011).

Finally, we would like to thank the colleagues and researchers with whom we have discussed and worked, over the years, on the field of topological condensed matter: Vítor R. Vieira, Eduardo Castro, Nikola Paunković, Pedro Ribeiro, Bruno Mera, Tilen Cadez, Linhu Li, Zlatko Tesanovic, José Lages, Henrik Johannesson, Rubem Mondaini, Marco Cardoso, Pedro Bicudo, Antonio Garcia-Garcia, Maxim Dzero, Stellan Ostlund, Masaki Tezuka, Norio Kawakami.

Lisbon, February 2019

Miguel A. N. Araújo[1,2] and Pedro D. Sacramento[1]
[1] *CeFEMA, Instituto Superior Técnico, Universidade de Lisboa*
[2] *Department of Physics, University of Évora*

Contents

Preface to the Portugese Edition   v

1. Basic notions on topology   1

     1.1   Loop at x 1
     1.2   Homotopy classes based at x 2
     1.3   The fundamental group π_1 2
     1.4   The second homotopy group π_2 4
     1.5   Examples of domains and groups 6
     1.6   Application to the Hamiltonians of fermionic systems . . . 6

2. Concepts   11

     2.1   Berry phase 11
     2.2   Berry phase effects 13
         2.2.1   Electric polarization of the unit cell 13
         2.2.2   Adiabatic current 14
         2.2.3   Anomalous velocity 16
     2.3   Discrete symmetries 17
     2.4   Topology of one-dimensional systems 23
         2.4.1   Shockley model and winding number 23
         2.4.2   Zak phase 25
         2.4.3   The number of edge states 26
         2.4.4   Higher dimensional Hamiltonian 26
     2.5   The two-level system 27
     2.6   Two-dimensional systems: the Chern number 29
     2.7   Calculating the Chern number from plaquettes 32
     2.8   Berry curvature as a sum over states 33

2.9   Edge states   34
2.10   Quantum transport by edge states   36
    2.10.1   Quantum transport in one dimension   36
    2.10.2   Hall conductance in two-dimensional systems . . .   37
    2.10.3   Quantum Hall effect   39
2.11   Dimensional reduction   40
2.12   Edge states in graphene   40
2.13   The calculation of edge states   43

3.   Topological insulators   47

3.1   The construction of the anomalous Hall insulator   47
3.2   Graphene and Haldane model   49
3.3   Edge states in Haldane model   50
3.4   The Chern insulator in a magnetic field   51
3.5   Kane-Mele topological insulator   53
3.6   The topological $\mathbb{Z}_2$ index   54
3.7   The three-dimensional topological insulator   57
3.8   Models for topological insulators   58
3.9   Higher order topological insulators   61
3.10   Symmetry classes of gapped Hamiltonians   61

4.   Topological superconductors   65

4.1   Bogoliubov-de Gennes equations   65
    4.1.1   Particle-hole symmetry   66
    4.1.2   Superconducting pairing   68
    4.1.3   The BCS wave function   71
    4.1.4   Majorana fermions   71
    4.1.5   The Nambu (or Balian-Werthammer) basis   72
4.2   One-dimensional Kitaev model   74
    4.2.1   Representation of the Kitaev model by Majorana
        fermions   75
    4.2.2   Fermionic parity of the groundstate   78
    4.2.3   Extended Kitaev model   80
    4.2.4   Shockley model expressed by Majorana fermions .   82
    4.2.5   SSH model with triplet pairing   83
4.3   Bound states in Josephson junctions   89
    4.3.1   Majorana states in a π junction   89
    4.3.2   Andreev bound states in a ϕ junction   91

4.4   Two-dimensional superconductors   93
       4.4.1   The spinless p+ip superconductor   93
       4.4.2   Dirac cone with s-wave superconductivity   96
       4.4.3   The $\mathbb{Z}_2$ superconductor   98
       4.4.4   Inclusion of pseudo-spin   98
       4.4.5   Examples of superconductors in a two-dimensional
                lattice   100
       4.4.6   Sato and Fujimoto model of a triplet
                superconductor   103
4.5   Superconductor with impurities   107
       4.5.1   Magnetic chain on a singlet superconductor   110
       4.5.2   Magnetic chain on a triplet superconductor   112
       4.5.3   Chern number in real space   114

5. Topological semimetals   121

5.1   Definition and symmetries   121
5.2   Type I Weyl points   122
       5.2.1   Sources and drains of Berry curvature   123
       5.2.2   Density of states   124
5.3   Surface states with "Fermi arcs"   124
5.4   Chiral anomaly   126
5.5   Perturbation of a Dirac point   128
5.6   Type II Weyl points   129
5.7   Nodal rings   130
       5.7.1   Topological invariant for nodal lines   131
       5.7.2   Drumhead edge states   133
5.8 $\mathbb{Z}_2$ nodal rings   134

6. Spin systems with topological properties   139

6.1   Representations of spin systems   139
6.2   Spin chains   142
       6.2.1   AKLT projection   143
       6.2.2   Berry phase   146
6.3   Topological defects   149
       6.3.1   Hedgehogs and skyrmions   150
       6.3.2   Vortices and Kosterlitz-Thouless transition   151
6.4   Duality and topology   156
       6.4.1   Inverse Jordan-Wigner transformation   156

    6.4.2   Fermionic representations of the one-dimensional
            Kitaev model 157

    6.4.3   Berry phase and change of representation 159

    6.4.4   Topology of the spin model in the fermionic
            representation 162

7.   Photonic systems with topological properties   167

    7.1   Topological phases in photonic systems 167

    7.2   Edge modes with time reversal symmetry breaking 168

        7.2.1   Waveguides 168

        7.2.2   Ferrite tubes 171

        7.2.3   Waves in a periodic system: photonic crystals . . 174

        7.2.4   TM modes in a periodic lattice 176

        7.2.5   Effective model for quadratic bands 177

        7.2.6   Experimental implementation 180

    7.3   Systems with time reversal symmetry 182

        7.3.1   Scattering of a particle by a potential 182

        7.3.2 S matrix for the scattering of electromagnetic
            waves 184

8.   Quantum information and topological systems   189

    8.1   Entanglement 189

        8.1.1   von Neumann entropy 191

        8.1.2   Relation with correlation functions 192

        8.1.3   Impurity in a conventional superconductor 194

    8.2   Entanglement spectrum 196

        8.2.1   Entanglement in real space 196

        8.2.2   Momentum space entanglement 198

    8.3   Fidelity 200

        8.3.1   Pure states 200

        8.3.2   Fidelity between partial states 201

        8.3.3   Two-level system 202

        8.3.4   States of a superconductor with magnetic
            impurities 203

        8.3.5   Fidelity spectrum and phase transitions
            in quantum systems 205

        8.3.6   Fidelity spectrum of a topological
            superconductor 206

8.3.7   Quantum phase transition in Kitaev model   209
8.3.8   Fidelity susceptibility   209
8.4   Non-abelian permutation of Majorana fermions   211
8.4.1   Products of Majorana fermions   212
8.4.2   Flux quantization and Majoranas
permutations   213

9.   Out of equilibrium topological systems   221

9.1   Sudden quantum transformations   221
9.1.1   Survival probability and Loschmidt echo   221
9.1.2   Energy non-conservation   224
9.1.3   Kitaev model: stability of edge states   227
9.1.4   Sato and Fujimoto model: stability of edge
modes   228
9.1.5   Evolution of the Chern numbers   229
9.2   Periodic perturbations: Floquet systems   232
9.2.1   Dirac cone under circularly polarized radiation . .   235
9.2.2   Magnus expansion   236
9.2.3   Invariants: frequency space formulation   237
9.2.4   Quasi-energy bands and creation of π modes . . .   240
9.2.5   Invariants: time formulation   242
9.2.6   Berry-Floquet phase   243
9.3   Instantaneous periodic pulses   244
9.3.1   Eigenvalues of the Floquet operator   245
9.3.2   Effective Hamiltonian   245

Appendix A   Physical realization of Kitaev model   251

Appendix B   Fermi surface topology   255

Index   261

Chapter 1

Basic notions on topology

The quantum nature of certain states of matter renders them robust to perturbations of their Hamiltonians. They exhibit properties that are, to some extent, independent of the details of the Hamiltonian. The mathematical properties of their wave functions are described by geometric concepts from Topology. Topological quantum states may appear in spin systems, bosonic (such as superfluids) and fermionic systems (such as electronic systems).

As we shall see in the following chapters, topological states of matter are described by topological invariants representing maps from the momentum space domain onto a target space of Hamiltonians. The equivalence classes of such maps form homotopy groups which depend not only on space dimensionality, but also on the physical symmetries present. This is because the latter impose constraints on the target space.

This chapter introduces the most basic concepts in Topology and the mathematical language that is commonly used in the description of topological quantum states. It is intended to be useful to the mathematically inclined reader who has an interest on the application of Topology to Hamiltonians of physical systems, and also to those whose main background is in Physics and need an introduction to the mathematical concepts of Topology.

1.1   Loop at x

Consider the continuous map $f{:}[0, 1] \to R$, where the codomain, R, can be, for instance, a subset of euclidean space, $\mathbb{R}^d$, and the function $f(z)$ is such that $f(0) = f(1) = x \in R$. More specifically, R could be, for instance, the plane of a paper sheet, and $f(z)$ would then represent a curve that starts and ends at point x. We write the set of all closed paths at x simply as

$$\{f(z) : f(0) = f(1) = x \in R\} . \tag{1.1}$$

1.2   Homotopy classes based at x

Consider now the possibility that two closed loops at x, f and g, can be continuously deformed into each other. Taking it more rigorously, such a deformation consists of a continuous family of paths, $h_t(z)$, with a parameter $t \in [0, 1]$, obeying

$$h_0(z) = f(z), \ h_1(z) = g(z). \tag{1.2}$$

If two closed loops, f and g, can be deformed onto each other, they are said to be *homotopicaly equivalent*. Such a relation between closed loops at x is reflexive, symmetric and transitive. Therefore, it is an equivalence relation. One can say, then, that all homotopicaly equivalent closed loops form a *homotopy class* based at x. And we denote by $[f]$ the class of homotopicaly equivalent loops to f.

The chosen codomain, R, is relevant to this discussion. If R is the plane defined by the paper sheet, we can then continuously deform any loop to a single point. In other words, the closed loop, f, is homotopicaly equivalent to the constant path, $e(z) = x$. Therefore, the class $[f]$ is that of the identity $[e]$: $[f] = [e]$. But now suppose there is a hole in the codomain R. Then, a closed loop, f, that goes around the hole can no longer belong in the identity class, $[f] \neq [e]$. In the case there are two or more holes in the codomain R, there will be even more homotopy classes.

1.3   The fundamental group π_1

In a set of homotopy classes, a multiplication operation among classes can be defined. The set of classes together with the multiplication operation defines a so-called *homotopy group*.

How can we then define a multiplication among homotopy classes? We start out by defining the product of two closed loops at x, f and g, as follows,

$$f \circ g(z) = \begin{cases} f(2z), & 0 \leq z \leq \frac{1}{2}, \\ g(2z - 1), & \frac{1}{2} \leq z \leq 1. \end{cases} \tag{1.3}$$

This means that we first perform the loop f with $z \in [0, \frac{1}{2}]$, and then perform the loop g with $z \in [\frac{1}{2}, 1]$. It is clear that $f \circ g$ is a closed path which, however, has the restriction $f \circ g(\frac{1}{2}) = x$. We now would like to relax this restriction. To that end, we shall continuously deform $f \circ g$ in such a way that $f \circ g(\frac{1}{2}) \neq x$ while still remaining a closed path based at x. By considering all such possible deformations we build up the homotopy class of $f \circ g$:

$$[f \circ g] = [f] \circ [g]. \tag{1.4}$$

We have thus defined the product of two classes in this way: multiply first two paths from $[f]$ and $[g]$, and then consider all possible continuous deformations that satisfy the condition $f \circ g(0) = f \circ g(1) = x$.

The homotopy group just defined is called *fundamental group* in the codomain R, based at x, and is denoted by $\pi_1(R, x)$. The subscript 1 refers to the map domain as $[0, 1] \to R$. In order to show that it really is a group, it is necessary to show that: *(i)* it satisfies closure; *(ii)* the multiplication is associative; *(iii)* there exists a neutral element, $[e]$; *(iv)* each element has an inverse, $[f]^{-1}$. The first three requirements are "intuitive" and we leave them as exercise to the reader. In order to prove *(iv)*, we first define the inverse of $f(z)$ as $f^{-1}(z) = f(1 - z)$. All possible continuous deformations of $f^{-1}(z)$ form the class $[f]^{-1}$.

If the codomain $R = \mathbb{R}$, the set of real numbers, it is then easy to see that any closed path at x can be continuously contracted to a constant path: $f(z) = x$, $z \in [0, 1]$. Therefore, $\pi_1(\mathbb{R}, x)$ contains only the element $[e]$. This same result holds true in the case of the plane $R = \mathbb{R}^2$ and, more generally, in the case of the d-dimensional space, $R = \mathbb{R}^d$.

Now let R be the closed circumference, $R = S^1$. The reader can check that each homotopy class is characterized by the number of turns around the circumference, $n \in \mathbb{Z}$, as one starts off from a point x and comes back to it. The number n is the so-called *winding number*. The sign of n is related to the clockwise/counterclockwise direction of those turns. The product of two classes, $[n]$ e $[m]$, is the class $[n + m]$, corresponding to $n + m$ turns. This means then that the group $\pi_1(S^1, x)$ is isomorphic to the group of relative integers with the addition operation. We write:

$$\pi_1(S^1, x) = \mathbb{Z}. \tag{1.5}$$

It is also easy to see that when R is the plane with one hole, the same group $\pi_1(S^1) = \mathbb{Z}$ is obtained, where each class is the number of turns performed around the hole. In other words, the group $\pi_1(\mathbb{R}^2 \setminus \{(0,0)\})$ is isomorphic to $\pi_1(S^1)$ and to $\mathbb{Z}$.

If the codomain R is the surface of a sphere, each closed path can be continuously deformed to a single point. Therefore, only the identity class, $[e]$, exists. If we further consider the isomorphism to the addition of integers, we can write:

$$\pi_1(S^2, x) = \{0\}. \tag{1.6}$$

In the examples studied above, the group is abelian. But if the codomain R is the plane with two holes, then the group based at point x is non-abelian. That this is so, can be checked as shown in figure 1.1, where the product of two loops around different holes is performed. In figure 1.1, the loop in class $[f \circ g]$ cannot be continuously deformed into the loop belonging to class $[g \circ f]$, because the homotopy is based at x.

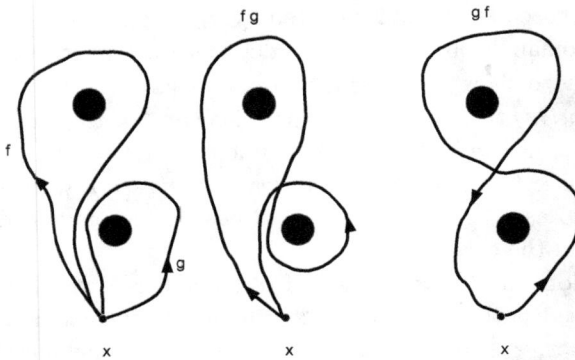

Fig. 1.1   Left: closed paths f and g. Center and right: loops in class $[f \circ g]$ and in class $[g \circ f]$, respectively.

1.4   The second homotopy group π_2

Let us consider, now, the map $f(u, v)$, where $0 \le u \le 1$, $0 \le v \le 1$. These are maps $[0, 1] \times [0, 1] \to R$. If we wish the set of departure to be the surface of a sphere (denoted as S^2 and called "2-sphere"), we then further impose the requirement that all the periphery of the domain $[0, 1] \times [0, 1]$ be mapped onto the same point, x:

$$f(0, v) = f(1, v) = f(u, 0) = f(u, 1) = x. \tag{1.7}$$

We can look upon $f(u, v)$ as a continuous family of paths $f_v(u)$ where the first ($v = 0$) and the last ($v = 1$) elements are the identity at x. The product of two maps is defined, in this case, as:

$$f \circ g = \begin{cases} f(2u, v), & 0 \le u \le \frac{1}{2}, \\ g(2u - 1, v), & \frac{1}{2} \le u \le 1. \end{cases} \tag{1.8}$$

The inverse of a map is defined as:

$$f^{-1}(u, v) = f(1 - u, v). \tag{1.9}$$

With the multiplication of homotopy classes defined previously, we can form the group $\pi_2(R)$. This group is always abelian. An interesting case is that of $\pi_2(S^2)$, the map of the sphere onto itself. As v goes over the interval $0 \to 1$, the family of curves $f_v(u)$ may cover the sphere a certain number of times, n, where $n \in \mathbb{Z}$. Figure 1.2 shows the case where the family of curves covers the sphere once. If we consider again an isomorphism to the group of relative integers with the addition operation, we obtain $\pi_2(S^2) = \mathbb{Z}$.

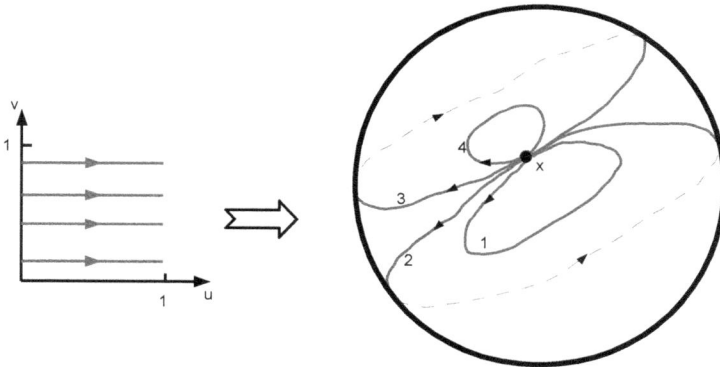

Fig. 1.2 The family of curves $f_v(u)$ covers the sphere once. Also represented are four curves in that family, for values $0 < v_1 < v_2 < v_3 < v_4 < 1$.

The structure of a group based at x does not depend on the point x. Consider another base point, $y \neq x$. As long as there is a path connecting[1] y to x, it can be shown that the homotopy groups based at x and y are isomorphic. It is then not necessary to identify the base point and we can henceforth simply write $\pi_1(X)$, or $\pi_2(X)$.

One can generalize the above ideas to the construction of other groups, $\pi_n(R)$. The maps $f(z_1, ..., z_n)$ satisfy the condition that when a variable on the sphere takes on the values $z_i = 0, 1$, then $f = x$. For $n \geq 1$ the group is abelian.

---

[1]The path $c(t) : [0, 1] \to X$ such that $c(0) = y$ and $c(1) = x$.

1.5   Examples of domains and groups

Consider the interval $[0,1]$. If one identifies the two endpoints, a circumference is obtained, denoted as S^1, the 1-sphere.

The domain $[0,1] \times [0,1]$ is a square in $\mathbb{R}^2$. If all the points (u,v) in the periphery are identified, then the surface of a sphere is obtained (the 2-sphere). If we instead identify opposite sides, $(0,v) \sim (1,v)$ and $(u,0) \sim (u,1)$, a torus is obtained, that we shall denote as $T^2 = S^1 \times S^1$. But if we instead make the identification $(0,v) \sim (1,1-v)$, then the Möbius ring is obtained.

We mentioned above the set of relative integers, $\mathbb{Z}$, with the addition operation. Another important group is the set $\{0,1\}$ with the addition modulo 2, defined as

$$0+0 = 0, \ 0+1 = 1, \ 1+1 = 0. \tag{1.10}$$

This group is referred to as $\mathbb{Z}_2$.

1.6   Application to the Hamiltonians of fermionic systems

In a crystal lattice, a fermion occupies states obeying Bloch theorem. As such, they are characterized by a wave vector, $\mathbf{k}$, in the Brillouin zone and an index identifying the energy band.

The Brillouin zone in n dimensions is equivalent to a torus, $\mathbf{k} \in T^n = S^1 \times S^1 \times \ldots$ (n terms). The Hamiltonian in momentum space is given by a hermitian matrix, $H(\mathbf{k})$, for each momentum $\mathbf{k}$. Therefore, it describes a map from the torus onto the set $\mathcal{H}$ of the hermitian matrices, i.e., $H : T^n \rightarrow \mathcal{H}$. A possible homotopy transformation consists of continuously deforming the map, H. The target domain, $\mathcal{H}$, may possibly be constrained by the physical system's symmetries (time reversal, parity, etc.). This will determine the relevant homotopy classes and groups.

A simple case occurs in one-dimensional systems ($k \in S^1$). In section 2.4, we shall study Hamiltonian 2×2 matrices that can be written in the form $\hat{H}(k) = \mathbf{h}(k) \cdot \boldsymbol{\tau}$, where $\boldsymbol{\tau}$ represents the three Pauli matrices. When chiral symmetry is imposed, the vector $\mathbf{h}(k)$ is constrained to live on a plane. If we exclude the origin ($\mathbf{h}(k) = 0$, which would be equivalent to closing the energy gap), we obtain the fundamental group, $\pi_1 = \mathbb{Z}$, as the relevant one for the problem. The homotopy classes are characterized by an element of $\mathbb{Z}$ (the topological index) measuring the number of times $\mathbf{h}(k)$ winds around the origin.

In the study of topological properties of insulators, the spectrum of $H(\mathbf{k})$ has a gap and all energy levels (or bands) below the gap are occupied. In a homotopy transformation, the Hamiltonian is modified continuously but the energy gap is never closed. A convenient approach is to replace the Hamiltonian matrix with another one which has the same eigenvectors, $|j, \mathbf{k}\rangle$, but where all the eigenenergies are set to -1, below the band gap, and to $+1$, above the gap. Let us suppose that there exist m occupied bands below the band gap, and n unoccupied bands above it. Let U denote the unitary matrix that diagonalizes the Hamiltonian, $UH(\mathbf{k})U^\dagger = diag(E_1, E_2, ...)$, where $U(\mathbf{k}) \in U(m+n)$, the set of unitary matrices with dimension $m+n$. Let us define the operator $Q(\mathbf{k})$ as

$$Q(\mathbf{k}) = 2 \sum_{j(occ)} |j, \mathbf{k}\rangle \langle j, \mathbf{k}| - 1, \tag{1.11}$$

where the summation is performed over the occupied bands j (below the energy gap). Because there are m occupied bands below the gap, the matrix $Q(\mathbf{k})$ takes the form, in the eigenbasis of the Hamiltonian,

$$Q(\mathbf{k}) = \begin{pmatrix} 1_m & \\ & -1_n \end{pmatrix} \equiv \Lambda, \tag{1.12}$$

where 1_m denotes an identity matrix of dimension m. Going back to the original basis, $Q(\mathbf{k}) = U^\dagger(\mathbf{k})\Lambda U(\mathbf{k})$. We then have $\text{Tr}[Q] = m - n$. Going from H to Q is a homotopy transformation. The only remaining homotopic transformation that can be done, i.e., without closing the band gap, is an arbitrary change of basis inside each of the subspaces above and below the gap[2]:

$$[U(m) \times U(n)] \, \Lambda \, [U^\dagger(m) \times U^\dagger(n)] = \Lambda. \tag{1.13}$$

Two Hamiltonians, H_1 and H_2, are homotopicaly equivalent if the unitary matrices that diagonalize them, U_1 and U_2, respectively, are related by

$$U_2 = U_1 [U(m) \times U(n)]. \tag{1.14}$$

Then, the homotopy classes are in one-to-one correspondence with the quotient space, or set of cosets, $U(m + n)/[U(m) \times U(n)] = G_{m,m+n}(\mathbb{C})$. This is the so-called *Grassmannian manyfold*. One can say that the Q matrices

---

[2]$U(m) \times U(n)$ denotes the group of pairs of unitary matrices, (u, v), where $\dim[u] = m$ and $\dim[v] = n$, with the cartesian product $(u_1, v_1) \cdot (u_2, v_2) = (u_1 u_2, v_1 v_2)$. But such a group is isomorphic to the group of block diagonal matrices, $diag(u, v)$, with the usual matrix product. It is this latter group that we have in mind in this discussion.

take on values in the Grassmannian. The Grassmannian itself is not a group because $U(m) \times U(n)$ is not a normal subgroup of $U(m+n)$.

When studying topological properties, one often considers the map from the Brillouin zone onto the set of matrices $Q(\mathbf{k})$, or Grassmannian manyfold, instead of the map H from the Brillouin zone onto the hermitian matrices, $H(\mathbf{k})$.

The symmetries imposed on a physical system (such as time reversal, for instance, or particle-hole) constrain the matrices Q. That is, they exclude some of the maps from the torus to the Grassmannian. The homotopy transformations change the matrix in the target space continuously, without closing the band gap, and must comply with the imposed symmetries. Chiral symmetry (to be studied later on), for instance, changes the target space in such a way that the latter is no longer the Grassmannian, but is, instead, the group of unitary matrices of dimension m. PT symmetry implies that $H(\mathbf{k})$ [and $Q(\mathbf{k})$, also] be real, so that the eigenstates are also real. Basis changes are then performed through orthogonal matrices [in the group $O(n)$] and the Grassmannian is then $O(m+n)/[O(m) \times O(n)] = G_{m,m+n}(\mathbb{R})$. In this case, the homotopy groups for are, for $m, n > 2$,

$$\pi_i \left[G_{m,m+n}(\mathbb{R}) \right] = \mathbb{Z}_2 \,, \qquad i = 1, 2 \,, \tag{1.15}$$

and also

$$\pi_1 \left[G_{1.2}(\mathbb{R}) \right] = \pi_2 \left[G_{2,4}(\mathbb{R}) \right] = \mathbb{Z} \,. \tag{1.16}$$

A *topological index* identifies a homotopy class, or an element of the homotopy group. Because the homotopy groups are isomorphic to the group of relative integers with an addition operation, each homotopy class, as well as the topological index, can be represented by an integer number.

We shall see that for the two-dimensional systems ($\mathbf{k} \in T^2$) to be studied in section 2.6, the relevant homotopy group is isomorphic to $\pi_2(S^2) = \pi_2 \left[G_{m,m+n}(\mathbb{C}) \right] = \mathbb{Z}$. The topological index is referred to as *Chern number* and is an integer describing the number of times that a map $S^2 \to S^2$ covers the sphere. If the system is three-dimensional, the homotopy group is $\pi_3 \left[G_{m,m+n}(\mathbb{C}) \right] = \{e\}$, isomorphic to the identity. The topology of such systems is therefore trivial, if no symmetries are imposed that constrain the maps onto the Grassmannian. If time-reversal symmetry for fermions is imposed, then the Q matrices are real and the relevant homotopy group is isomorphic to $\mathbb{Z}_2$ (topological insulators).

Concerning chiral symmetry, we here state an important theorem. As we shall see in chapter 2, the Hamiltonian matrices with chiral symmetry can always be written in the form of two off-diagonal blocks. The off-diagonal blocks of the corresponding Q matrices are then unitary matrices of dimension m. Therefore, the map is onto $U(m)$, not the Grassmannian. The relevant homotopy group, in the case of d spatial dimensions, is $\pi_d[U(m)]$, and the latter is given by *Bott periodicity theorem*:

$$m \geq \frac{d+1}{2} \;\Rightarrow\; \pi_d[U(m)]) = \pi_d[SU(m)] = \begin{cases} \{e\} & d \text{ even} \\ \mathbb{Z} & d \text{ odd.} \end{cases} \tag{1.17}$$

Further reading:

Concepts on geometry and topology can be found in:

- N. D. Mermin, *Reviews of Modern Physics* **51**, 591 (1979).
- M. Nakahara, *Geometry, Topology and Physics*, CRC Press (2003).

The application of homotopy groups to Hamiltonians can be found in section III of the paper:

- Schnyder et al., *Physical Review* B **78**, 195125 (2008).

Chapter 2

Concepts

2.1 Berry phase

Let the Hamiltonian, $\hat{H}(\mathbf{R})$, depend on a set of parameters, $\mathbf{R}$, and assume that the latter vary slowly in time, $\mathbf{R}(t)$. At every instant, t, $\hat{H}$ has a spectrum:

$$\hat{H}(\mathbf{R})|\psi_n(\mathbf{R})\rangle = E_n(\mathbf{R})|\psi_n(\mathbf{R})\rangle. \tag{2.1}$$

The wave function, $|\psi(t)\rangle$, evolves from an eigenstate at $t = 0$, $|\psi_n[\mathbf{R}(0)]\rangle$:

$$|\psi(t)\rangle = e^{-\frac{i}{\hbar}\int^t E_n[\mathbf{R}']dt'}e^{i\gamma}|\psi_n[\mathbf{R}(t)]\rangle, \tag{2.2}$$

where, in addition to the dynamical phase in the first exponential, the Berry phase, γ, also appears. Schrödinger equation reads:

$$i\hbar\frac{\partial}{\partial t}|\psi(t)\rangle = \hat{H}(\mathbf{R})|\psi(t)\rangle, \tag{2.3}$$

and it yields the time evolution of (2.2):

$$i\hbar\frac{\partial}{\partial t}|\psi(t)\rangle =$$

$$e^{-\frac{i}{\hbar}\int^t E_n(\mathbf{R}')dt'}e^{i\gamma}\left[\left(E_n(\mathbf{R}) - \hbar\frac{\partial\gamma}{\partial t}\right)|\psi_n[\mathbf{R}(t)]\rangle + i\hbar\dot{\mathbf{R}}\cdot\frac{\partial}{\partial\mathbf{R}}|\psi_n[\mathbf{R}(t)]\rangle\right]$$

$$= E_n(\mathbf{R})e^{-\frac{i}{\hbar}\int^t E_n(\mathbf{R}')dt'}e^{i\gamma}|\psi_n[\mathbf{R}(t)]\rangle. \tag{2.4}$$

This implies, after multiplying both sides by $\langle\psi_n[\mathbf{R}(t)]|$,

$$\frac{\partial\gamma}{\partial t} = i\langle\psi_n[\mathbf{R}(t)]|\frac{\partial}{\partial\mathbf{R}}|\psi_n[\mathbf{R}(t)]\rangle \cdot \dot{\mathbf{R}}, \tag{2.5}$$

The Berry phase resulting from a slow variation of the Hamiltonian parameters, $\delta\mathbf{R}$, is given by $\delta\gamma = i\langle\psi_n(\mathbf{R})|\frac{\partial}{\partial\mathbf{R}}|\psi_n(\mathbf{R})\rangle\cdot\delta\mathbf{R}$. When the parameters vary along a closed path, the resulting accumulated Berry phase is

$$\gamma = i\oint\langle\psi_n(\mathbf{R})|\frac{\partial}{\partial\mathbf{R}}|\psi_n(\mathbf{R})\rangle \cdot \delta\mathbf{R}. \tag{2.6}$$

The integrand in (2.6) defines the vector $\mathcal{A}_n(\mathbf{R})$, known as *Berry connection*:

$$\mathcal{A}_n(\mathbf{R}) = i\langle\psi_n(\mathbf{R})|\,\frac{\partial}{\partial\mathbf{R}}\,|\psi_n(\mathbf{R})\rangle\,, \qquad (2.7)$$

and it is real because we assume the state to be normalized, $\langle\psi_n(\mathbf{R})|\psi_n(\mathbf{R})\rangle = 1$. Hence,

$$\frac{\partial}{\partial\mathbf{R}}\langle\psi_n|\psi_n\rangle = 0 \Rightarrow \left\langle\frac{\partial}{\partial\mathbf{R}}\psi_n|\psi_n\right\rangle + \left\langle\psi_n|\frac{\partial}{\partial\mathbf{R}}\psi_n\right\rangle = 0$$

$$\Rightarrow Re\left\langle\psi_n|\frac{\partial}{\partial\mathbf{R}}\psi_n\right\rangle = 0\,.$$

For a complex scalar, we always have $\psi^*\nabla\psi - \psi\nabla\psi^* = 2i|\psi|^2\nabla arg(\psi) = 2\psi^*\nabla\psi$. If the wave function is multidimensional (a column vector with several entries), then

$$\langle\psi|\nabla\psi\rangle = i\sum_j |\psi_j|^2\nabla arg(\psi_j)\,, \qquad (2.8)$$

where j runs over the components of $|\psi\rangle$.

We can see this in another way by discretizing the derivative and noting that

$$\langle\psi_n(\mathbf{R})|\psi_n(\mathbf{R}+d\mathbf{R})\rangle - \langle\psi_n(\mathbf{R})|\psi_n(\mathbf{R})\rangle = 1 - i\alpha - 1 = -i\alpha, \quad \text{where} \quad \alpha \ll 1.$$

If we discretize the integral (2.6), we get

$$-i\gamma = \langle\psi(\mathbf{R})|\psi(\mathbf{R}+d\mathbf{R})\rangle - \langle\psi(\mathbf{R})|\psi(\mathbf{R})\rangle$$
$$+ \langle\psi(\mathbf{R}+d\mathbf{R})|\psi(\mathbf{R}+2d\mathbf{R})\rangle - \langle\psi(\mathbf{R}+d\mathbf{R})|\psi(\mathbf{R}+d\mathbf{R})\rangle + ...$$
$$= -i\alpha_1 - i\alpha_2 - .... \qquad (2.9)$$

The vector *Berry curvature* is defined as

$$\mathbf{\Omega} = \nabla_{\mathbf{R}} \times \mathcal{A}_n(\mathbf{R})\,, \qquad (2.10)$$

and is clearly gauge invariant[1]. Therefore, the curvature is analogous to a magnetic field, except for the fact that it is associated to the Berry vector potential. Using the definition (2.10), we can check, for an energy level n, that

$$(\mathbf{\Omega})_\xi = \epsilon_{\mu\nu\xi}\,i\left\langle\frac{\partial\psi_n}{\partial R^\mu}|\frac{\partial\psi_n}{\partial R^\nu}\right\rangle. \qquad (2.11)$$

The Berry curvature can also be defined as

$$\Omega_{\mu\nu} = i\left[\left\langle\frac{\partial\psi_n}{\partial R^\mu}|\frac{\partial\psi_n}{\partial R^\nu}\right\rangle - \left\langle\frac{\partial\psi_n}{\partial R^\nu}|\frac{\partial\psi_n}{\partial R^\mu}\right\rangle\right] = (\mathbf{\Omega})_\xi\epsilon_{\mu\nu\xi}\,. \qquad (2.12)$$

---

[1]The transformation $|\psi(\mathbf{R})\rangle \to e^{-i\phi(\mathbf{R})}|\psi(\mathbf{R})\rangle$ implies $\mathcal{A} \to \mathcal{A} + \nabla\phi$ and leaves $\mathbf{\Omega}$ invariant.

2.2 Berry phase effects

The energy spectrum of an electron in a crystal lattice obeys Bloch theorem. The Berry phase applied to Bloch states has interesting effects on the electronic properties. In what follows, we shall describe some of those effects.

2.2.1 *Electric polarization of the unit cell*

The Berry connection is related to the polarization of Wannier states of an energy band. We first consider the one-dimensional case. Let $\phi(x - n)$ denote a Wannier state localized at site n of a chain with N sites. We assume that $N \to \infty$. The electron coordinate, x, is continuous and we calculate the electric polarization of orbital ϕ as

$$P = e \int dx \; x |\phi(x)|^2 \, . \tag{2.13}$$

A Bloch wave with wave vector k can be written as

$$\psi_k(x) = \frac{1}{\sqrt{N}} \sum_n e^{ikn} \phi(x - n) = e^{ikx} u_k(x) \, , \tag{2.14}$$

where the Bloch periodic function, $u_k(x)$, takes the form:

$$u_k(x) = \frac{1}{\sqrt{N}} \sum_n e^{ik(n-x)} \phi(x - n) \, . \tag{2.15}$$

Now, we can see that

$$u_k^*(x) i \partial_k u_k(x) = \frac{1}{N} \sum_{n,m} e^{ik(n-m)} (x - n) \phi(x - n) \phi^*(x - m) \, . \tag{2.16}$$

Summing over all momenta, k, in the Brillouin zone,

$$\frac{1}{N} \sum_k u_k^*(x) i \partial_k u_k(x) = \frac{1}{N} \sum_n (x - n) |\phi(x - n)|^2 \, , \tag{2.17}$$

where the relation $\sum_k e^{ik(n-m)} = N\delta_{n,m}$ has been used. We now integrate expression (2.17) over x. From a comparison with equation (2.13) and converting the sum into an integral, we conclude that:

$$P = e \int dx \int_{-\pi}^{\pi} \frac{dk}{2\pi} u_k^*(x) i \partial_k u_k(x) \, . \tag{2.18}$$

If we now see the integration over x as a dot product,

$$\int dx \; u_k^*(x) i \partial_k u_k(x) = \langle u_k | i \partial_k | u_k \rangle = \mathcal{A}_k \, , \tag{2.19}$$

we can write

$$P = \frac{e}{2\pi} \int_{-\pi}^{\pi} \mathcal{A}_k dk \,, \tag{2.20}$$

where the integral defines the so-called *Zak phase*. Multiplying the latter by $e/(2\pi)$, we obtain the electric polarization per unit cell, P.

The reader can generalize (2.18) to three-dimensional space by proceeding in an analogous fashion. The polarization along the cartesian axis $j(= 1, 2, 3)$ is given by

$$P_j = e \int d^3r \int \frac{d^3k}{(2\pi)^3} u_{\mathbf{k}}^*(\mathbf{r}) i \partial_{k_j} u_{\mathbf{k}}(\mathbf{r}) \,. \tag{2.21}$$

2.2.2 *Adiabatic current*

In section 2.1, we have considered a slow change of the Hamiltonian and introduced the Berry phase for the wave function. However, expression (2.2), which results from the evolution from an eigenstate at $t = 0$, is only approximate. This is because at any instant $t > 0$, $|\psi(t)\rangle$ has a finite overlap with all eigenstates of $\hat{H}[\mathbf{R}(t)]$. In what follows, we shall go beyond the approximation (2.2).

Consider, then, that the Hamiltonian $\hat{H}(t)$ has a slow time evolution and its spectrum, at each instant,

$$\hat{H}(t)|\phi_n(t)\rangle = E_n(t)|\phi_n(t)\rangle \,, \tag{2.22}$$

where the time dependence of $\hat{H}$ may either be explicit, or through the time dependence of the parameters $\mathbf{R}(t)$. The wave function evolves in time according to

$$|\psi(t)\rangle = \sum_j e^{-\frac{i}{\hbar} \int^t E_j(t')dt'} a_j(t)|\phi_j(t)\rangle \,, \tag{2.23}$$

and we shall assume that the initial conditions are $a_j(0) = \delta_{j,m}$. This means an evolution from the state $\phi_m(0)$. Introducing this into the time dependent Schrödinger equation,

$$i\hbar \frac{\partial}{\partial t}|\psi(t)\rangle = \sum_j e^{-\frac{i}{\hbar} \int^t E_j(t')dt'} [\, E_j(t)a_j(t) + i\hbar\dot{a}_j(t) + a_j(t)i\hbar\partial_t]\, |\phi_j(t)\rangle$$

$$= \hat{H}(t) \sum_j e^{-\frac{i}{\hbar} \int^t E_j(t')dt'} a_j(t)|\phi_j(t)\rangle \,, \tag{2.24}$$

we obtain, after multiplying by $\langle\phi_n(t)|$,

$$\partial_t a_n(t) = - \sum_j e^{-\frac{i}{\hbar} \int^t [E_j(t') - E_n(t')]dt'} a_j(t)\langle\phi_n(t)|\partial_t\phi_j(t)\rangle \,. \tag{2.25}$$

We now make some approximations. If the time dependence of $\hat{H}$ results from the slowly varying parameters, $\mathbf{R}(t)$, then the time derivatives of ϕ_j and E_j are proportional to $\partial_t \mathbf{R} \to 0$. Hence, the right-hand-side of (2.25) tends to zero. Since $a_j(0) = \delta_{j,m}$, we can then assume that $a_j \ll a_m$ for $j \neq m$, and that the modulus of a_m is close to unity. So, $a_m \approx e^{i\gamma}$. Under these assumptions, we may consider that the summation on the right-hand-side of (2.25) only contains the term $j = m$:

$$\partial_t a_m(t) \approx -a_m(t)\langle\phi_m(t)|\partial_t\phi_m(t)\rangle \; ; \tag{2.26}$$

$$\partial_t a_n(t) \approx -e^{-\frac{i}{\hbar}\int^t [E_m(t')-E_n(t')]dt'} a_m(t)\langle\phi_n(t)|\partial_t\phi_m(t)\rangle, \; n \neq m \tag{2.27}$$

The integration of (2.26) leads us back to the Berry phase, γ, from section 2.1. We are left with the time integration of equation (2.27). In doing so, we can introduce one further approximation. We assume that the term $a_m(t)\langle\phi_n(t)|\partial_t\phi_m(t)\rangle$ varies so slowly in time that we can take it as constant and place it outside the time integral. We are then left with the integration of the exponentials of the energies:

$$a_n(t) \approx -i\hbar \frac{\langle\phi_n(t)|\partial_t\phi_m(t)\rangle}{E_m(t) - E_n(t)} e^{i\gamma(t)} e^{-\frac{i}{\hbar}\int^t [E_m(t')-E_n(t')]dt'}, \; n \neq m \,. \tag{2.28}$$

The wave function can be written as:

$$|\psi(t)\rangle = e^{i\gamma(t)} e^{-\frac{i}{\hbar}\int^t E_m(t')dt'} \left[|\phi_m(t)\rangle - i\hbar \sum_{j(\neq m)} \frac{\langle\phi_j(t)|\partial_t\phi_m(t)\rangle}{E_m(t) - E_j(t)} |\phi_j(t)\rangle \right], \tag{2.29}$$

which now includes the terms that had been neglected in section 2.1. We can consider the electron in a crystal lattice and the functions $|\phi_j(t)\rangle = |u_j(\mathbf{k}, t)\rangle$ which give the instantaneous spectrum of the Hamiltonian $\hat{H}(\mathbf{k}, t)$, where j denotes the band index. Equation (2.29) yields the function $|\psi_m(\mathbf{k}, t)\rangle$ which results from the time evolution from $|u_m(\mathbf{k}, t)\rangle$. At any instant t, the velocity operator

$$\hat{v}(\mathbf{k}, t) = \frac{1}{\hbar} \partial_{\mathbf{k}} \hat{H}(\mathbf{k}, t) \,. \tag{2.30}$$

The velocity in a state $\mathbf{k}$ that has adiabatically evolved from $|u_m(\mathbf{k}, 0)\rangle$, is given by (2.29):

$$\langle\psi_m(\mathbf{k}, t)|\hat{v}(\mathbf{k}, t)|\psi_m(\mathbf{k}, t)\rangle = \langle u_m(\mathbf{k}, t)|\hat{v}(\mathbf{k}, t)|u_m(\mathbf{k}, t)\rangle$$

$$-i\hbar \sum_{j(\neq m)} \frac{\langle u_m(\mathbf{k}, t)|\hat{v}(\mathbf{k}, t)|u_j(\mathbf{k}, t)\rangle\langle u_j(\mathbf{k}, t)|\partial_t u_m(\mathbf{k}, t)\rangle - c.c.}{E_m(t) - E_j(t)} \,. \tag{2.31}$$

The term in the summation can be rewritten in a more convenient form if we make use the result, for $j \neq m$,

$$\hbar \langle u_m | \hat{\boldsymbol{v}}(\mathbf{k},t) | u_j \rangle = (E_j - E_m) \langle u_m | \partial_\mathbf{k} u_j \rangle = (E_m - E_j) \langle \partial_\mathbf{k} u_m | u_j \rangle, \quad (2.32)$$

(here we have simplified the notation somewhat) which can be proven by taking the derivative of the time-independent Schrödinger equation with respect to $\mathbf{k}$:

$$\hat{H}(\mathbf{k}) | u_j(\mathbf{k}) \rangle = E_j(\mathbf{k}) | u_j(\mathbf{k}) \rangle$$

$$\Rightarrow \frac{\partial \hat{H}}{\partial \mathbf{k}} | u_j \rangle + \hat{H} | \partial_\mathbf{k} u_j \rangle = \frac{\partial E_j}{\partial \mathbf{k}} | u_j \rangle + E_j | \partial_\mathbf{k} u_j \rangle. \quad (2.33)$$

Multiplying both sides of (2.33) by $\langle u_m |$, where $m \neq j$, the first equality in (2.32) is obtained. The second equality in (2.32) follows from the normalization of the states $| u_n \rangle$. Inserting the result (2.32) in (2.31), we obtain

$$\langle \psi_m(\mathbf{k},t) | \hat{\boldsymbol{v}}(\mathbf{k},t) | \psi_m(\mathbf{k},t) \rangle = \langle u_m(\mathbf{k},t) | \hat{\boldsymbol{v}}(\mathbf{k},t) | u_m(\mathbf{k},t) \rangle$$
$$- i \, [\langle \partial_\mathbf{k} u_m(\mathbf{k},t) | \partial_t u_m(\mathbf{k},t) \rangle - c.c \,]. \quad (2.34)$$

In the second term we recognize a Berry curvature, $\Omega^{(m)}_{k_j t}$, where $j = x, y, z$. We may rewrite (2.34) as

$$\langle \psi_m(\mathbf{k},t) | \hat{\boldsymbol{v}}(\mathbf{k},t) | \psi_m(\mathbf{k},t) \rangle = \boldsymbol{v}_m(\mathbf{k},t) - \Omega^{(m)}_{\mathbf{k}t} \quad (2.35)$$

In order to calculate the energy density current it is necessary to integrate over the states $\mathbf{k}$ and over the occupied bands,

$$\boldsymbol{j} = e \int \frac{d^d k}{(2\pi)^d} \left[\boldsymbol{v}_m(\mathbf{k},t) - \Omega^{(m)}_{\mathbf{k}t} \right] = -e \int \frac{d^d k}{(2\pi)^d} \Omega^{(m)}_{\mathbf{k}t}. \quad (2.36)$$

The result (2.36) is referred to as *adiabatic current*. It is the result of a slow evolution of the Hamiltonian with Bloch waves with curvature $\Omega_{\mathbf{k}t}$. Note that the integral in (2.36) is not the same as that for the Chern number of a two-dimensional system[2] because Ω has a subscript t instead of the spatial index k_j.

2.2.3 *Anomalous velocity*

An electric field, $\boldsymbol{E}$, may be described by means of a vector potential, $\boldsymbol{A}(t)$, such that $\boldsymbol{E} = -\partial_t \boldsymbol{A}$. The Hamiltonian,

$$\hat{H}(\mathbf{k},t) = \hat{H}[\mathbf{k} - \frac{e}{\hbar} \boldsymbol{A}(t)], \quad (2.37)$$

---

[2]The Chern number will be defined in section 2.6.

changes slowly in time if the field is weak. The above velocity operator,

$$\hat{\boldsymbol{v}}(\mathbf{k}, t) = \frac{1}{\hbar} \partial_{\mathbf{k}} \hat{H} [\mathbf{k} - \frac{e}{\hbar} \boldsymbol{A}(t)] , \qquad (2.38)$$

and the time derivatives in (2.34) may be calculated from the chain rule as

$$|\partial_t u_m(\mathbf{k}, t)\rangle = \frac{e}{\hbar} \boldsymbol{E} \cdot \partial_{\mathbf{k}} |u_m(\mathbf{k}, t)\rangle . \qquad (2.39)$$

Equation (2.34), or (2.35), can be more clearly written by explicitly writing, for the cartesian components,

$$\langle v_i(t)\rangle = (\boldsymbol{v}_m)_i - \frac{ie}{\hbar} E_j \left[\langle \partial_{k_i} u_m | \partial_{k_j} u_m \rangle - c.c \right] . \qquad (2.40)$$

We may now invoke the definition of the curvature vector,

$$(\boldsymbol{\Omega})_\mu \epsilon_{ij\mu} = \Omega_{ij}^{(m)} = i \langle \partial_{k_i} u_m | \partial_{k_j} u_m \rangle - c.c , \qquad (2.41)$$

and rewrite (2.40) as

$$\langle v_i(t)\rangle = (\boldsymbol{v}_m)_i - \frac{ie}{\hbar} (\boldsymbol{E} \times \boldsymbol{\Omega})_i ,$$

or, going back to vectorial notation,

$$\langle \boldsymbol{v}(\mathbf{k}, t)\rangle = \boldsymbol{v}_m(\mathbf{k}, t) - \frac{e}{\hbar} \boldsymbol{E} \times \boldsymbol{\Omega}^{(m)} . \qquad (2.42)$$

The second term goes under the designation *anomalous velocity* and is perpendicular to the electric field. The Berry curvature may produce a Hall current, perpendicular to the applied field, after the integration over occupied states is performed.

2.3 Discrete symmetries

In this section, we review the most relevant discrete symmetries for the topology of fermionic systems. In many cases, the topological properties of a Hamiltonian are protected by, at least, one symmetry. This means that, if perturbations are added to the Hamiltonian, its topological properties will remain as long as that symmetry is not broken.

The symmetries present will determine the type of topological index a Hamiltonian can have (a Chern number or other index). We may here invoke the concept of homotopy group from chapter 1. A Hamiltonian represents a map from points $\mathbf{k}$ onto Hamiltonian matrices (the codomain of the map). If it satisfies a set of symmetries, then the target domain is constrained. The homotopy group may be isomorphic to $\mathbb{Z}$ or $\mathbb{Z}_2$.

We first clarify a point concerning spatial Fourier transforms. In what follows, when we mention the symmetry of a Hamiltonian or wave function, we write the latter as H or ψ, having in mind that they are expressed in real space, $\mathbf{r}$, possibly including other degrees of freedom (spin, for instance) that we denote by α. Therefore, we see H as having matrix elements between states $|\mathbf{r}\alpha\rangle$. If a system is invariant with respect to spatial translations, $H_{\mathbf{r}'\alpha',\mathbf{r}\alpha}$ only depends on $\mathbf{r}' - \mathbf{r}$ and we may then go over to Bloch waves with momentum $\mathbf{k}$ through the Fourier transformation:

$$\frac{1}{N}\sum_{\mathbf{r}',\mathbf{r}} H_{\mathbf{r}'\alpha',\mathbf{r}\alpha} e^{i\mathbf{k}\cdot(\mathbf{r}-\mathbf{r}')} = H_{\alpha',\alpha}(\mathbf{k}). \qquad (2.43)$$

The Fourier transform of H^T, for instance, can be calculated as

$$\frac{1}{N}\sum_{\mathbf{r}',\mathbf{r}} H^T_{\mathbf{r}'\alpha',\mathbf{r}\alpha} e^{i\mathbf{k}\cdot(\mathbf{r}-\mathbf{r}')} = \frac{1}{N}\sum_{\mathbf{r}',\mathbf{r}} H_{\mathbf{r}\alpha,\mathbf{r}'\alpha'} e^{i\mathbf{k}\cdot(\mathbf{r}-\mathbf{r}')} = H_{\alpha\alpha'}(-\mathbf{k}). \qquad (2.44)$$

One can easily derive the following rules for the Fourier transform along momentum $\mathbf{k}$:

$$H \to H(\mathbf{k}), \ H^T \to H^T(-\mathbf{k}), \ H^* \to H^*(-\mathbf{k}), \ H^\dagger \to H^\dagger(\mathbf{k}). \qquad (2.45)$$

We also adopt the following notation for the Pauli matrices: we use the symbol $\boldsymbol{\tau}$ when acting on the sub-lattice degree of freedom (if there are two orbitals per primitive cell) and we shall use the symbol $\boldsymbol{\sigma}$ when acting on the physical spin.

We now turn to the discussion of the main symmetries. An important point that should be kept in mind is that there does not *a priori* exist an operator associated with each particular symmetry. Given a Hamiltonian, we must search for an operator that satisfies the condition defining the symmetry. If that operator does not exist, we then conclude that that symmetry is not present.

- *Time-reversal symmetry* is defined by the condition: if ψ satisfies the Schrödinger equation, then $T\psi^*$ satisfies the Schrödinger equation with the substitution $t \to -t$. Here, T is a unitary matrix still to be found.

One can always write

$$i\hbar\partial_t\psi = H\psi \ \Leftrightarrow \ -i\hbar\partial_t\psi^* = H^*\psi^*$$
$$\Leftrightarrow \ -i\hbar\partial_t(T\psi^*) = TH^*T^\dagger(T\psi^*). \qquad (2.46)$$

The time reversal operation, $\mathcal{T}$, is then defined as:

$$\psi \to T\psi^*, \qquad (2.47)$$

$$H \to TH^*T^\dagger, \qquad (2.48)$$

where it is assumed that ψ e H are expressed in real space. But if $T\psi^*$ satisfies the Schrödinger equation with $t \to -t$, then $-i\hbar\partial_t(T\psi^*) = H(T\psi^*)$. From equation (2.46) we get

$$TH^*T^\dagger = H\,, \qquad TH^*(-\mathbf{k})T^\dagger = H(\mathbf{k})\,. \qquad (2.49)$$

In the absence of spin, T is a symmetric matrix.

Consider the following example:

$H(\mathbf{k}) = (\cos k_x + \cos k_y)\,\tau_x + \sin k_z\tau_z.$

In this case, $T = \tau_x$ implements time reversal symmetry. Indeed,

$$\tau_x H^*(-\mathbf{k})\tau_x = (\cos k_x + \cos k_y)\,\tau_x + \sin k_z\tau_z = H(\mathbf{k})\,,$$

and the time-reversed state of $\psi(\mathbf{k})$ is the state $\tau_x\psi^*(-\mathbf{k})$.

Considering the physical spin, T must be an anti-symmetric matrix. This leads us to the result:

$$\text{(spin)}: T = \sigma_y\,. \qquad (2.50)$$

The notation $\mathcal{T} = i\sigma_y K$ is often used, where K denotes complex conjugation.

The operator $\mathcal{T}$, either with or without physical spin, is antiunitary[3] and for spin $\frac{1}{2}$ we have $\mathcal{T}^2 = -1$. This implies that the two states, ψ and $\mathcal{T}\psi$, are orthogonal and have the same energy, making up a Kramers pair. For momentum $\mathbf{k} = 0$, the two eigenstates ψ and $\mathcal{T}\psi$ are two orthogonal spinors. The same holds for other $\mathcal{T}$-invariant momenta (at the corners of the Brillouin zone, for instance).

- *space inversion* symmetry or *parity* is defined by the condition: if ψ satisfies Schrödinger equation, then $P\psi(-\mathbf{r})$ also satisfies, where P is a suitable matrix. Inverting the coordinates in Schrödinger equation, including inside the matrix H, one can always write

$$H\psi(-\mathbf{r}) = i\hbar\partial_t\psi(-\mathbf{r}) \iff PHP^\dagger P\psi(-\mathbf{r}) = i\hbar\partial_t P\psi(-\mathbf{r})\,, \quad (2.51)$$

and, therefore, the symmetry imposes that, in real space, $PH(-\mathbf{r}, -\mathbf{r}')P^\dagger = H(\mathbf{r}, \mathbf{r}')$. In momentum space we write

$$PH(-\mathbf{k})P^\dagger = H(\mathbf{k})\,. \qquad (2.52)$$

---

[3]An operator $\mathcal{O}$ is said to be antiunitary if and only if $\mathcal{O}\psi \cdot \mathcal{O}\phi = (\psi \cdot \phi)^*$.

Let us consider some examples:

$H(\mathbf{k}) = \mathbf{k} \cdot \boldsymbol{\tau}$: here, it is necessary that $-\mathbf{k} \cdot P\boldsymbol{\tau}P^\dagger = \mathbf{k} \cdot \boldsymbol{\tau}$, and there is no P, therefore, parity does not exist.

Weyl points: $H(\mathbf{K} + \mathbf{q}) = \mathbf{q} \cdot \boldsymbol{\tau} = H(-\mathbf{K} - \mathbf{q})$. Here, $P = 1$. So, the inversion operation acts only on the space coordinates.

Nodal line: $H(\mathbf{k}) = \left(\sqrt{k_x^2 + k_y^2} - k_0\right)\tau_z + k_z\tau_x$. We have $H(-\mathbf{k}) = \left(\sqrt{k_x^2 + k_y^2} - k_0\right)\tau_z - k_z\tau_x$. In this case, $P = \tau_z$ implements parity.

Graphene: In the vicinity[4] of cone $\mathbf{K}$ we have $H(\mathbf{K} + \mathbf{q}) = q_x\tau_x + q_y\tau_y$, for small $\mathbf{q}$, and close to the cone $-\mathbf{K}$ we have $H(-\mathbf{K} + \mathbf{q}) = -q_x\tau_x + q_y\tau_y$. Here, $P = \tau_x$ implements parity according to equation (2.52). The addition of a constant term, $m\tau_z$, to the Hamiltonian breaks parity. Such a term is called "Dirac mass".

- *Particle-hole or charge conjugation symmetry* is defined by the following condition: there is a unitary matrix, C, such that $C\psi^*$ satisfies Schrödinger equation. This implies $CH^*C^\dagger = -H$. The operation of charge conjugation or particle-hole is defined as:

$$C : H \to CH^*C^\dagger. \qquad (2.53)$$

  In the above example for graphene, for instance, $C = \tau_z$ implements the symmetry. The latter is broken by a constant Dirac mass term. If, however, the Dirac mass depends on momentum and takes on symmetric values at cones $\mathbf{K}$ and $-\mathbf{K}$, then the symmetry is preserved.

  Charge conjugation for spin $\frac{1}{2}$ particles is mostly considered in the context of superconductivity and applied to Bogolubov-de Gennes matrices, as we shall discuss later on. In that case, the matrix C will be identified with the first Pauli matrix acting in particle-hole space.

- *Chiral* or *sublattice symmetry* is defined as

$$\exists_{\mathcal{O}} : \mathcal{O}H\mathcal{O}^\dagger = -H, \; \mathcal{O}^2 = 1. \qquad (2.54)$$

  Example: for massless graphene, $\mathcal{O} = \tau_z$. A Dirac mass term breaks chirality.

---

[4]The Hamiltonian for graphene will be introduced in section 3.2. For now, we take it as given and study its symmetry.

This symmetry implies that the Hamiltonian can be brought to off-diagonal form. Indeed, expressing $\mathcal{O}$ in the basis of its eigenvectors,

$$\mathcal{O} = \begin{pmatrix} 1 & 0 \\ 0 & -1 \end{pmatrix} = \tau_z \implies H = \begin{pmatrix} 0 & D \\ D^\dagger & 0 \end{pmatrix}. \qquad (2.55)$$

Now assume that there is a unitary matrix, U, that rotates $\hat{H}$ to the off-diagonal form:

$$UHU^\dagger = \begin{pmatrix} 0 & D \\ D^\dagger & 0 \end{pmatrix}.$$

Then, we can write

$$\tau_z UHU^\dagger \tau_z = - \begin{pmatrix} 0 & D \\ D^\dagger & 0 \end{pmatrix} = -UHU^\dagger \Leftrightarrow U^\dagger \tau_z UHU^\dagger \tau_z U = -H\,,$$

which means that the definition (2.54) is satisfied for $\mathcal{O} = U^\dagger \tau_z U$.

In the case $\hat{H}(\mathbf{k}) = \mathbf{h}(\mathbf{k}) \cdot \boldsymbol{\tau}$, the above SU(2) rotation corresponds to a SO(3) rotation of the vector $\mathbf{h}$ so as to place it in the OXY plane and $D = h_x - ih_y$.

Putting it in more general terms, chiral symmetry means that $\mathbf{h}(\mathbf{k})$ lives on a plane in (h_x, h_y, h_z) space, containing the origin. If the plane does not contain the origin, then the distance from the origin to the plane is a Dirac mass that breaks chirality.

Chiral symmetry must be present whenever the Hamiltonian enjoys the symmetries $\mathcal{T}$ and $\mathcal{C}$. Starting with the Hamiltonian H and applying the symmetry operators in succession, we get

$$H \xrightarrow{\mathcal{T}} TH^*T^\dagger = H \xrightarrow{\mathcal{C}} CH^*C^\dagger = -H\,,$$

which means that we have rotated H into $-H$ through the operator $\mathcal{O} = \mathcal{CT}$. Chiral symmetry may also exist alone, without any of the symmetries $\mathcal{T}$ and $\mathcal{C}$.

- $\mathcal{PT}$ *symmetry*: composition of parity and time reversal may exist as a symmetry by itself when neither parity nor time reversal are separately present. The composition of the above symmetries yields $H(\mathbf{k}) \to PTH^*(\mathbf{k})T^\dagger P^\dagger$. We can then introduce the definition as follows: there exists a unitary matrix, A, such that

$$\mathcal{PT}: H(\mathbf{k}) = AH^*(\mathbf{k})A^\dagger. \qquad (2.56)$$

The $\mathcal{PT}$ operator is antiunitary.

Taking $\hat{H} = \mathbf{h} \cdot \tau$, we can check that one of the components of $\mathbf{h}$ has to be eliminated. The following possibilities arise:
$$A = \tau_x \,, h_z = 0; \quad A = \tau_z \,, h_x = 0; \quad A = 1 \,, h_y = 0.$$
In any of the above cases, we have $(\mathcal{PT})^2 = 1$.

For Hamiltonians $H(\mathbf{k}) = h_{\mu\nu}(\mathbf{k})\tau_\mu\sigma_\nu$, the condition (2.56) may be satisfied with $(\mathcal{PT})^2 = \pm 1$. For spin $\frac{1}{2}$ systems it is often required that $(\mathcal{PT})^2 = -1$.

- *Reflection symmetry.* The *reflection in a plane* perpendicular to the j axis is defined as:

$$\mathcal{M}_j\psi(\mathbf{r}) = Ri\sigma_j\psi\,(x_j \to -x_j)\,, \tag{2.57}$$

where σ_j acts on the physical spin and the matrix R on pseudo-spin. In this expression, the operator σ_j arises from the composition of a π rotation around the j axis, in spin space, followed by an inversion. The reflection applied to the spin Hamiltonian, $\boldsymbol{h}(\mathbf{k}) \cdot \boldsymbol{\sigma}$, yields

$$\boldsymbol{h}(\mathbf{k}) \cdot \boldsymbol{\sigma} \to \sigma_j\boldsymbol{h}(-k_j) \cdot \boldsymbol{\sigma}\sigma_j \tag{2.58}$$

$$\Rightarrow \quad \mathcal{M}_j : h_i(\mathbf{k}) \to (2\delta_{ij} - 1)\, h_i(-k_j) \tag{2.59}$$

(no summation over repeated indices implied).

Example: the spinless nodal line in $H(\mathbf{k}) = \left(m - Bk^2\right)\tau_x + vk_z\tau_z$ is protected by reflection symmetry along the Oz axis, where $R = \tau_x$:

$$\tau_x H(k_x, k_y, -k_z)\tau_x = H(\mathbf{k})\,.$$

Exercise 2.1

Consider the Hamiltonian matrix for spinless fermions:

$$H(\mathbf{k}) = (\cos k_x + \cos k_y - m)\, \tau_x + \sin k_z\tau_z\,.$$

(1) Show that it enjoys the following symmetries: chiral, time reversal, space inversion, reflection and particle-hole.
(2) Consider now that the Pauli matrices act on spin and investigate the symmetries again.
(3) Let

$$H(\mathbf{k}) = (\cos k_x + \cos k_y - m)\, \tau_x\sigma_0 + \sin k_z\tau_z\sigma_0 + \sum_{i=0}^{3} M_i(\mathbf{k})\tau_y\sigma_i\,.$$

Check that the term M_0 breaks $\mathcal{PT}$ symmetry.

2.4 Topology of one-dimensional systems

The Shockley model may be seen as the fundamental model for the study of topological properties of fermions in one dimension. It allows us to understand how localized edge states appear, in the most simple way. The latter are probably the most important manifestation of topological properties. The model is also important for the study of higher dimensional systems, as we shall see later, when we talk about "dimensional reduction".

In the topology of one-dimensional systems, chiral symmetry plays a central role.

2.4.1 *Shockley model and winding number*

Consider a one-dimensional tight-binding model where the matrix elements connecting the lattice sites alternate between two values, t_0 and t_1. The primitive cell then has two sites (a e b). See figure 2.1.

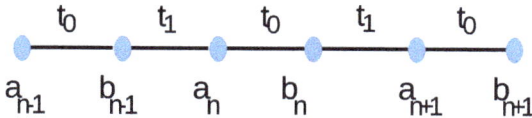

Fig. 2.1   One-dimensional system with alternate hopping amplitudes.

The wave function has amplitudes (a_n, b_n) in the n-th cell. The Hamiltonian's matrix elements read:

$$\langle a, n|\hat{H}|b, n\rangle = t_0 , \qquad \langle b, n|\hat{H}|a, n + 1\rangle = t_1 . \tag{2.60}$$

For an eigenstate with energy E, we can write:

$$\psi(n) = \begin{pmatrix} a_n \\ b_n \end{pmatrix} , \qquad \begin{array}{l} Ea_n = t_0 b_n + t_1^* b_{n-1} \\ Eb_n = t_1 a_{n+1} + t_0^* a_n \end{array} . \tag{2.61}$$

We consider first a semi-infinite chain, $n \geq 0$. The above system of equations allows a solution with $E = 0$ as long as:

$$a_n = \left(-\frac{t_0^*}{t_1}\right)^n a_0 , \qquad b_n = 0 , \tag{2.62}$$

which can only exist if $|t_1| > |t_0|$. This means that *the chain starts with the weak link*. This state is localized at the edge. When the crystal ends with a weak link, an edge state appears.

We may also consider terminating the crystal in the cell $n = L$. We then obtain, for $n \leq L$:

$$a_n = 0, \qquad b_n = \left(-\frac{t_0}{t_1^*}\right)^{L-n} b_L, \qquad (2.63)$$

so, the state with $E = 0$ lives on sublattice b. In a very long finite chain, $0 \leq n \leq L \to \infty$, there exist two edge states living on different sublattices.

We note a subtle point here: once the amplitudes t_0, t_1 are chosen, it is the choice of the primitive cell that will determine whether the chain ends with a weak link, because a crystal always begins with the first cell. Had we chosen the cell to be ba, with the link t_1, there would be no edge states, because the first/last cell would be a strong link.

How can we here identify a topological invariant? The invariant is defined for an infinite system, that is, when equations (2.61) hold for all $n \in \mathbb{Z}$. In that case, the wave function is a Bloch wave:

$$\psi(n) = e^{ikn}\begin{pmatrix} a \\ b \end{pmatrix}, \qquad E\begin{pmatrix} a \\ b \end{pmatrix} = H(k)\begin{pmatrix} a \\ b \end{pmatrix}, \qquad (2.64)$$

where the Hamiltonian matrix:

$$H(k) = \begin{pmatrix} 0 & t_0 + t_1^* e^{-ik} \\ t_0^* + t_1 e^{ik} & 0 \end{pmatrix}, \qquad (2.65)$$

has eigenenergies $E(k) = \pm|t_0 + t_1^* e^{-ik}|$. The condition $|t_1| > |t_0|$ has the following geometric interpretation. As k takes on values $0 \to 2\pi$, the complex number $t_0 + t_1^* e^{-ik}$ makes one turn around the origin in the Argand plane. Hence we can define a topological concept: the number of times that the matrix element $H_{12}(k)$ travels counterclockwise around the origin of the complex plane, as k runs over all its domain. If $|t_1| < |t_0|$ this winding number is zero. Actually, it is the phase of the complex number $z = t_0 + t_1 e^{-ik}$ that accumulates a variation equal to a multiple of 2π. We can then write the topological invariant, known as "winding number" as:

$$W = \frac{1}{2\pi i}\int_0^{2\pi} dk \frac{d\log(t_0 + t_1^* e^{-ik})}{dk}. \qquad (2.66)$$

At the transition between the two regimes (or topological phases), the trajectory of $H_{12}(k)$ intercepts the origin, closing the energy gap of the Hamiltonian $H(k)$.

The Hamiltonian (2.65) has chiral symmetry because it anticommutes with the third Pauli matrix, τ_z. What is the effect of chiral symmetry breaking? This can be achieved by introducing a perturbation proportional to τ_z (we call it Dirac mass). Let then $\epsilon\tau_z$ be such a perturbation. Because the wave function (2.62) or (2.63) occupies only one sublattice, the edge states still exist but their energy is shifted by $\pm\epsilon \neq 0$. Chiral symmetry breaking by a Dirac mass lifts the degeneracy between the sublattices and, as a result, between the edge states.

Summarizing:

- the existence of a winding number, W, determines the existence of edge states;
- chiral symmetry guarantees that the latter have zero energy;
- chiral symmetry breaking is due to a Dirac mass term that shifts the edge states' energies.

Exercise 2.2

Consider the Hamiltonian matrix (Kitaev model):
$$H(k) = \Delta \sin(k)\tau_y + [\mu - w\cos(k)]\,\tau_z\,.$$
(1) Which Pauli matrix, $\mathcal{O}$, implements chiral symmetry?
(2) Apply the same procedure as in equation (2.55): find the unitary matrix, U, that rotates the chiral operator, $\mathcal{O}$, into the third Pauli matrix: $U\mathcal{O}U^\dagger = \tau_z$.
(3) Rotate the Hamiltonian, $UHU^\dagger$, and apply Shockley's criterion in order to obtain the index W. In a topological phase, what is the relation between w and μ?

2.4.2 *Zak phase*

Let us consider the Hamiltonian matrix in off-diagonal form, with matrix element $H_{12}(k)$, and let $\phi(k)$ denote the argument of the complex number $H_{12}(k)$. The lowest energy level's wave function can be written as:
$$\psi(k) = \frac{1}{\sqrt{2}}\begin{pmatrix} -e^{i\phi(k)} \\ 1 \end{pmatrix}\,. \tag{2.67}$$
We then obtain the Berry connection as
$$\mathcal{A}(k) = i\langle\psi(k)|\partial_k\psi(k)\rangle = -\frac{1}{2}\frac{\partial\phi(k)}{\partial k}\,. \tag{2.68}$$
The Berry phase (2.6) is given by
$$\gamma = \int_{-\pi}^{\pi}\mathcal{A}(k)dk = -\pi W\,, \tag{2.69}$$
and it is quantized in the presence of chiral symmetry (W is integer). The Berry phase for one-dimensional systems is called *Zak phase*.

2.4.3   *The number of edge states*

The modulus of the index W is the number of edge states. This can be easily seen in a simple extension of the Shockley model.

We can write down a Hamiltonian where $W = 2$:

$$H(k) = \begin{pmatrix} 0 & t_0 + t_2 e^{-2ik} \\ (t_0 + t_2 e^{-2ik})^* & 0 \end{pmatrix}. \tag{2.70}$$

The equations for the wave function amplitudes in real space take the form

$$\begin{aligned} Ea_n &= t_0 b_n + t_2^* b_{n-2}, \\ Eb_n &= t_2 a_{n+2} + t_0^* a_n. \end{aligned} \tag{2.71}$$

One can easily check, by inspection, that (2.71) describes two decoupled chains: one made of the cells $(a_n b_n)$ with even n, and the other with odd n. Each of these chains admits an edge state, i.e., a solution of the form (2.63).

2.4.4   *Higher dimensional Hamiltonian*

If the Hamiltonian's dimension is higher than 2, chirality still allows us to bring it to block off-diagonal form,

$$H(k) = \begin{pmatrix} 0 & q(k) \\ q^\dagger(k) & 0 \end{pmatrix}, \tag{2.72}$$

where $q(k)$ is a unitary matrix. Bott periodicity theorem tells us that there is an integer index, W. In this case, W is the winding number of the function given by the determinant of q. The determinant of a unitary matrix is a complex number with modulus 1, therefore, we can write

$$\det(q) = e^{i\phi}.$$

Thus,

$$W = \frac{1}{2\pi} \int d\phi = \frac{1}{2i\pi} \int_0^{2\pi} \frac{d\left[\log \det(q)\right]}{dk} dk. \tag{2.73}$$

We may write, in general, for the unitary matrix q obtained from a hermitian matrix a:

$$q = e^{ia} \;\Rightarrow\; \det(q) = e^{i\mathrm{Tr}(a)}, \qquad \phi = \mathrm{Tr}(a). \tag{2.74}$$

Let λ_j label the eigenvalues of matrix a. Then,

$$\begin{aligned} \mathrm{Tr}\left\{q^\dagger dq\right\} &= \mathrm{Tr}\left\{e^{-ia} d\left(e^{ia}\right)\right\} \\ &= \sum_j e^{-i\lambda_j} d\left(e^{i\lambda_j}\right) = \sum_j i\, d\lambda_j = i\, d\phi. \end{aligned} \tag{2.75}$$

(Hellmann-Feynman theorem has been used in going from the first to the second line). Hence we obtain the expression

$$W = W[q] = \frac{1}{2i\pi} \int_0^{2\pi} \text{Tr} \left\{ q^\dagger \frac{dq}{dk} \right\} dk \,, \tag{2.76}$$

which is useful for the calculation of the index W in (2.73).

2.5 The two-level system

The two-level system is the most simple and fundamental model allowing the understanding of topological properties of fermionic bands. Hence its importance. A two-level (or two-band) Hamiltonian takes often the form:

$$\hat{H} = \mathbf{h} \cdot \boldsymbol{\tau} + h_0 \tau_0, \tag{2.77}$$

where $\boldsymbol{\tau}$ denotes the three Pauli matrices in orbital space in each cell, and τ_0 denotes the identity matrix. The vector $\mathbf{h}$ has spherical coordinates (h, θ, ϕ) and is a function of several parameters, $\mathbf{R}$. The eigenfunctions of (2.77) read:

$$|\psi_-\rangle = \begin{pmatrix} \sin \frac{\theta}{2} \\ -e^{i\phi} \cos \frac{\theta}{2} \end{pmatrix} \quad \text{and} \quad |\psi_+\rangle = \begin{pmatrix} \cos \frac{\theta}{2} \\ e^{i\phi} \sin \frac{\theta}{2} \end{pmatrix}. \tag{2.78}$$

The Berry connection for each level is given by:

$$\mathcal{A}_- = i\langle \psi_- | \nabla_\mathbf{R} | \psi_- \rangle = -\cos^2 \frac{\theta}{2} \nabla_\mathbf{R} \phi \,, \tag{2.79}$$

$$\mathcal{A}_+ = i\langle \psi_+ | \nabla_\mathbf{R} | \psi_+ \rangle = -\sin^2 \frac{\theta}{2} \nabla_\mathbf{R} \phi \,, \tag{2.80}$$

which is just equation (2.8) for this case. These expressions depend on the phase of the wave function (2.78), and are not, therefore, gauge invariant. If the parameters that $\mathbf{h}$ depends on are the spherical angles themselves, $(\mathbf{R} = \theta, \phi)$, then the Berry connection has two components:

$$\mathcal{A}_\theta = 0 \,, \mathcal{A}_\phi = -\cos^2 \frac{\theta}{2} \qquad \text{for level - ,} \tag{2.81}$$

$$\mathcal{A}_\theta = 0 \,, \mathcal{A}_\phi = -\sin^2 \frac{\theta}{2} \qquad \text{for level + .} \tag{2.82}$$

Applying the definition of curvature to level $|\psi_\pm\rangle$ above, we obtain:

$$\Omega_{\mu\nu}^\pm = \pm \frac{1}{2} \left[\frac{\partial \cos \theta}{\partial R^\mu} \frac{\partial \phi}{\partial R^\nu} - \frac{\partial \cos \theta}{\partial R^\nu} \frac{\partial \phi}{\partial R^\mu} \right]. \tag{2.83}$$

In the case (2.81) we would get $\Omega_{\theta\phi} = \frac{1}{2} \sin \theta$.

Let us now consider the case where the cartesian components of $\boldsymbol{h}$ are the parameters $\boldsymbol{R} = (h_x, h_y, h_z)$. Since $\mathcal{A}_\theta = i\langle\psi_-|\frac{\partial}{\partial\theta}|\psi_-\rangle = i\langle\psi_-|\frac{\partial}{\partial\boldsymbol{h}}|\psi_-\rangle \cdot \frac{\partial\boldsymbol{h}}{\partial\theta}$, we can see that, for the lower level,

$$\mathcal{A}_\theta = h\mathcal{A} \cdot \boldsymbol{u}_\theta = 0\,, \tag{2.84}$$

$$\mathcal{A}_\phi = h\sin\theta\mathcal{A} \cdot \boldsymbol{u}_\phi = -\cos^2\frac{\theta}{2}\,. \tag{2.85}$$

On the other hand, $\mathcal{A}_h = i\langle\psi_-|\frac{\partial}{\partial h}|\psi_-\rangle = 0$, so that the vector $\mathcal{A}$ has only a single component, along $\boldsymbol{u}_\phi$. A simple calculation yields:

$$\mathcal{A} = -\frac{1}{2h}\cotg\frac{\theta}{2}\,\boldsymbol{u}_\phi \qquad \text{(level -)}\,. \tag{2.86}$$

This is similar to the vector potential for a magnetic monopole. Taking the curl in $\boldsymbol{h}$ space,

$$\nabla_{\boldsymbol{h}} \times \mathcal{A} = \boldsymbol{\Omega}^- = \frac{1}{2h^2}\hat{\boldsymbol{h}}\,. \tag{2.87}$$

We might also have used the expressions relating spherical to cartesian coordinates[5] in expression (2.83) and obtain $\boldsymbol{\Omega}^\pm = \mp\boldsymbol{h}/(2h)^3$. Making the analogy with a magnetic field, this describes a monopole. The total flux through the sphere is $\mp 2\pi$.

The Hamiltonian often describes two Bloch bands in $\boldsymbol{k}$ space. Then, $\boldsymbol{h}$ is a function of Bloch momentum, $\boldsymbol{k}$. The Berry phase is given by the line integral of the connection:

$$\oint_{\boldsymbol{k}\,\text{space}} i\langle\psi|\frac{\partial}{\partial k_j}|\psi\rangle dk_j = \oint_{\boldsymbol{h}\,\text{space}} i\langle\psi|\frac{\partial}{\partial h_j}|\psi\rangle dh_j\,, \tag{2.88}$$

because $\boldsymbol{h}$ is a function of $\boldsymbol{k}$. From Stokes theorem, it is equal to the flux of its curl:

$$\oint_{\boldsymbol{h}\,\text{space}} i\langle\psi|\frac{\partial}{\partial h_j}|\psi\rangle dh_j = \int_{\boldsymbol{h}\,\text{space}} \boldsymbol{\Omega} \cdot d\boldsymbol{S}$$

$$= \int_{\boldsymbol{h}\,\text{space}} \frac{\boldsymbol{h}}{2h^3} \cdot d\boldsymbol{S} = \frac{1}{2}\int \sin\theta\,d\phi d\theta\,, \tag{2.89}$$

5

$$\frac{\partial\cos\theta}{\partial x} = \frac{\partial}{\partial x}(\frac{z}{r}) = -\frac{xz}{r^3};\quad \frac{\partial\cos\theta}{\partial y} = -\frac{yz}{r^3};\quad \frac{\partial\cos\theta}{\partial z} = \frac{x^2+y^2}{r^3};$$

$$\frac{\partial\phi}{\partial x} = \frac{\partial}{\partial x}arctg(\frac{y}{x}) = -\frac{y}{x^2+y^2};\quad \frac{\partial\phi}{\partial y} = \frac{x}{x^2+y^2};\quad \frac{\partial\phi}{\partial z} = 0.$$

where the flux through a sector of the sphere $h = const.$ has been taken. This line integral is the solid angle in h space (see figure 2.2), which can be written as:

$$\frac{1}{2} \int \sin\theta \, d\phi d\theta = \frac{1}{2} \int \frac{\partial \hat{h}}{\partial \theta} \times \frac{\partial \hat{h}}{\partial \phi} \cdot \hat{h} \, d\phi d\theta \,, \qquad (2.90)$$

where the unit vector $\hat{h} = h/|h|$ has been used.

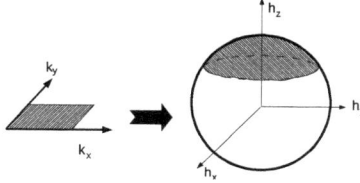

Fig. 2.2   The flux of Ω through the area $\int dk_x dk_y$ corresponds to a flux of Ω through a spherical sector.

2.6   Two-dimensional systems: the Chern number

We have seen that the winding number characterizes the topology of one-dimensional systems. Likewise, the Chern number is the most important topological invariant in two-dimensional systems. It is also related to the number of edge states. The edge states manifest themselves, for instance, through the quantum Hall effect.

Suppose that h is a function of (k_x, k_y). Then,

$$\frac{\partial \hat{h}}{\partial \theta} \times \frac{\partial \hat{h}}{\partial \phi} \cdot \hat{h} = \frac{\partial \hat{h}}{\partial k_x} \times \frac{\partial \hat{h}}{\partial k_y} \cdot \hat{h} \left(\frac{\partial k_x}{\partial \theta} \frac{\partial k_y}{\partial \phi} - \frac{\partial k_x}{\partial \phi} \frac{\partial k_y}{\partial \theta} \right). \qquad (2.91)$$

The expression in brackets is the determinant of the Jacobian for the transformation of the element of area:

$$dk_x dk_y = \left| \frac{\partial k_x}{\partial \theta} \frac{\partial k_y}{\partial \phi} - \frac{\partial k_x}{\partial \phi} \frac{\partial k_y}{\partial \theta} \right| d\phi d\theta \,.$$

Thus, equation (2.90) may be rewritten as[6]:

$$\frac{1}{2} \int \sin\theta \, d\phi d\theta = \frac{1}{2} \int \frac{\partial \hat{h}}{\partial \theta} \times \frac{\partial \hat{h}}{\partial \phi} \cdot \hat{h} \, d\phi d\theta = \frac{1}{2} \int \frac{\partial \hat{h}}{\partial k_x} \times \frac{\partial \hat{h}}{\partial k_y} \cdot \hat{h} \, dk_x dk_y \,.$$

$$(2.92)$$

---

[6]There is here a problem with the sign of the determinant of the Jacobian, which can be negative or positive. The same applies to equation (2.93).

Alternatively, one could have made use of (2.83), which already contains the determinant of the Jacobian for the transformation from spherical coordinates to (k_x, k_y). We have:

$$
\int \Omega^-_{k_x,k_y} dk_x \; dk_y = \frac{1}{2} \int dk_x \; dk_y \left[\frac{\partial \cos\theta}{\partial k_x} \frac{\partial \phi}{\partial k_y} - \frac{\partial \cos\theta}{\partial k_y} \frac{\partial \phi}{\partial k_x} \right]
$$

$$
= \frac{1}{2} \int \sin\theta \; d\phi d\theta \,. \tag{2.93}
$$

The Chern invariant is defined as the Berry phase obtained from the line integral of the connection around the Brillouin zone, or from the integral of the curvature over the Brillouin zone,

$$
C = \frac{1}{2\pi} \oint_{ZB} \mathcal{A} \cdot d\mathbf{k} = \frac{1}{2\pi} \int dk_x \; dk_y \; \Omega_{k_x,k_y} \,, \tag{2.94}
$$

where Stokes theorem has been used.

Another expression for C can be obtained from equation (2.92) for the flux of Berry curvature :

$$
C = \frac{1}{4\pi} \int dk_x \; dk_y \; \frac{\partial \hat{\boldsymbol{h}}}{\partial k_x} \times \frac{\partial \hat{\boldsymbol{h}}}{\partial k_y} \cdot \hat{\boldsymbol{h}} = \frac{1}{4\pi} \int dk_x \; dk_y \; \frac{1}{h^2} \frac{\partial \boldsymbol{h}}{\partial k_x} \times \frac{\partial \boldsymbol{h}}{\partial k_y} \cdot \hat{\boldsymbol{h}} \,. \tag{2.95}
$$

It is then understood that C counts the number of times that the integration domain $\int dk_x \; dk_y$ covers the sphere in $\boldsymbol{h}$ space (see figure 2.2).

In the same way that the Shockley model served as the basic model for the study of topology of one-dimensional systems, the Dirac cone serves as the fundamental model for the understanding of the topological invariant (Chern number) in two-dimensional systems.

Dirac cone: Let the Hamiltonian

$$
\hat{H} = (k_x, k_y, h_z) \cdot \boldsymbol{\tau} \,, \tag{2.96}
$$

where $h_z = const$ is a Dirac mass term. Then, the Berry curvature, $\boldsymbol{\Omega}$,

$$
\Omega^-_{k_x,k_y} = \frac{h_z}{2k^3} = (\boldsymbol{\Omega}^-)_z \,. \tag{2.97}
$$

If we wish to integrate the Berry curvature over the Brillouin zone (k_x, k_y),

$$
\int \Omega^-_{k_x,k_y} dk_x \; dk_y = \frac{h_z}{2} \int \frac{dk_y \; dk_y}{(k_x^2 + k_y^2 + h_z^2)^{3/2}}
$$

$$
= \frac{h_z}{2} \int_0^\infty \frac{2\pi r dr}{(r^2 + h_z^2)^{3/2}}
$$

$$
= \frac{\pi h_z}{2} \int_0^\infty \frac{dt}{(t + h_z^2)^{3/2}}
$$

$$
= \pi \cdot \mathrm{sgn}\,(h_z) \,, \tag{2.98}
$$

where the domain of integration $\int dk_x \, dk_y$ has been extended to infinity. As we should expect, the flux of the monopole through a half-sphere is π. A Dirac cone gives a half-integer contribution to the Chern number. Since the latter must be integer, there must be an even number of Dirac points in the Brillouin zone. This is guaranteed by the *fermion doubling theorem* by Nielsen and Ninomiya, which we shall return to in chapter 3.

Symmetry transformations of the Berry phase

What is the effect of $\mathcal{T}$ on the Berry curvature in equation (2.95) ? The transformation $\boldsymbol{k} \longrightarrow -\boldsymbol{k}$ implies $\partial \boldsymbol{h}(-\boldsymbol{k})/\partial k_j = -\partial \boldsymbol{h}(\boldsymbol{k})/\partial k_j$, so, the product of derivatives does not change sign. After the rotation by T, only one component of $\boldsymbol{h}$ changes sign. Hence, we get for the Berry curvature,

$$\mathcal{T}: \frac{1}{h^2} \frac{\partial \boldsymbol{h}}{\partial k_x} \times \frac{\partial \boldsymbol{h}}{\partial k_y} \cdot \hat{\boldsymbol{h}} \longrightarrow -\frac{1}{h^2} \frac{\partial \boldsymbol{h}}{\partial k_x} \times \frac{\partial \boldsymbol{h}}{\partial k_y} \cdot \hat{\boldsymbol{h}}. \qquad (2.99)$$

We conclude that time reversal inverts the Chern number, $\mathcal{T}: C \longrightarrow -C$. Because the operation $\mathcal{T}$ changes the sign of the Berry curvature, a $\mathcal{T}$-invariant system has $C = -C = 0$. Even so, we shall see in chapter 3 that there exists, in this case, a topological $\mathbb{Z}_2$ number, which can take on the values $\nu = 0$ (trivial) or 1 (topological).

The inversion operation on Hamiltonian (2.77), in momentum space, projects the Berry curvature at point $\boldsymbol{k}$ onto the curvature at point $-\boldsymbol{k}$. Because the Chern number involves an integration over all of the Brillouin zone, its value is not changed under the inversion operation.

The Berry curvature in equation (2.41) vanishes in the presence of time reversal and space inversion symmetry, therefore, the anomalous velocity is not present either. Indeed, time reversal transforms $u_m(\mathbf{k})$ into $u_m^*(-\mathbf{k})$, therefore, $\Omega_{ij}(\mathbf{k})$ into $\Omega_{ji}(-\mathbf{k}) = -\Omega_{ij}(\mathbf{k})$. On the other hand, space inversion symmetry transforms $\Omega_{ij}(\mathbf{k})$ into $\Omega_{ij}(-\mathbf{k})$. The two symmetries combined yield zero curvature.

Under a symmetry transformation, $\hat{O}$, a set of functions $\psi(\mathbf{k})$ is changed for another set, $\hat{O}\psi(\mathbf{k})$. This implies the replacement of a Berry connection by another one. Note that this is not equivalent to a gauge transformation of the type $\mathcal{A} \to \mathcal{A} + \nabla\chi$.

Considering the Hamiltonian (2.77) without the spin degree of freedom, and the time reversal operation, $\mathcal{T}: \psi(\mathbf{k}) \to \psi^*(-\mathbf{k})$. Then,

$$-i\mathcal{A}(\mathbf{k}) = \psi^\dagger(\mathbf{k})\nabla\psi(\mathbf{k}) \to \psi^T(-\mathbf{k})\nabla_k\psi^*(-\mathbf{k}) = -\psi(-\mathbf{k})^T\nabla_{-k}\psi^*(-\mathbf{k})$$
$$= -\left[\psi(-\mathbf{k})^\dagger\nabla\psi(-\mathbf{k})\right]^* = -i\mathcal{A}(-\mathbf{k}).$$

Thus, $\mathcal{T} : \mathcal{A}(\mathbf{k}) \to \mathcal{A}(-\mathbf{k})$. The Berry phase obtained from the line integral along a path $\mathcal{C}$ transforms as

$$\gamma = \oint_{\mathcal{C}} \mathcal{A}(\mathbf{k}) \cdot d\mathbf{k} \to \oint_{\mathcal{C}} \mathcal{A}(-\mathbf{k}) \cdot d\mathbf{k} = -\oint_{\mathcal{C}} \mathcal{A}(-\mathbf{k}) \cdot (-d\mathbf{k}) = -\oint_{-\mathcal{C}} \mathcal{A}(\mathbf{k}) \cdot d\mathbf{k}.$$

The inversion $\mathbf{k} \to -\mathbf{k} \Rightarrow \mathcal{C} \to -\mathcal{C}$. A line element $d\mathbf{k}$ along $\mathcal{C}$ corresponds to a line element $-d\mathbf{k}$ along $-\mathcal{C}$, both line elements being performed in the same direction. Thus,

$$\mathcal{T} : \mathcal{A}(\mathbf{k}) \to \mathcal{A}(-\mathbf{k}), \quad \gamma(\mathcal{C}) \to -\gamma(-\mathcal{C}). \qquad (2.100)$$

This result holds if we include spin space because σ_y is unitary.

We now consider parity, $\psi(\mathbf{k}) \to P\psi(-\mathbf{k})$. Then,

$$-i\mathcal{A}(\mathbf{k}) = \psi^{\dagger}(\mathbf{k})\nabla\psi(\mathbf{k}) \to \psi^{\dagger}(-\mathbf{k})P^{\dagger}P\nabla_k\psi(-\mathbf{k})$$
$$= -\psi^{\dagger}(-\mathbf{k})\nabla\psi(-\mathbf{k}) = i\mathcal{A}(-\mathbf{k}).$$

Therefore,

$$\mathcal{P} : \mathcal{A}(\mathbf{k}) \to -\mathcal{A}(-\mathbf{k}), \quad \gamma(\mathcal{C}) \to \gamma(-\mathcal{C}). \qquad (2.101)$$

From the composition of the two symmetries above we obtain

$$\mathcal{PT} : \mathcal{A}(\mathbf{k}) \to -\mathcal{A}(\mathbf{k}), \quad \gamma(\mathcal{C}) \to -\gamma(\mathcal{C}). \qquad (2.102)$$

2.7   Calculating the Chern number from plaquettes

Looking at the wave functions (2.78), we see that they are not single-valued in all space. Indeed, ψ_- is indeterminate at the north pole ($\theta = 0$) because the angle ϕ is not determined there. For an analogous reason, ψ_+ is indeterminate at the south pole. One could attempt to solve the problem by multiplying any of those wave functions by $e^{-i\phi}$, which means adopting a different gauge. Then again, that would only transfer the indeterminacy to the opposite poles.

When the Chern number is nonzero, all the sphere is covered and, from the previous paragraph, we conclude that *it is not possible to adopt a gauge where the wave function is continuous and single-valued all over the Brillouin zone.*

Consider, then, the insulator with states in the lower band given by ψ_-, which are occupied. At a point $\mathbf{k}^*$ there is a north pole, $\boldsymbol{h}(\mathbf{k}^*) = (0, 0, h_z > 0)$. In the vicinity of $\mathbf{k}^*$, h_x and h_y vanish linearly, producing a Dirac cone. In going around that point, ψ_- accumulates a phase 2π coming from the factor $e^{i\phi}$ in the wave function, the same one that caused the indeterminacy. We may then think about calculating the Chern number from the line integral (2.94) around the Brillouin zone by decomposing the latter into plaquettes and summing over all of them:

$$C = \frac{1}{2\pi} \oint_{ZB} \mathcal{A} \cdot d\mathbf{k} = \frac{1}{2\pi} \sum_{\Box} \oint_{\Box} \mathcal{A} \cdot d\mathbf{k}. \qquad (2.103)$$

Here, $\Box$ denotes a plaquette. The line integrals cancel along the sides of neighboring plaquettes. But there is a plaquette that contains the point $\mathbf{k}^*$. For that one, the line integral yields $\pm 2\pi$. If there are $|C|$ such points in the Brillouin zone, the summation (2.103) yields the Chern number. This allows the calculation of C in a numerically very efficient way.

2.8 Berry curvature as a sum over states

The Berry curvature, $\Omega_{\mu\nu}^{(n)}$, may be written as a sum over states which does not involve derivatives of the wave function. After taking the derivative of the Schrödinger equation,

$$\hat{H}(\mathbf{R})|\psi_n(\mathbf{R})\rangle = E_n(\mathbf{R})|\psi_n(\mathbf{R})\rangle, \qquad (2.104)$$

with respect to $\mathbf{R}$, we can then either left-multiply by $\langle\psi_n(\mathbf{R})|$, thereby obtaining Hellman-Feynman theorem,

$$\langle\psi_n(\mathbf{R})|\frac{\partial\hat{H}}{\partial R^\nu}|\psi_n(\mathbf{R})\rangle = \frac{\partial E_n(\mathbf{R})}{\partial R^\nu}, \qquad (2.105)$$

or we can left-multiply by $\langle\psi_{n'}(\mathbf{R})|$, obtaining

$$\langle\psi_{n'}(\mathbf{R})|\frac{\partial\hat{H}}{\partial R^\nu}|\psi_n(\mathbf{R})\rangle = (E_n - E_{n'})\langle\psi_{n'}|\frac{\partial\psi_n}{\partial R^\nu}\rangle, \qquad n \neq n'. \qquad (2.106)$$

This allows us to write:

$$|\frac{\partial\psi_n}{\partial R^\mu}\rangle = \sum_{n'(\neq n)} \frac{1}{E_n - E_{n'}}\langle\psi_{n'}|\frac{\partial\hat{H}}{\partial R^\mu}|\psi_n\rangle \, |\psi_{n'}\rangle + \alpha_{n,\mu}|\psi_n\rangle, \quad (2.107)$$

$$|\frac{\partial\psi_n}{\partial R^\nu}\rangle = \sum_{n'(\neq n)} \frac{1}{E_n - E_{n'}}\langle\psi_{n'}|\frac{\partial\hat{H}}{\partial R^\nu}|\psi_n\rangle \, |\psi_{n'}\rangle + \alpha_{n,\nu}|\psi_n\rangle, \quad (2.108)$$

where α is pure imaginary, so $\alpha^* = -\alpha$. This property of α stems from the derivative $\partial/\partial\mathbf{R}$ of $\langle\psi_n(\mathbf{R}) \mid \psi_n(\mathbf{R})\rangle$ being zero. Inserting (2.107)-(2.108) into (2.12) and simplifying the notation, $|n\rangle \equiv |\psi_n\rangle$, we obtain:

$$\Omega_{\mu\nu}^{(n)} = i \sum_{n'(\neq n)} \frac{\langle n|\frac{\partial\hat{H}}{\partial R^\mu}|n'\rangle\langle n'|\frac{\partial\hat{H}}{\partial R^\nu}|n\rangle - \langle n|\frac{\partial\hat{H}}{\partial R^\nu}|n'\rangle\langle n'|\frac{\partial\hat{H}}{\partial R^\mu}|n\rangle}{(E_n - E_{n'})^2}. \qquad (2.109)$$

From Kubo formula it can be verified that the Hall conductivity for a filled band, n, is

$$\sigma_{yx} = \frac{e^2}{h} \int \Omega_{k_x,k_y}^{(n)} dk_x \, dk_y = \frac{e^2}{h} C^{(n)} \,. \qquad (2.110)$$

Instead of resorting to Kubo formula, we shall rederive this result in section 2.10.2 in a way that allows a better understanding of its physical meaning.

2.9   Edge states

The Chern number, C, of a Hamiltonian is a bulk property of a system. It is calculated in momentum space, $\mathbf{k}$, as if the system was infinite. If such a system is in contact with another topologically different region (the vacuum, for instance), then the difference in their in topological properties manifests itself in the appearance of localized states at the boundary between them. This phenomenon is referred to as *bulk-boundary correspondence principle.*

We now consider again the Dirac cone model (2.96). We see that changing the value of C in equation (2.98) implies changing the value of the mass, h_z. When making such a continuous change of h_z, a point must be reached where $h_z = 0$, which implies closing the band gap.

From this example we see that the Chern number remains constant if the band gap is not closed. One may modify Hamiltonian parameters continuously in order to change C. But the gap closes in the process, and opens up again with a new value of C. This means that **a topological transition requires closing the band gap**. [7]

If two regions with different Chern numbers are in contact, then there appear localized zero energy states at the boundary between those regions. The rigorous proof of this result is sophisticated. Nevertheless, we can understand it as follows: going from one region to the other is equivalent to a topological transition which necessarily implies closing the band gap. The localized state at the boundary corresponds to the closed gap and has, then, zero energy.

We can sate another important result as follows: the difference between the Chern numbers of two regions in contact is equal to the number of localized states at the boundary between those regions.

---

[7]For interacting systems (not treated in this book), the effective Hamiltonian parameters are obtained from the minimization of the free energy. This can yield a discontinuous variation of one or more parameters, namely, the Dirac mass. A first order topological transition between topological phases may then occur.

The bulk-boundary correspondence principle may be used to define what is meant by a topological insulating system: a system that is insulating in the bulk but exhibits a gapless (metallic) spectrum of surface states. These surface states cannot be removed by local single-particle perturbations without breaking a symmetry.

Chirality of the edge states and Jackiw-Rebbi model

The term *chirality* refers, in this context, to the direction of propagation of the edge states. It should not be confused with the chiral symmetry of a Hamiltonian (i.e, the Hamiltonian's chirality). The chirality of edge states is related to the sign of the Chern number, as shown in figure 2.3.

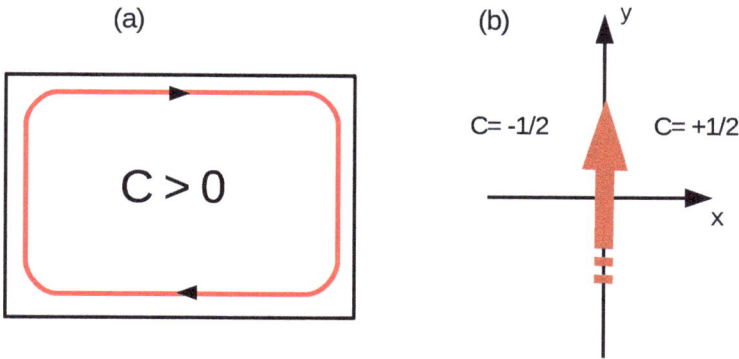

Fig. 2.3   Chirality (direction of propagation) of edge states of a two-dimensional topological system: *(a)* a finite system along x and y is surrounded by the vacuum; *(b)* model (2.111)-(2.112). The arrow in red represents an edge state.

We shall not present here the general proof of this result. But we can solve a model by Jackiw and Rebbi (1976), whereby the edge states can be explicitly obtained. We take the two regions as the half-planes $x < 0$ and $x > 0$, and Dirac cones as Hamiltonians in each region, albeit with symmetrical masses:

$$\hat{H} = \hbar v_F \left(k_x \tau_x + k_y \tau_y \right) - m\tau_z \,, \qquad x < 0 \,, \qquad (2.111)$$

$$\hat{H} = \hbar v_F \left(k_x \tau_x + k_y \tau_y \right) + m\tau_z \,, \qquad x > 0 \,. \qquad (2.112)$$

Taking the Dirac mass $m > 0$ in the half-plane $x < 0$, we get $C = -1/2$, whereas in the half-plane $x > 0$ we get $C = 1/2$. The difference between the Chern numbers on the right- and left-hand side is $+1$. Therefore, there will be a zero energy localized state at the boundary $x = 0$. What is its direction of propagation? We make the replacement $k_j \to -i\partial_j$ in the Hamiltonian and try a wave function of the form

$$\psi(\mathbf{r}) = e^{iky} e^{-\rho|x|} \begin{pmatrix} a \\ b \end{pmatrix}, \tag{2.113}$$

in Schrödinger equation. We obtain

$$a = \begin{cases} \dfrac{-i(E+m)}{\hbar v_F (\rho+k)} b \,, \text{ for } x > 0 \,, \\[3mm] \dfrac{-i(E-m)}{\hbar v_F (-\rho+k)} b \,, \text{ for } x < 0 \,. \end{cases} \tag{2.114}$$

$$E^2 = m^2 + \hbar^2 v_F^2 (k^2 - \rho^2) \,. \tag{2.115}$$

The only way both conditions (2.114) can be satisfied is by using the dispersion $E = \hbar v_F k$. We obtain, then,

$$\psi(\mathbf{r}) = e^{iky} e^{-\frac{m}{\hbar v_F}|x|} \begin{pmatrix} -i \\ 1 \end{pmatrix}. \tag{2.116}$$

The group velocity is, therefore, positive along the y axis. This identifies the chirality as shown in figure 2.3(b).

2.10   Quantum transport by edge states

2.10.1   *Quantum transport in one dimension*

Consider a one-dimensional conductor with length $L \to \infty$, where the dispersion relation $\varepsilon(\mathbf{k}) = \hbar^2 k^2 / (2m)$. The electron group velocity is then $v = \hbar k/m$ and the momentum is quantized as $k = 2\pi n/L$, with $n \in \mathbb{Z}$. The number of Bloch waves in the momentum range dk, or energy range dE, is:

$$dn = \frac{L}{2\pi} dk = \frac{L}{2\pi} \frac{1}{\left(\frac{\partial E}{\partial k}\right)} dE = \frac{L}{hv} dE \,. \tag{2.117}$$

How does the system respond to an electric field? The effect of a battery is to create different chemical potentials at the left- and right-hand ends of the wire. If the chemical potential on the left is higher, then the role of the battery is to inject charges that move to the right (right-movers) in the branch $k > 0$. This means that the Fermi momentum is higher on the left than on the right end of the wire. We neglect, for the time being, the spin degeneracy. Let $\mu_{R/L}$ denote the chemical potentials on the right/left side. The number of injected electrons with $k > 0$ in the energy range $\mu_L - \mu_R$

is then $dn = L(\mu_L - \mu_R)/(hv)$ and the energy current density, j, is given by:

$$j = ev\frac{dn}{L} = \frac{e}{h}(\mu_L - \mu_R). \qquad (2.118)$$

In order to make contact with Ohm's law, we take into account that, for one-dimensional systems, the current density is equal to the actual current, $j = I$. On the other hand, Ohm's law can be written as $I = GV = V/R$, where G is the conductance (the inverse of resistance) and V is the electric potential difference, or voltage. We identify $\mu_L - \mu_R = eV$ and obtain

$$G = \frac{e^2}{h}, \qquad (2.119)$$

which implies that the conductance is quantized.

- If we consider Ohm's law in microscopic form, $j = \sigma E$, we see that the relation between electric field and chemical potential is $E = -\frac{1}{e}\nabla\mu$, for the situation considered above.
- In the case above, the electrons in excess on the left will propagate ballistically to the right. If we use Fick's law relating the particle current density to the density gradient,

$$j_P = -D \cdot (-\frac{dn}{L^2}) = \frac{j}{e}, \qquad (2.120)$$

  we obtain the diffusion coefficient for ballistic transport in one dimension: $D = vL$.
- In the above analysis, we only counted Bloch waves. Other degrees of freedom may be taken into account. For instance, by multiplying the result (2.119) by the spin degeneracy, or by the number of transverse channels.

2.10.2   Hall conductance in two-dimensional systems

We now take, as a model for a two-dimensional system, a ribbon with length $L_x \to \infty$ and finite width, L_y. Edge states (effectively one-dimensional) propagate along the edges $y = 0$ and $y = L_y$ in opposite directions. We may also assume that the Bloch waves have positive (negative) velocity along the edge $y = L_y$ ($y = 0$). This geometry is depicted in figure 2.4. We are now in a position to apply the above ideas on quantum transport to these two one-dimensional systems. The only difference to the case discussed above is that the right- and left-movers are physically separated, since they lie on opposite edges of the ribbon.

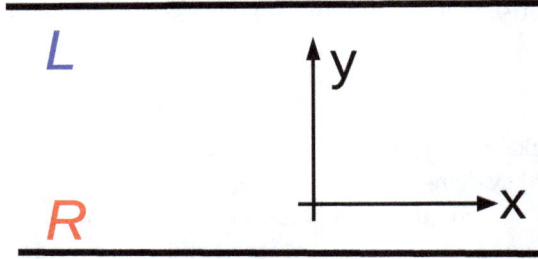

Fig. 2.4   Geometry of the Hall bar. The sample edges $y = 0$ and $y = L_y$ are labeled as R and L, respectively.

The battery injects right-movers on the left-hand side, which must propagate along the edge $y = L_y$ in the x direction. The battery also injects left-movers on the right-hand side, which propagate along the edge $y = 0$ in the negative x direction. This brings about the following consequences:

(*i*) μ_L and μ_R are the chemical potentials at the edges $y = L_y$ and $y = 0$, respectively;

(*ii*) if $\mu_L > \mu_R$, there will be an excess of electrons at $y = L_y$ and a potential difference along y;

(*iii*) the current is then related to a transverse potential difference, along y, instead of along x;

(*iv*) the charge density at the edges is related to an electric field along the y axis. The current density, j_x, is the same as that in expression (2.118):

$$j_x = ev \left(\frac{dn}{L_x} \right) = \frac{e}{h}(\mu_L - \mu_R).$$ (2.121)

The microscopic Ohm's law then reads $j_x = \sigma_{xy} E_y$, where, in the case, $j_x < 0$ e $E_y > 0$. We identify $\mu_L - \mu_R = -eV_y$ and obtain[8]

$$j_x = -\frac{e^2}{h}V_y = \sigma_{xy}V_y \Rightarrow \sigma_{yx} = \frac{e^2}{h}.$$ (2.122)

Note that the conductivity coincides with the conductance, in two-dimensional systems.

We can here relate the result (2.122) to the Chern number of a topological system. Indeed, since C is the number of edge states, we simply

---

[8]It is easy to understand that $\sigma_{yx} = -\sigma_{xy}$. If a field E_x originates a transverse current, $j_y > 0$, then, by rotating this geometry by $90°$, we find that a field E_y originates a current $j_x < 0$ with the same modulus. The axis perpendicular to the electric field allows two possible propagating directions and the appearance of a transverse current (which has "chosen" a direction) implies that time reversal symmetry is broken.

have to multiply the result (2.122) by C in order to understand the result obtained in (2.110).

2.10.3 *Quantum Hall effect*

Note that the Hall conductivity (2.122) takes on a value that is a fundamental constant. Taking into account the possible existence of various channels, the Hall conductance then becomes quantized.

In order for this situation to hold true, there must be no mechanism causing reflection of the electronic states along the boundary.

As is well known, a potential barrier causes the reflection/transmission of an otherwise free electron in one dimension. However, there are situations where the reflection of an edge state in a two-dimensional system is forbidden, simply because the "backward" moving state does not even exist. One such case is due to the Chern invariant, which makes the edge states propagate in a definite direction. Another case is that of an applied magnetic field, which gives rise to Landau levels having chiral edge states (this can actually be described through the Chern invariant as well). What happens, then, if an impurity perturbation is introduced at the edge? The electronic state is modified in such a way that it "goes around" the impurity while propagating in the same direction. No reflection occurs.

The above discussion applies to the transport of *charge*. However, there is also a spin quantum Hall effect. Let us assume that spin ↑ and spin ↓ electrons run along the edge in opposite directions. If the chirality of the edge states depends on spin, we say that there is *helicity*. The edge states are said to be *helical*. If the charge current density of ↑ spin electrons is given by (2.122), then the charge current density of ↓ spin electrons must be opposite and, as a result, no net transport of charge exists. Nevertheless, there is *transport of spin* and we may define the spin current density by modifying equation (2.121):

$$j_x^\uparrow = \frac{\hbar}{2} v_\uparrow \left(\frac{dn}{L_x} \right), \qquad j_x^\downarrow = -\frac{\hbar}{2} v_\downarrow \left(\frac{dn}{L_x} \right). \qquad (2.123)$$

In other words, each electron carries the spin $\pm\hbar/2$ instead of the charge e. Because the ↑ and ↓ spin electrons have opposite velocities, $v_\uparrow = -v_\downarrow$, hence, opposite charge currents, the spin currents add up as:

$$j_x^\uparrow + j_x^\downarrow = -\frac{e}{2\pi} V_y \equiv \sigma_{xy}^s V_y \; \Rightarrow \; \sigma_{yx}^s = \frac{e}{2\pi}. \qquad (2.124)$$

One may need to multiply this result by the number of channels, if necessary. The result (2.124) describes the quantum spin Hall effect.

Note that the role of a battery, introducing the chemical potentials $\mu_{L,R}$ at the two ends, is to inject right-movers. But because of helicity, electrons with opposite spins are injected from opposite ends of the Hall bar. One of the edges of the bar will then have an excess of $\uparrow$ spin, while the other edge will have an excess of $\downarrow$ spin: the battery causes opposite spin densities on opposite edges of a Hall bar.

2.11   Dimensional reduction

How can one relate the winding number of a one-dimensional system to the Chern number of a two-dimensional system? Let us assume the chemical potential to be located in the middle of the energy gap between two bands, as would be the case of an infinite insulator. We can take the chemical potential to be zero. If the Hamiltonian of a two-dimensional system, $H(k_x, k_y)$, has a finite Chern number, then there must be edge states in a semi-infinite system. If the edge is, for instance, $x = 0$, then the states are localized at $x = 0$ and run along the y axis. The dispersion relation for the edge states has to cross the energy gap, making up a branch that connects the conduction to the valence band. And it must intercept the chemical potential (zero energy) somewhere. Take k_y as a fixed parameter so that the Hamiltonian depends only on k_x: $H(k_x, k_y) = H_{k_y}(k_x)$. Now $H_{k_y}(k_x)$ may be seen as describing a one-dimensional chain along x. If a winding number exists, $W \neq 0$, as $k_x : 0 \to 2\pi$, then there will be a localized state at a $x = 0$ boundary. The existence of a zero energy state comes from the fact that there is some value of k_y for which the Hamiltonian $H_{k_y}(k_x)$ has chiral symmetry. Away from that value of k_y, the chirality is broken by a Dirac mass term. The Dirac mass, which is a function of k_y, yields the dispersion relation for the edge states running along the boundary.

Consider again the situation depicted in figure 2.4: the momentum k_x is fixed and the Hamiltonian, $H_{k_x}(k_y)$, has an index W which guarantees the existence of edge states.

2.12   Edge states in graphene

Graphene is made of carbon atoms sitting at the sites of a honeycomb lattice. The honeycomb lattice is not a Bravais lattice. It can be described as a triangular lattice with a basis of two sites (A e B). We may also see it as two interpenetrating triangular lattices. The closest neighbors to an atom A are B atoms and vice-versa. Because of this property, the lattice is said to be *bipartite*.

The three vectors joining a site B to its three A neighbors are:

$$\boldsymbol{\delta}_1 = \hat{x}, \qquad \boldsymbol{\delta}_2 = (-\frac{1}{2}, \frac{\sqrt{3}}{2}), \qquad \boldsymbol{\delta}_3 = (-\frac{1}{2}, -\frac{\sqrt{3}}{2}). \qquad (2.125)$$

The underlying triangular Bravais lattice is generated by the primitive vectors:

$$\mathbf{a}_1 = \boldsymbol{\delta}_1 - \boldsymbol{\delta}_3, \qquad \mathbf{a}_2 = \boldsymbol{\delta}_2 - \boldsymbol{\delta}_3.$$

The choice of primitive cell implicitly determines the boundary of the system, because a crystal starts with a primitive cell. The boundary of a graphene half-plane can be of the following three types: armchair, zigzag and dangling, as shown in figure 2.5.

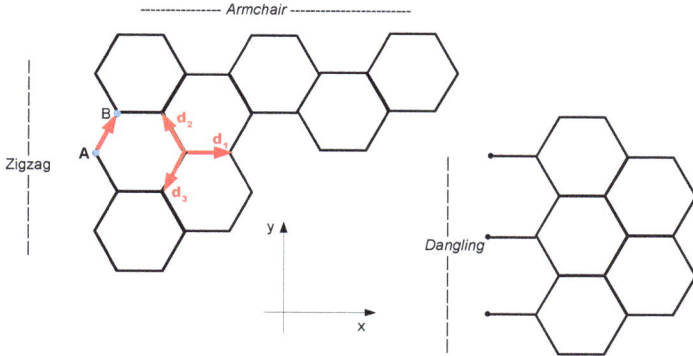

Fig. 2.5 The possible boundaries of a graphene sheet and their designations normally found in the literature.

We now analyze, as an example, the zigzag boundary. Inside the primitive cell, atom A sits at position $\mathbf{R}$ and atom B at position $\mathbf{R} - \boldsymbol{\delta}_3$. Using the primitive vectors $\mathbf{a}_1$ and $\mathbf{a}_2$, this AB segment is repeated along all of the y axis, as well as along the positive x semi-axis. Hence, a semi-infinite crystal is generated with the zigzag boundary along y.

The Bloch wave function takes the form:

$$|\psi\rangle = \sum_{\mathbf{R}} (A|\mathbf{R}\rangle + B|\mathbf{R} - \boldsymbol{\delta}_3\rangle) \, e^{i\mathbf{k}\cdot\mathbf{R}}. \qquad (2.126)$$

The two amplitudes, A and B, apply to each sublattice. We write Schrödinger equation in the form $E\langle\mathbf{R}|\psi\rangle = \langle\mathbf{R}|\hat{H}|\psi\rangle$. If we set the site energies to zero, the Hamiltonian has no diagonal matrix elements. We also set the hopping amplitude between nearest neighbors to unity, so,

$\langle \mathbf{R}|\hat{H}|\mathbf{R} \pm \boldsymbol{\delta}_j \rangle = 1$, and obtain the relation between amplitudes A and B as:

$$EA = B\left[1 + e^{-i\sqrt{3}k_y} + e^{-i\frac{\sqrt{3}}{2}k_y}e^{-i\frac{3}{2}k_x}\right]. \tag{2.127}$$

The term in square brackets is the $(1,2)$ matrix element of the off-diagonal Hamiltonian matrix, $H(\mathbf{k})$, in the sublattice basis. A 2×2 off-diagonal Hamiltonian matrix enjoys, of course, chiral symmetry. The analysis presented in subsection 2.4.1 applies here, then.

Because the crystal is semi-infinite along x, we can take a fixed k_y and make the identification with Shockley model (2.65):

$$t_0 = 1 + e^{-i\sqrt{3}k_y}, \qquad t_1 = e^{-i\frac{\sqrt{3}}{2}k_y}, \qquad k = \frac{3}{2}k_x. \tag{2.128}$$

Note that by fixing k_y we are making use of dimensional reduction. Hence, we know that a zero energy edge state exists if:

$$|t_0| < |t_1| \; \Leftrightarrow \; |1 + e^{-i\sqrt{3}k_y}|\langle 1 \; \Leftrightarrow \; |k_y|\rangle \frac{2\pi}{3\sqrt{3}}. \tag{2.129}$$

An interesting reasoning helps us understand this result. All along the zigzag boundary, the A sites are connected by the vector $\mathbf{a}_2 = (0, \sqrt{3})$. And the exponential

$$e^{i\mathbf{k}\cdot\mathbf{R}} = e^{i\sqrt{3}k_y m} \qquad m \in \mathbb{Z}.$$

Now consider a zero energy state. It is easy to see that the sum of the amplitudes of $|\psi\rangle$ on the neighboring sites of any given site must vanish. Let us write down Schrödinger equation in real space for the B sites shown in the figure:

$$\begin{aligned} e^{i\sqrt{3}k_y(m+1)} + e^{i\sqrt{3}k_y m} + x &= 0, \\ e^{i\sqrt{3}k_y(m-1)} + e^{i\sqrt{3}k_y m} + y &= 0, \\ x + y + z &= 0, \end{aligned} \tag{2.130}$$

whereby we can obtain the wave function's amplitude

$$z = 4e^{i\sqrt{3}k_y m}\cos^2\left(\frac{\sqrt{3}}{2}k_y\right).$$

This is the first A site that we encounter as we start out from the boundary and move along the x axis. In order for the state to decay exponentially into the bulk, it is necessary that $|z| < 1$, hence the result (2.129) is recovered.

Exercise 2.3

In order to study the dangling boundary, define the primitive cell with the A site at $\mathbf{R}$ and the B site at $\mathbf{R} + \boldsymbol{\delta}_1$. Then obtain the condition $|k_y| < 2\pi/(3\sqrt{3})$ for the zero energy edge states.

Exercise 2.4

Make an analogous reasoning to that in equation (2.130) and obtain the condition $|k_y| < 2\pi/(2\sqrt{3})$ for the edge states in the dangling boundary.

Exercise 2.5

And now take k_x as fixed while k_y is allowed to vary. Check that the winding number $W = 0$ and, therefore, no edge states exist. Identify the boundary type.

2.13 The calculation of edge states

We now explain a technique that allows us to find whether edge states exist, and, in affirmative case, their dispersion relation. The advantage of this method is its wide applicability to any spatial dimensionality, or Hamiltonian matrix dimension (the Hamiltonian matrix is not restricted to be 2×2). The main idea of the technique is based on the considerations made at the end of subsection 2.4.1.

Let us consider a Hamiltonian matrix of the form:

$$\hat{H}(\mathbf{k}) = \boldsymbol{h}(\mathbf{k}) \cdot \boldsymbol{\Gamma}\,, \qquad (2.131)$$

where $\boldsymbol{\Gamma} = (..., \Gamma_i, ...)$ denotes a set of hermitian matrices which satisfy the anticommutation relations of a Clifford algebra,

$$\{\Gamma_i, \Gamma_j\} = 2\delta_{ij}I\,, \qquad (2.132)$$

where I is the identity and the multidimensional vector $\boldsymbol{h} = (..., h_i, ...)$ is a function of the momentum, $\mathbf{k}$.

One can easily see that the trace of any Γ_i matrix is zero. And the trace of $\hat{H}(\mathbf{k})$ is, therefore, also zero. Indeed, it follows from (2.132) that for any matrix, $\Gamma_j^2 = I$ holds. On the other hand, if we take any two matrices, $\Gamma_i \neq \Gamma_j$, we have:

$$\Gamma_j \Gamma_i \Gamma_j = -\Gamma_i\,. \qquad (2.133)$$

Taking the trace on both sides of (2.133) and using the cyclic property of the trace on the left-hand side, we obtain:

$$\mathrm{Tr}\{\Gamma_j \Gamma_i \Gamma_j\} = \mathrm{Tr}\{\Gamma_i\} = -\mathrm{Tr}\{\Gamma_i\} = 0\,. \qquad (2.134)$$

The trace of the Hamiltonian also vanishes, by virtue of (2.134). Taking the square of both sides of (2.131), we obtain $\hat{H}^2(\mathbf{k}) = \boldsymbol{h}^2(\mathbf{k})I$. It follows, then, that the eigenenergies $E(\mathbf{k}) = \pm|\boldsymbol{h}(\mathbf{k})|$.

Consider a semi-infinite system, in any spatial dimension and let $k_\perp$ be the momentum component normal to the boundary. The momentum $\mathbf{k} = (\mathbf{k}_\parallel, k_\perp)$. We can then isolate the part of $\boldsymbol{h}(\mathbf{k})$ that depends on $k_\perp$ as:

$$\boldsymbol{h}(\mathbf{k}) = \boldsymbol{b}_0 + 2\boldsymbol{b}_r \cos k_\perp + 2\boldsymbol{b}_i \sin k_\perp \,, \qquad (2.135)$$

where $\boldsymbol{b}_0, \boldsymbol{b}_r, \boldsymbol{b}_i$ may be functions of $\mathbf{k}_\parallel$. Considering that $k_\perp \in [-\pi, \pi[$ and keeping $\mathbf{k}_\parallel$ constant, the vector $\boldsymbol{h}(k_\perp)$ traces out an ellipse centered at the point $\boldsymbol{b}_0$. The ellipse's plane is defined by the vectors $\boldsymbol{b}_r$, $\boldsymbol{b}_i$, and it may contain the origin or not. We may then split $\boldsymbol{b}_0$ into the components parallel and perpendicular to the ellipse's plane:

$$\boldsymbol{b}_0 = \boldsymbol{b}_\parallel + \boldsymbol{b}_\perp \,, \qquad (2.136)$$

where $\boldsymbol{b}_\parallel$ must be a linear combination of $\boldsymbol{b}_r$ and $\boldsymbol{b}_i$. And $\boldsymbol{b}_\perp$ is normal to these two vectors. The modulus of $\boldsymbol{b}_\perp$ is the distance from the ellipse's plane to the origin. We are now in a position to state the following result concerning the edge states:

- *Existence condition*: edge states exist if the projection of the origin onto the plane of the ellipse is contained in the latter. Equivalently, the edge states exist if the ellipse $\boldsymbol{b}_\parallel + 2\boldsymbol{b}_r \cos k_\perp + 2\boldsymbol{b}_i \sin k_\perp$ contains the origin.
- *The energy* of the edge states is given by $E_s(\mathbf{k}_\parallel) = \pm|\boldsymbol{b}_\perp|$. This is the distance from the origin to plane of the ellipse.

We can make an analogy between these results and the considerations presented in the last paragraph of subsection 2.4.1. By tracing out the ellipse as $k_\perp : -\pi \to \pi$, we obtain a winding number that determines the index W. And $\boldsymbol{b}_\perp \cdot \boldsymbol{\Gamma}$ plays the role of the Dirac mass that breaks chiral symmetry.

Further reading:

On the definition of Hamiltonian's symmetries:

- Schnyder et al., *Physical Review* B **78**, 195125 (2008).
- V. Gurarie, *Physical Review* B **83**, 085426 (2011).

On the reflection operator:

- J. C. Y. Teo, L. Fu, and C. L. Kane, *Physical Review* B **78**, 045426 (2008).

On the Shockley model and winding number:

- S. S. Pershoguba and V. M. Yakovenko, *Physical Review* B **86**, 075304 (2012).

On the Berry phase and its application to transport:

- D. Xiao, M.-C. Chang and Q. Niu, *Reviews of Modern Physics* **82**, 1959 (2010).

On quantum transport in low dimensional systems:

- M. Di Ventra, *Electrical Transport in Nanoscale Systems*, Cambridge University Press, 1st edition (2008).
- S. Datta, *Electronic Transport in Mesoscopic Systems*, Cambridge University Press (1997).

On the relation between Chern number and Hall conductivity:

- D. J. Thouless et. al., *Physical Review Letters* **49**, 405 (1982).

On the calculation of the Chern number from the plaquette method:

- Y. Hatsugai, T. Fukui, and H. Aoki, *Physical Review* B **74**, 205414 (2006).
- T. Fukui, Y. Hatsugai, and H. Suzuki, *Journal of the Physical Society of Japan* **74**, 1674 (2005).

On the Dirac Hamiltonian's edge states in two spatial dimensions:

- R. Jackiw, and C. Rebbi, *Physical Review* D **13**, 3398 (1976).

On the calculation of edge states and results for graphene:

- R. S. K. Mong, and V, Shivamoggi, *Physical Review* B **83**, 125109 (2011).
- R. Seshadri, *et. al.*, *Physical Review* B **93**, 035431 (2016).

Chapter 3

Topological insulators

The interest on topological properties of matter has been spurred by the realization that topological insulating materials are different from ordinary insulators: they are insulating in the bulk and metallic on the surface. The surface charge and spin currents are robust and do not dissipate energy. This is of high interest for potential applications in electronics and spintronics.

In what follows, we use concepts introduced in the previous chapters, namely, the Chern number and the Dirac cone, in order to study the models for topological insulators that are more fundamental and more frequently mentioned in the literature. Once the logic behind those models is understood, we explain, through examples, how new models having the desired topological properties can be constructed.

3.1 The construction of the anomalous Hall insulator

The anomalous Hall insulator is a two-dimensional band insulator of spinless fermions with a finite Chern number. In order to write down the corresponding model, we need to take into account that the system must have two sublattices, at least. The Hamiltonian can be written as $\hat{H} = \boldsymbol{h}(\mathbf{k}) \cdot \boldsymbol{\tau}$ with Pauli matrices τ_μ acting in sublattice space, which is also dubbed "pseudo-spin". The wave vector $\boldsymbol{k} = (k_x, k_y)$ belongs in the Brillouin zone.

There are points in the Brillouin zone where the band gap must close when a topological transition (change of Chern number) occurs. Close to the transition, the Hamiltonian in the vicinity of such points takes the form of a Dirac cone.

More specifically, let us assume that close to a point K in the Brillouin zone the Hamiltonian may be linearized as

$$\hat{H} \approx (\hbar v_F q_x, \hbar v_F q_y, h_z) \cdot \boldsymbol{\tau} \,, \tag{3.1}$$

where $q_{x(y)}$ denote small deviations from the point K. Close to the transition, the small gap is the Dirac mass, h_z. The Chern invariant for the lower band is given by

$$C = \frac{1}{4\pi} \int dk_x \, dk_y \, \frac{\partial \hat{\boldsymbol{h}}}{\partial k_x} \times \frac{\partial \hat{\boldsymbol{h}}}{\partial k_y} \cdot \hat{\boldsymbol{h}} \,. \tag{3.2}$$

The calculation of the Dirac point's contribution, ΔC, to the Chern number, C, has been presented in equation (2.98): $\Delta C = \frac{1}{2}\text{sgn}\,[h_z\,(K)]$. We here used the notation $\text{sgn}[x]$ to denote the sign of x.

Time reversal symmetry requires $h_{x(z)}$ to be even functions of momentum, and h_y to be odd. Therefore, there exists another Dirac point at K', which is the time-reversed state of K, so $K' = -K$. The Hamiltonian can be linearized in the vicinity of point K' as

$$\hat{H} \approx (-\hbar v_F k_x, \hbar v_F k_y, h_z) \cdot \boldsymbol{\tau} \,, \tag{3.3}$$

which then gives a contribution $-\frac{1}{2}\text{sgn}\,[h_z\,(K')]$ to the Chern invariant. Time-reversal symmetry requires that the Dirac masses be equal, $h_z\,(K) = h_z\,(K')$. So, the total Chern number $C = 0$. The insulator is thus trivial in the presence of time reversal symmetry.

Time-reversal symmetry requires Dirac cones to occur in pairs. This is the content of the *fermion doubling theorem* by Nielsen and Ninomiya.

We had already seen, in chapter 2, that time reversal symmetry must be broken in order to obtain $C \neq 0$. The application of a uniform magnetic field indeed breaks the symmetry but does not lead to the anomalous Hall insulator. Haldane's proposal, in 1988, is the model that we shall present in section 3.2. It assume symmetrical Dirac masses for the Dirac cones,

$$h_z\,(K) = -h_z\,(K') \,, \tag{3.4}$$

so, the Chern invariant in the lower band is $C = \text{sgn}[h_z\,(K)]$. This is the essence of Haldane's anomalous Hall insulator. The Dirac masses determine the sign of C and, consequently, the edge states' chirality.

One can obtain models with $|C| > 1$ from Hamiltonians containing $|C|$ pairs of Dirac cones. At the Dirac points, the masses are treated in the same way as above.

3.2 Graphene and Haldane model

In 1988, Haldane introduced the best known model for an anomalous Hall insulator. It was conceived for the honeycomb lattice, which is graphene's lattice.

Consider a A site at position $\mathbf{R}$. Let $t = 1$ be the hopping amplitude to the neighboring B sites at positions $\mathbf{R} - \boldsymbol{\delta}_3$, $\mathbf{R} - \boldsymbol{\delta}_1$, $\mathbf{R} - \boldsymbol{\delta}_2$, as shown in figure 2.5.

We may take, as primitive cell, the one made up of atoms A and B at positions $\mathbf{R}$ and $\mathbf{R} - \boldsymbol{\delta}_3$, respectively. This is an appropriate choice for studying the zigzag boundary.

We write the Hamiltonian matrix as $H(\mathbf{k}) = \mathbf{h}(\mathbf{k}) \cdot \boldsymbol{\tau}$, where $\boldsymbol{\tau}$ acts in sublattice space (A and B). We have

$$h_x - ih_y = 1 + e^{-i\sqrt{3}k_y} + e^{-i\frac{\sqrt{3}}{2}k_y} e^{-i\frac{3}{2}k_x}, \qquad h_z = 0. \qquad (3.5)$$

Exercise: Check that inversion symmetry (or parity) is present in the cases: $h_z = 0$; constant $h_z \neq 0$.

At the points where $h_x = h_y = h_z = 0$ there are Dirac cones. Two equivalent Dirac points are:

$$\boldsymbol{K} = \frac{2\pi}{3}(1, \frac{1}{\sqrt{3}}), \qquad \boldsymbol{K}' = \frac{2\pi}{3}(1, -\frac{1}{\sqrt{3}}), \qquad (3.6)$$

and the coordinates for other equivalent cones can be obtained from $120°$ rotations. Graphene is half-filled, therefore, the Fermi energy is zero. The Fermi surface is just the Dirac points. With the aim of obtaining a low energy effective Hamiltonian, we linearize $h_x - ih_y$ in equation (3.5) in the vicinity of the Dirac points. For instance, for the point $\boldsymbol{K}$,

$$h_x - ih_y \approx \frac{3}{2}e^{i\pi/6}(q_x + iq_y), \qquad (3.7)$$

where $\mathbf{q}$ denotes a small deviation from the point $\boldsymbol{K}$. We may further simplify the expression (3.7) by making a suitable rotation of the cartesian axes through the angle $\pi/6$:

$$\begin{pmatrix} q_x \\ q_y \end{pmatrix} = \begin{pmatrix} \cos\frac{\pi}{6} & \sin\frac{\pi}{6} \\ -\sin\frac{\pi}{6} & \cos\frac{\pi}{6} \end{pmatrix} \begin{pmatrix} q_{x'} \\ q_{y'} \end{pmatrix}. \qquad (3.8)$$

Then, the linearized expressions (3.5) and (3.7), near cone $\boldsymbol{K}$, take the form

$$h_x - ih_y \approx \frac{3}{2}(q_{x'} + iq_{y'}). \qquad (3.9)$$

We now discuss the symmetry. For $h_z = 0$, graphene enjoys chiral symmetry, parity (implemented by the matrix $P = \tau_x$), and time reversal symmetry. The physical meaning of parity is easy to understand if we put the origin at the center of a hexagon (real lattice). The displacement of an atom from $\mathbf{r}$ to $-\mathbf{r}$ implies a change of sublattice. Hence, the matrix $P = \tau_x$.

Haldane modified graphene's Hamiltonian in such a way as to obtain finite and symmetrical masses, h_z, at the Dirac points, as described in (3.4). A way to achieve this is to introduce an intra-sublattice hopping matrix element as

$$\langle \mathbf{R}|\hat{H}|\mathbf{R} + \boldsymbol{\delta}_2 - \boldsymbol{\delta}_3 \rangle = \langle \mathbf{R}|\hat{H}|\mathbf{R} + \boldsymbol{\delta}_1 - \boldsymbol{\delta}_2 \rangle = \langle \mathbf{R}|\hat{H}|\mathbf{R} + \boldsymbol{\delta}_2 - \boldsymbol{\delta}_1 \rangle \equiv -im \,,$$

$$\langle \mathbf{R}|\hat{H}|\mathbf{R} + \boldsymbol{\delta}_2 - \boldsymbol{\delta}_1 \rangle = \langle \mathbf{R}|\hat{H}|\mathbf{R} + \boldsymbol{\delta}_1 - \boldsymbol{\delta}_3 \rangle = \langle \mathbf{R}|\hat{H}|\mathbf{R} + \boldsymbol{\delta}_3 - \boldsymbol{\delta}_2 \rangle \equiv im \,.$$

$$(3.10)$$

Pictorially, these hopping terms can be described as follows: the electron's hopping amplitude between next-nearest neighbors is $\pm im$, depending on whether it deviates to the right or left. Going over to momentum space, we obtain,

$$h_z = 4m \sin\left(\frac{\sqrt{3}}{2}k_y\right)\left[\cos\left(\frac{3}{2}k_x\right) - \cos\left(\frac{\sqrt{3}}{2}k_y\right)\right], \qquad (3.11)$$

At points $\boldsymbol{K}, \boldsymbol{K}'$ it obeys the relation (3.4).

Exercise 3.1

(1) Verify that inversion symmetry exists.
(2) Linearize the Hamiltonian close to the Dirac points.

3.3   Edge states in Haldane model

We here make use of the technique explained in section 2.13. Let us consider, then, the Hamiltonian given in equations (3.5) and (3.11):

$$h_x = 1 + \cos(\sqrt{3}k_y) + \cos\left(\frac{\sqrt{3}}{2}k_y\right)\cos\left(\frac{3}{2}k_x\right) - \sin\left(\frac{\sqrt{3}}{2}k_y\right)\sin\left(\frac{3}{2}k_x\right),$$

$$h_y = \sin(\sqrt{3}k_y) + \sin\left(\frac{\sqrt{3}}{2}k_y\right)\cos\left(\frac{3}{2}k_x\right) + \cos\left(\frac{\sqrt{3}}{2}k_y\right)\sin\left(\frac{3}{2}k_x\right),$$

$$h_z = 4\sin\left(\frac{\sqrt{3}}{2}k_y\right)\left[\cos\left(\frac{3}{2}k_x\right) - \cos\left(\frac{\sqrt{3}}{2}k_y\right)\right]. \qquad (3.12)$$

We now fix the value of k_y in order to study the semi-infinite system $x \geq 0$. This means that we put $k_\perp = k_x$ and $k_\parallel = k_y$. We can write, in a similar way to what we did in (2.135),

$$h(\mathbf{k}) = \boldsymbol{b}_0 + 2\boldsymbol{b}_r \cos(\frac{3}{2}k_x) + 2\boldsymbol{b}_i \sin(\frac{3}{2}k_x) \,, \qquad (3.13)$$

where:

$$\boldsymbol{b}_0 = \left(1 + \cos(\sqrt{3}k_y), \sin(\sqrt{3}k_y), -4\sin(\frac{\sqrt{3}}{2}k_y)\cos(\frac{\sqrt{3}}{2}k_y) \right) \,, (3.14)$$

$$2\boldsymbol{b}_r = \left(\cos(\frac{\sqrt{3}}{2}k_y), \sin(\frac{\sqrt{3}}{2}k_y), 4\sin(\frac{\sqrt{3}}{2}k_y) \right) \,, \qquad (3.15)$$

$$2\boldsymbol{b}_i = \left(-\sin(\frac{\sqrt{3}}{2}k_y), \cos(\frac{\sqrt{3}}{2}k_y), 0 \right) \,. \qquad (3.16)$$

We now make the decomposition $\boldsymbol{b}_0 = \boldsymbol{b}_\parallel + \boldsymbol{b}_\perp$. Because $\boldsymbol{b}_r, \boldsymbol{b}_i$ are orthogonal, we can write

$$\boldsymbol{b}_\parallel = \left(\hat{\boldsymbol{b}}_r \cdot \boldsymbol{b}_0 \right) \hat{\boldsymbol{b}}_r + \left(\hat{\boldsymbol{b}}_i \cdot \boldsymbol{b}_0 \right) \hat{\boldsymbol{b}}_i \,, \qquad (3.17)$$

$$\boldsymbol{b}_\perp = \boldsymbol{b}_0 - \boldsymbol{b}_\parallel \,. \qquad (3.18)$$

The condition for the edge states existence is $|\boldsymbol{b}_\perp| < |2\boldsymbol{b}_r|$ and a bit of further algebraic manipulation leads us to the result:

$$2 \left| \cos\left(\frac{\sqrt{3}}{2}k_y\right) \left[1 - 8\,\text{sen}^2\left(\frac{\sqrt{3}}{2}k_y\right) \right] \right| < 1 + 16\,\text{sen}^2\left(\frac{\sqrt{3}}{2}k_y\right) \,. \quad (3.19)$$

In such a region of k_y, the edge states dispersion is given by $E_s = \pm|\boldsymbol{b}_\perp|$, or,

$$E_s = \pm 6 \frac{|\sin(\sqrt{3}k_y)|}{\sqrt{1 + 16\sin^2\left(\frac{\sqrt{3}}{2}k_y\right)}} \,. \qquad (3.20)$$

The reader can analyze other boundary types and the compare his results to those presented in the paper by R. Seshadri, K. Sengupta, and D. Sen, *Physical Review* B **93**, 035431 (2016).

3.4   The Chern insulator in a magnetic field

The effect of a vector potential, $\boldsymbol{A}$, on the anomalous Hall insulator may be studied, in the case of a weak magnetic field (this means a small flux per unit cell), by doing the minimal coupling $-i\hbar\nabla \to -i\hbar\nabla - e\boldsymbol{A}$. Let us assume the magnetic field to be perpendicular to the plane of the system, $\boldsymbol{B} = B\hat{z} = \nabla \times \boldsymbol{A}$. The Hamiltonian matrix (3.1) in the vicinity of a Dirac point at $\boldsymbol{K}$ takes the form:

$$\hat{H} = \begin{pmatrix} h_z & \hat{O} \\ \hat{O}^\dagger & -h_z \end{pmatrix}, \tag{3.21}$$

where the off-diagonal element, $\hat{O}$, obeys the commutation relation $\left[\hat{O}, \hat{O}^\dagger\right] = -2\hbar v_F^2 eB$, independently of the chosen gauge. The operator $\hat{O}$ obeys the same algebra as the excitation operator for a harmonic oscillator. Suppose that $\left[\hat{O}, \hat{O}^\dagger\right] \equiv c > 0$. One can find a set of functions $|\phi_n\rangle$ such that:

$$\hat{O}|\phi_n\rangle = nc|\phi_{n-1}\rangle \qquad n > 0, \tag{3.22}$$

$$\hat{O}|\phi_0\rangle = 0, \tag{3.23}$$

$$|\phi_n\rangle = \left(\hat{O}^\dagger\right)^n |\phi_0\rangle, \tag{3.24}$$

$$\langle\phi_n|\phi_n\rangle = c^n\, n!, \tag{3.25}$$

where ϕ_n are the wave functions for the one-dimensional harmonic oscillator. With the aim of diagonalizing the matrix (3.21), we seek eigenvectors with energy E of the form

$$\begin{pmatrix} (E + h_z)\,\phi_{n-1} \\ \phi_n \end{pmatrix} \qquad n > 0, \tag{3.26}$$

which have energy $E = \pm\sqrt{h_z^2 + nc}$. There is also the level with $n = 0$:

$$\begin{pmatrix} 0 \\ \phi_0 \end{pmatrix}, \tag{3.27}$$

that has energy

$$E_0 = -h_z. \tag{3.28}$$

The above column vectors have to be normalized taking into account equation (3.25). Taking into account that there are positive and negative energies, we generalize to $n \in \mathbb{Z}$ and write the dispersion relation as:

$$E_n = \text{sgn}[n]\sqrt{h_z^2 + |n| \cdot 2\hbar v_F^2\, |eB|} \qquad \text{for } n \neq 0, \tag{3.29}$$

$$E_0 = -h_z. \tag{3.30}$$

These are the Landau levels for a Dirac cone. Each level is degenerate: there exist $|eB|/h$ states per unit area. They are symmetrically distributed for $n \neq 0$.

Exercise 3.2

Consider now the case $\left[\hat{O}, \hat{O}^\dagger\right] \equiv c < 0$. Check that we have to swap the column vector components in (3.26) and (3.27), and make the replacement $h_z \to -h_z$ in (3.26). Show that $E_0 = h_z$.

Following an analogous procedure for the cone $\boldsymbol{K'}$, we use the matrix (3.3) and obtain the commutator $\left[\hat{O}, \hat{O}^\dagger\right] = 2\hbar v_F^2 eB$, which has the opposite sign to the one above. In the topological case, the mass h_z is the same for both cones. Therefore, the $n = 0$ levels have symmetrical energies on the two cones.

Concerning the occupation of the electronic states, the Chern insulator is half filled. This implies that the $n < 0$ levels are occupied. As to the 0 level, we may first consider that it is half filled in each cone, separately. But because of the energy splitting , $E_0 = \pm h_z$, thermal equilibrium implies that the electrons migrate to one of the cones, thereby filling up the 0 level. The 0 level of the other cone is then left empty.

In the topological case, we must invert the sign of the Dirac mass in one of the cones, in accordance with (3.4). In this case, the 0 levels of both cones have the same energy. Taking into account the Chern number, C, that existed in the absence of the field, we can write $E_0 = -|h_z| \operatorname{sgn}[C \cdot B]$. Note that the degeneracy of the 0 level implies that the latter is half-filled in both cones: the weak magnetic field *turns the Chern insulator into a metallic system*.

The Hall conductivity of fermions in a Dirac cone has been studied in the context of graphene. The results can be summarized as follows. The value of σ_{yx} is quantized only when the Landau levels are filled. As the chemical potential increases, each filled Landau level of a single cone gives a contribution $\operatorname{sgn}[B]\frac{1}{2}e^2/h$ to the value of σ_{yx}. If the $n = 0$ level is empty, then $\sigma_{yx} = -\operatorname{sgn}[B]\frac{1}{2}e^2/h$ for a single Dirac cone.

In the topological case we obtain $\sigma_{yx} = 0$ because one of the 0 levels is filled and the other is empty.

3.5   Kane-Mele topological insulator

The model by Kane and Mele is a modification to Haldane model: the spin degree of freedom is introduced in such a way that time reversal symmetry is restored.

We here mention that the term *topological insulator*, as often used in the literature, refers to two kinds of systems: the ones that enjoy time reversal symmetry and are characterized by a $\mathbb{Z}_2$ topological index, and of which the Kane-Mele model is an example; the systems with chiral symmetry, characterized by a relative integer index (an element of $\mathbb{Z}$), which is the winding number, W, and have been dubbed "chiral topological insulators". The term does not usually refer to a system with a Chern number: the latter is commonly referred to as anomalous Hall (or Chern) insulator.

Kane and Mele considered graphene and argued that the terms (3.10) describe the effect of spin-orbit coupling, where the spin s_z component is conserved. Spin-orbit coupling is time-reversal invariant and is proportional to s_z, therefore, the sign of the terms (3.10) is the same as that of s_z. So, the model can be written as

$$\hat{H}(\mathbf{k}) = h_x \tau_x + h_y \tau_y + h_z \tau_z \sigma_z , \qquad (3.31)$$

where the Pauli matrices σ_μ act on the spin degree of freedom, the components $h_{x,y}$ of the $\mathbf{h}(\mathbf{k})$ vector are given by (3.5), and h_z is given by (3.11).

The mental picture of Kane-Mele model is simple: electrons with spin $\uparrow$ (or $\downarrow$) have a Chern number $C = 1$ (or -1). These electronic states with opposite spins are related by time reversal. They constitute, therefore, a Kramers pair. The edge states have opposite chiralities, connected to the spin projection value. Because the edge states' chiralities are connected to the spin values, it is said that there is *helicity*. Note that this concept is used in Condensed Matter with a different meaning than that in Particle Physics.

The model also describes a quantum *spin-Hall insulator* as described at the end of section 2.10.

3.6   The topological $\mathbb{Z}_2$ index

Edge states that map onto each other under the time reversal operation are robust to any perturbation, $\hat{V}$, that conserves time reversal symmetry:

$$\hat{V} = \mathcal{T}\hat{V}\mathcal{T}^{-1} , \qquad (3.32)$$

where the time reversal operator is antiunitary and satisfies

$$\mathcal{T}^2 = -1 ,$$

because it operates on the physical $\frac{1}{2}$ spin. One may prove that such a perturbation has zero matrix element between the Kramer pair states ψ and $\mathcal{T}\psi$. Indeed, such a matrix element satisfies

$$\langle \psi | \hat{V} | \mathcal{T}\psi \rangle = \langle \mathcal{T}\psi | \hat{V} | \psi \rangle^* \qquad \text{because } \hat{V} \text{ is hermitian.} \qquad (3.33)$$

But, on the other hand, we can write:

$$\langle \psi | \hat{V} | \mathcal{T}\psi \rangle = \langle \mathcal{T}\psi | \mathcal{T}\hat{V} | \mathcal{T}\psi \rangle^* \qquad \text{because } \mathcal{T} \text{ is antiunitary,}$$
$$= \langle \mathcal{T}\psi | \mathcal{T}\hat{V}\mathcal{T}^{-1}\mathcal{T} | \mathcal{T}\psi \rangle^*$$
$$= \langle \mathcal{T}\psi | \hat{V} | \mathcal{T}^2\psi \rangle^* \qquad \text{from (3.32),}$$
$$= -\langle \mathcal{T}\psi | \hat{V} | \psi \rangle^* . \tag{3.34}$$

By comparing (3.33) with (3.34), we conclude that $\langle \psi | \hat{V} | \mathcal{T}\psi \rangle = 0$.

We just proved that *one Kramers pair* of states is robust to a time-reversal symmetric perturbation. Suppose that two states, ψ_1 and ψ_2, propagate along the boundary. We may think of them as k-states with dispersion relations that cross the band gap, connecting the lower and upper band. Their Kramers partners, $\mathcal{T}\psi_1$ and $\mathcal{T}\psi_2$ run along the boundary in the opposite direction. A perturbation $\hat{V}$ satisfying $\langle \psi_1 | \hat{V} | \mathcal{T}\psi_2 \rangle = -\langle \psi_2 | \hat{V} | \mathcal{T}\psi_1 \rangle \neq 0$, conserves time reversal symmetry. Yet, it opens a gap in the spectrum. Because an incident electron in state ψ_1 (or ψ_2) may be back-scattered to the state $\mathcal{T}\psi_2$ (or $\mathcal{T}\psi_1$). Hence, two pairs of states propagating long the boundary are not robust to a *single-particle* perturbation. The corresponding quantum spin-Hall insulator may then become a trivial insulator.

The $\mathbb{Z}_2$ topological classification assumes the protection of an odd number of edge state Kramers pairs by time reversal symmetry. An even number of Kramers pairs is unstable and eventually becomes non-topological. If an odd number of such pairs exists before the perturbation, then there will be one pair that survives a time-reversal invariant perturbation.

The topological index, ν, may take on two values, $\nu = 0$ (trivial insulator), or $\nu = 1$ (topological insulator), corresponding to the above even/odd situations, respectively. For the boundary between two regions with N_K Kramers pairs we can write:

$$N_K = \Delta\nu \qquad \text{modulo } 2 \,, \tag{3.35}$$

where $\Delta\nu$ denoted the difference between the topological indices of the two regions.

The Kane-Mele insulator preserves the spin component s_z. If the fermions with each spin component have Chern numbers $C_\uparrow$ and $C_\downarrow$, time reversal symmetry implies that $C_\uparrow = -C_\downarrow$. The topological index may then be obtained as

$$\nu = (C_\uparrow - C_\downarrow)/2 \qquad \text{modulo } 2 \,, \tag{3.36}$$

or, in other words, ν is the parity of the Chern number C_σ.

If s_z is not a good quantum number (for that to happen a spin-orbit coupling term containing the operator $\hat{s}_x$ or $\hat{s}_y$ suffices), then, equation (3.36) no longer makes sense and the ν index has to be calculated in a different way. There exist several formulations in the literature. We here describe one of them.

Let $|u_j(\mathbf{k})\rangle$ denote the eigenvectors of a Hamiltonian matrix of dimension n, $H(\mathbf{k})$, where $j = 1, ..., n$. The matrix elements of the time reversal operator, $\mathcal{T}$, between eigenvectors shall be denoted as:

$$w_{ij}(\mathbf{k}) \equiv \langle u_i(\mathbf{k})|\mathcal{T}|u_j(-\mathbf{k})\rangle. \tag{3.37}$$

Making use of the antinunitarity of $\mathcal{T}$, and of the property $\mathcal{T}^2 = -1$, one can check that

$$w_{ij}(\mathbf{k}) = \langle \mathcal{T}u_i(\mathbf{k})|\mathcal{T}^2|u_j(-\mathbf{k})\rangle^* = -\langle u_j(-\mathbf{k})|\mathcal{T}|u_i(\mathbf{k})\rangle = -w_{ji}(-\mathbf{k}), \tag{3.38}$$

which means that $w(\mathbf{k}) = -w^T(-\mathbf{k})$. In the Brillouin zone there are p points such that $\mathbf{k}$ and $-\mathbf{k}$ denote the same state:

(1) in one dimension, these would be the points $k = 0$, $k = \pi$, hence, $p = 2$;
(2) in two dimensions, in the square lattice, for instance, these would be the points $(0, 0)$, $(\pi, 0)$, $(0, \pi)$ e (π, π), hence, $p = 4$;
(3) in the three-dimensional cubic lattice, for instance, there are $p = 8$ points.

We shall denote any of such points by Λ_a ($a = 1, ..., p$). Then the matrix $w(\Lambda_a)$ is anti-symmetric. Note that the determinant of an anti-symmetric matrix is the square of its Pfaffian. Let δ_a denote the sign of the Pfaffian:

$$\delta_a = \frac{\text{Pf}[w(\Lambda_a)]}{\sqrt{\text{Det}[w(\Lambda_a)]}}. \tag{3.39}$$

Then, ν may be obtained from:

$$(-1)^\nu = \Pi_{a=1}^p \delta_a. \tag{3.40}$$

It is important the the vectors $|u(\mathbf{k})\rangle$ are continuous over the Brillouin zone.

When inversion symmetry is also present, there is a simpler expression for ν. At each point Λ_a, the eigenvalue of the parity operator is ± 1 for an occupied state. Both states of a Kramers pair have the same parity eigenvalue. Let $\xi_m(\Lambda_a)$ be the eigenvalue of the m-th Kramers pair occupied at point Λ_a. We can then take the product of the eigenvalues for the occupied states:

$$\delta(\Lambda_a) = \Pi_m \xi_m. \tag{3.41}$$

The topological index, ν, is defined as:

$$(-1)^\nu = \Pi_{a=1}^p \delta(\Lambda_a). \tag{3.42}$$

3.7 The three-dimensional topological insulator

The three-dimensional topological insulator is characterized by four $\mathbb{Z}_2$ indices, $(\nu_0; \nu_1, \nu_2, \nu_3)$, where ν_0 is the so-called *strong* topological index given by expression (3.40) or (3.42). The remaining *weak* indices are given by the same expressions as in the two-dimensional case, because only the points Λ_a in the plane $k_i = \pi$ are considered when ν_i is calculated:

$$(-1)^{\nu_i} = \Pi_{\Lambda_a \in k_i = \pi} \delta(\Lambda_a) \qquad i = 1, 2, 3. \tag{3.43}$$

Figure 3.1 shows examples of topological indices. The signs $\pm$ at the vertices represent the values of $\delta(\Lambda_a) = \pm 1$. The index ν_0 determines whether the topological insulator is strong or weak.

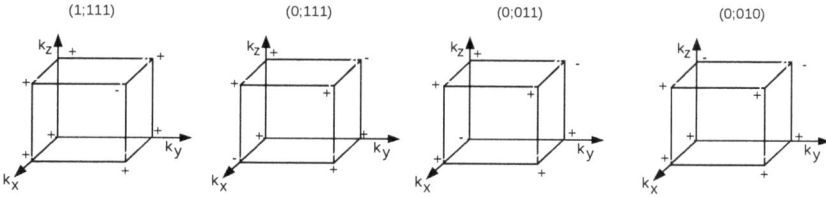

Fig. 3.1 Examples of $\delta(\Lambda_a)$ values in the Brillouin zone of a cubic lattice and the corresponding topological indices $(\nu_0; \nu_1 \nu_2 \nu_3)$.

One may understand the meaning of the weak indices as follows. Consider a superstructure made up of two-dimensional spin-Hall insulating layers, stacked along the z axis. If there is no inter-layer coupling, the layers are independent and the index $\nu_3 = 1$ is obtained from the product $\delta(\Lambda_a)$ for $k_z = 0$ or π. Then $\nu_0 = 0 = \nu_1 = \nu_2$, so, we have $(0; 001)$. This remains so when the inter-layer coupling is turned on. On the other hand, the edge states exist only on the surfaces perpendicular to the x or y axis (or to the xy plane). Hence, a surface perpendicular to the i axis, and for which $\nu_i = 0$, will have surface states.

Such an insulator is "weak" because the surface states my be localized by disorder. Indeed, suppose that the number of layers is even. We have seen that time reversal symmetry does not prevent that a perturbation couples different Kramers pairs. One pair would survive if there was an odd number of layers. On a surface where states exist, there appear Dirac cones at the points Λ_a of the corresponding two-dimensional Brillouin zone.

The surface Fermi surface contains an even number of Kramers degenerate Dirac points.

A strong insulator cannot be obtained by stacking layers of two-dimensional insulators. In other words, it does not descend from the two-dimensional spin-Hall insulator. On a surface of the strong insulator there are surface states, and the surface Fermi surface contains an odd number of Kramers degenerate Dirac cones. For example, if a (doubly degenerate) Dirac cone exists on a surface of a sample, then there will be a Dirac cone on the opposite surface.

3.8　Models for topological insulators

The Haldane and Kane-Mele models were devised for the honeycomb lattice. We shall now see how other models can be devised for the square lattice.

Let $H(\mathbf{k}) = \boldsymbol{h}(\mathbf{k}) \cdot \boldsymbol{\tau}$. In order for $C = \pm 1$ we need the vector $\boldsymbol{h}(\mathbf{k})$ to visit once the north pole and once the south pole. This happens at points $\mathbf{k}$ such that $h_x(\mathbf{k}) = h_y(\mathbf{k}) = 0$ and $h_z(\mathbf{k}) \neq 0$. The latter is just the Dirac mass at $\mathbf{k}$.

Example with C=1: Haldane mechanism for the spinless square lattice.

$$h_x = \cos k_x + \cos k_y + \mu \,,$$
$$h_y = \sin (k_x + k_y) \,,$$
$$h_z = -\sin k_x \,. \tag{3.44}$$

The condition $h_x = h_y = 0$ is satisfied at two symmetrical points on the line $k_y = -k_x$. For instance, if $\mu = 0$, such points would be $(\frac{\pi}{2}, -\frac{\pi}{2})$ and $(-\frac{\pi}{2}, \frac{\pi}{2})$. On the other hand, the functions h_x e h_y satisfy time reversal symmetry. Therefore, we must choose symmetrical Dirac masses at those points (hence visiting the north and south poles). The function $h_z = \sin k_x$ is odd and satisfies such a requirement. Note that h_z breaks time reversal symmetry, a necessary condition for having $C \neq 0$. The choice of the $-$ sign in (3.44) allows us to obtain the desired sign of C: $C = +1$.

The topological insulator can be obtained simply by introducing spin in such a way that opposite spin particles have symmetrical Chern numbers. This is analogous to Kane-Mele model. A simple way to do it is to put $h_z = \sigma_z \sin k_x$.

Exercise: Show that if $h_z = \sigma_z \sin k_x$, then time reversal symmetry is restored.

Example with $|C|=2$: Let $H(\mathbf{k}) = \boldsymbol{h}(\mathbf{k}) \cdot \boldsymbol{\tau}$, with

$$h_x = (\cos k_x + \cos k_y) \,,$$
$$h_y = (\cos k_x - \cos k_y) \,,$$
$$h_z = t \sin k_x \sin k_y \,. \tag{3.45}$$

The condition for visiting the north and south poles, $h_x = h_y = 0$, is satisfied at the points

$$(\frac{\pi}{2}, \frac{\pi}{2}), \qquad (\frac{\pi}{2}, -\frac{\pi}{2}), \qquad (-\frac{\pi}{2}, \frac{\pi}{2}), \qquad (-\frac{\pi}{2}, -\frac{\pi}{2}) \,.$$

The value of $\frac{\partial \boldsymbol{h}}{\partial k_x} \times \frac{\partial \boldsymbol{h}}{\partial k_y} \cdot \hat{\boldsymbol{z}} = -2 \sin k_x \sin k_y$. In two such points we have $h_z = t$ and in the remaining two we have $h_z = -t$. Therefore, each pole is visited twice in the Brillouin zone, hence $C = -2\mathrm{sgn}(t)$. The bands of the Hamiltonian $\hat{H} = \boldsymbol{h} \cdot \boldsymbol{\sigma}$ are $E = \pm|\boldsymbol{h}(\boldsymbol{k})|$ and they do not touch because $\boldsymbol{h} \neq 0$.

The integrand function in (2.95) has the same value at all north and south poles visited. It does not suffice to visit the poles. It is also necessary that the sign of the integrand function be always the same there, so that the integral does not vanish.

We consider now a model where a topological transition occurs by allowing a parameter to vary. The idea is that by varying the parameter, one of the poles is no longer visited.

Transition $C = 2 \to C = 1$: If in the previous example with $|C| = 2$ we modify h_z as:

$$h_z = 4t_2 \sin k_x \sin k_y + 2t_1 (\sin k_x + \sin k_y) + \delta \,, \tag{3.46}$$

one can check that

$$t_2 > t_1 - \delta/4 \;\Rightarrow\; C = -2 \,,$$
$$t_2 < t_1 - \delta \;\Rightarrow\; C = -1 \,. \tag{3.47}$$

The transition occurs because the point $(-\frac{\pi}{2}, -\frac{\pi}{2})$ is a north pole for $C = -2$ and becomes a south pole for $C = -1$.

Therefore, for $C = -1$ the north pole was visited once only. In order for a pole to be no longer visited, h_z has to change sign at that point. But because $h_{x(y)} = 0$ at that point, then $\boldsymbol{h} = 0$ at the transition, thus closing the band gap. Again, we see that a topological transition implies closing the gap.

Exercise 3.3

Dimensional reduction. Consider $H(\mathbf{k}) = \boldsymbol{h}(\mathbf{k}) \cdot \boldsymbol{\tau}$, with

$$h_x = (\cos k_x + \cos k_y) \,,$$
$$h_y = (\cos k_x - \cos k_y) \,,$$
$$h_z = 4t_2 \sin k_x \sin k_y + 2t_1 (\sin k_x + \sin k_y) \,. \qquad (3.48)$$

(1) Verify that when $k_x = \pm\pi/2$, the Hamiltonian $H_{k_x}(k_y)$ has chiral symmetry.
(2) Verify that $H(k_y)$ can be brought to off-diagonal form by making a rotation through the matrix U given by:

$$U = \frac{1}{\sqrt{2}} \begin{pmatrix} 1 & e^{-i\pi/4} \\ -e^{i\pi/4} & 1 \end{pmatrix} \,, \qquad U H U^\dagger = \begin{pmatrix} 0 & D(k_y) \\ D^\dagger(k_y) & 0 \end{pmatrix} \,.$$

(3) Check that at point $k_x = \pi/2$ the winding number of $D(k_y)$ is always 1. But that at point $k_x = -\pi/2$, the winding number of $D(k_y)$ can be either 0 or -1, depending on whether the Chern number $C = -1$ or $C = -2$, respectively.

Exercise 3.4

Transition $C = 1 \to C = 0$. Consider $H(\mathbf{k}) = \boldsymbol{h}(\mathbf{k}) \cdot \boldsymbol{\tau}$, with

$$h_x = (\cos k_x + \cos k_y) \,,$$
$$h_y = (\cos k_x - \cos k_y) \,,$$
$$h_z = t_1 (\sin k_x + \sin k_y) + t_2 [\cos(2k_x) + \cos(2k_y)] \,. \qquad (3.49)$$

(1) Find the points where $h_x = h_y = 0$.
(2) What is the value of C in the regimes $|t_2| > |t_1|$ and $|t_2| < |t_1|$?
(3) Which terms break time reversal symmetry?

Exercise 3.5

Transition to a topological insulator. Let $H(\mathbf{k}) = \boldsymbol{h}(\mathbf{k}) \cdot \boldsymbol{\tau}$, with

$$h_x = \sigma_z \sin k_y \,,$$
$$h_y = -\sin k_x \,,$$
$$h_z = 4t_2 \cos k_x \cos k_y + 2t_1 (\cos k_x + \cos k_y) \,. \qquad (3.50)$$

Verify whether time reversal symmetry is present. Identify the points where the north and south poles are visited. Study the regimes $|t_1| > |t_2|$ and $|t_1| < |t_2|$. In which one of them do we obtain a topological insulator?

3.9 Higher order topological insulators

In the cases studied so far, a d-dimensional topological system has surface states in $d - 1$ dimensions.

If the edge states' dimensionality is $d - 2$, the insulator is said to be of second order. That is the case of a two-dimensional system whose edge states are localized at the corners; or of a cube with states localized at the edges. If the cube had states localized at the vertices, the insulator would be of third order.

We can construct an example for a three-dimensional system by making use of chiral symmetry and obtain a second order insulator. Consider the Hamiltonian:

$$H_1(\mathbf{k}) = \sin k_z \tau_y + [\cos k_z + m(k_x, k_y)] \tau_x . \tag{3.51}$$

There exists a winding number for $k_z \in [0, 2\pi[$ as long as the condition $|m(k_x, k_y)| < 1$ is satisfied. If the system is finite in the z direction and infinite along x and y, then there are zero energy localized states on the surfaces. Such a three-dimensional system has been named *chiral topological insulator*, and does not need to be time-reversal invariant, in contrast to Kane-Mele's topological insulator. The existence of the winding number guarantees the existence of the surface states.

One may confer a non-vanishing dispersion to the surface states by adding to H_1 a term, $H_2(k_x, k_y)$, that anticommutes with it. $H_2(k_x, k_y)$ will then behave as a chiral symmetry breaking mass term. Obviously, $H_2 \propto \tau_z$,

$$H_2(\mathbf{k}) = \boldsymbol{h}(k_x, k_y) \cdot \boldsymbol{\sigma} \tau_z , \tag{3.52}$$

$$H = H_1 + H_2 , \tag{3.53}$$

where $\boldsymbol{\sigma}$ represents a new degree of freedom such as spin or pseudo-spin. Then, $\boldsymbol{h} \cdot \boldsymbol{\sigma}$ is effectively a surface Hamiltonian. Suppose now that the latter has a Chern number. If the system is cut in such a way that it becomes finite along the x and/or y axes, then there will be localized states at the edges of the three-dimensional solid.

3.10 Symmetry classes of gapped Hamiltonians

The relevant symmetries for the classification of insulators and gapped superconductors are time reversal ($\mathcal{T}$), particle-hole ($\mathcal{C}$) and chiral (S). They also determine which type of topological index ($\mathbb{Z}$ or $\mathbb{Z}_2$) a Hamiltonian can have, for a system in any space dimensionality.

The square of a symmetry operator, like $\mathcal{C}$ or $\mathcal{T}$, can be ± 1. Or the symmetry may be absent. One often writes $\mathcal{C}^2 = 0, \pm 1$, and similarly for $\mathcal{T}$, where 0 means that the symmetry is not present. Hence, nine possibilities emerge. But there is also the possibility that chirality (S) is present while $\mathcal{T}$ and $\mathcal{C}$ are absent. There are, then, 10 possibilities corresponding to 10 symmetry classes.

The ten symmetry classes are identified in table 3.1, after the Altland-Zirnbauer classification. The table does not tell us, however, how a given topological index can be calculated. The last four classes apply to Bogoliubov-de Gennes Hamiltonians (superconducting systems), which we shall study in the next chapter.

Class	$\mathcal{T}^2$	$\mathcal{C}^2$	S	d=1	d=2	d=3
A (unitary)	0	0	0	-	$\mathbb{Z}$	-
AI (orthogonal)	1	0	0	-	-	-
AII (simpletic)	-1	0	0	-	$\mathbb{Z}_2$	$\mathbb{Z}_2$
AIII (chiral unitary)	0	0	1	$\mathbb{Z}$	-	$\mathbb{Z}$
BDI (chiral orthogonal)	1	1	1	$\mathbb{Z}$	-	-
CII (chiral simpletic)	-1	-1	1	$\mathbb{Z}$	-	$\mathbb{Z}_2$
D	0	1	0	$\mathbb{Z}_2$	$\mathbb{Z}$	-
C	0	-1	0	-	$\mathbb{Z}$	-
DIII	-1	1	1	$\mathbb{Z}_2$	$\mathbb{Z}_2$	$\mathbb{Z}$
CI	1	-1	1	-	-	$\mathbb{Z}$

Further reading:

On the fermion doubling theorem:

- H. B. Nielsen, and M. Ninomiya, *Physics Letters* B **105**, 219 (1981).
  H. B. Nielsen, and M. Ninomiya, *Nuclear Physics* B **185**, 20 (1981); **193**, 173 (1981).

Haldane and Kane-Mele models were introduced in:

- F. D. M. Haldane, *Physical Review Letters* **61**, 2015 (1988).
- C. L. Kane, and E. J. Mele, *Physical Review Letters* **95**, 146802 (2005).
- C. L. Kane, and E. J. Mele, *Physical Review Letters* **95**, 226801 (2005).

A review on band topology and models:

- M. Z. Hasan, and C. L. Kane, *Reviews of Modern Physics* **82**, 3045 (2010).
- X.-L- Qi, and S.-C. Zhang, *Reviews of Modern Physics* **83**, (2011).

On the effect of an applied magnetic field to Chern and topological insulators:

- N. Goldman, W. Beugeling, and C. Morais Smith, *Europhysics Letters* **97** 23003 (2012); *Physical Review* B **86**, 075118 (2012).
- M. A. N. Araújo, and E. V. Castro, *J. Physics: Condensed Matter* **26**, 075501 (2014).

The chiral topological insulator is discussed in:

- P. Hosur, S. Ryu, and A. Vishwanath, *Physical Review* B **81**, 045120 (2010).

On higher order topological insulators:

- F. Zhang, C. L. Kane, and E. J. Mele, *Physical Review Letters* **110**, 046404 (2013).
- W. A. Benalcazar et al., *Science* **357**, 61 (2017).
- F. Schindler et al., *Science Advances* **4**, no. 6, eaat0346 (2018).
- J. Langbehn et al., *Physical Review Letters* **119**, 246401 (2017).
- S. Imhof et al., *Nature Physics* **14**, 925 (2018).
- L. Li et al., Physical Review B **96**, 235424 (2017).

There are topological systems outside Condensed Matter, also. One instance are the so-called "topoelectric circuits". Some examples can be found in:

- V. V. Albert, L. I. Glazman, and L. Jiang, *Physical Review Letters* **114**, 173902 (2015).
- J. Ningyuan et al., *Physical Review* X **5**, 021031 (2015).
- C. H. Lee et al., *Communications Physics* **1**, 39 (2018).

Chapter 4

Topological superconductors

4.1 Bogoliubov-de Gennes equations

Let us consider a system of electrons described by a Bloch Hamiltonian matrix which we shall denote as $\Xi_{\alpha\beta}(\mathbf{k})$, and which describes the normal metal phase. The indices α, β denote spin projections. We refer to this Hamiltonian term as the "kinetic energy". The field operators for destruction and annihilation are denoted by $\hat{a}^\dagger_{\mathbf{k}\alpha}$ and $\hat{a}_{\mathbf{k}\beta}$, respectively. In mean field theory, superconducting pairing involves the simultaneous creation or annihilation of two particles. We write the effective Hamiltonian for a superconductor as

$$\hat{H} = \sum_{\mathbf{k},\alpha,\beta} \left[\Xi_{\alpha\beta}(\mathbf{k})\hat{a}^\dagger_{\mathbf{k}\alpha}\hat{a}_{\mathbf{k}\beta} + \frac{1}{2}\Delta_{\alpha\beta}(\mathbf{k})\hat{a}^\dagger_{\mathbf{k}\alpha}\hat{a}^\dagger_{-\mathbf{k}\beta} + \frac{1}{2}\Delta^*_{\alpha\beta}(\mathbf{k})\hat{a}_{-\mathbf{k}\beta}\hat{a}_{\mathbf{k}\alpha} \right].$$

(4.1)

Introducing the column vector $a_k = (\hat{a}_{\mathbf{k}\uparrow}\hat{a}_{\mathbf{k}\downarrow}\hat{a}^\dagger_{-\mathbf{k}\uparrow}\hat{a}^\dagger_{-\mathbf{k}\downarrow})^T$, we may rewrite the Hamiltonian (4.1) as

$$\hat{H} = \frac{1}{2}\sum_{\mathbf{k}} a^\dagger_k \begin{pmatrix} \Xi(\mathbf{k}) & \Delta(\mathbf{k}) \\ \Delta^\dagger(\mathbf{k}) & -\Xi^*(-\mathbf{k}) \end{pmatrix} a_k,$$

(4.2)

where the block $\Xi(\mathbf{k})$ is hermitian. We shall see in detail in section 4.1.2 how the term $\Delta_{\alpha\beta}$ is related to electron-electron interactions. Our aim now is to write the Bogoliubov-de Gennes (BdG) equations. To that end, we first calculate the commutators

$$\left[\hat{a}_{\mathbf{k}\mu}, \hat{H}\right] = \Xi_{\mu\beta}(\mathbf{k})\hat{a}_{\mathbf{k}\beta} + \Delta_{\mu\beta}(\mathbf{k})\hat{a}^\dagger_{-\mathbf{k}\beta},$$

(4.3)

$$\left[\hat{a}^\dagger_{-\mathbf{k}\mu}, \hat{H}\right] = -\Xi_{\beta\mu}(-\mathbf{k})\hat{a}^\dagger_{-\mathbf{k}\beta} + \Delta^*_{\beta\mu}(\mathbf{k})\hat{a}_{\mathbf{k}\beta}.$$

(4.4)

Note that $\Delta^*_{\beta\mu}(\mathbf{k}) = \Delta^\dagger_{\mu\beta}(\mathbf{k})$.

If the lementary excitation operators are defined as

$$\hat{\gamma}(\mathbf{k}) = \sum_{\alpha} \left[u_\alpha^*(\mathbf{k})\hat{a}_{\mathbf{k}\alpha} + v_\alpha^*(\mathbf{k})\hat{a}_{-\mathbf{k}\alpha}^\dagger \right] , \tag{4.5}$$

$$\hat{\gamma}^\dagger(-\mathbf{k}) = \sum_{\alpha} \left[u_\alpha(-\mathbf{k})\hat{a}_{-\mathbf{k}\alpha}^\dagger + v_\alpha(-\mathbf{k})\hat{a}_{\mathbf{k}\alpha} \right] , \tag{4.6}$$

then, we impose that the destruction operator for an excitation of energy E must obey the condition

$$\left[\hat{\gamma}, \hat{H} \right] = E\hat{\gamma} ,$$

and we obtain the BdG equations:

$$E u_\mu(\mathbf{k}) = \sum_{\alpha} \left[\Xi_{\mu\alpha} u_\alpha(\mathbf{k}) + \Delta_{\mu\alpha}(\mathbf{k}) v_\alpha(\mathbf{k}) \right] , \tag{4.7}$$

$$E v_\mu(\mathbf{k}) = \sum_{\alpha} \left[\Delta_{\mu\alpha}^\dagger(\mathbf{k}) u_\alpha(\mathbf{k}) - \Xi_{\mu\alpha}^*(-\mathbf{k}) v_\alpha(\mathbf{k}) \right] . \tag{4.8}$$

4.1.1 *Particle-hole symmetry*

We write, in general, the Hamiltonian in real space in the so-called *particle-hole basis*, $\left(c_\uparrow c_\downarrow c_\uparrow^\dagger c_\downarrow^\dagger \right)$. The Hamiltonian matrix, H, is then:

$$\hat{H} = \left(c_\uparrow^\dagger c_\downarrow^\dagger c_\uparrow c_\downarrow \right) \begin{pmatrix} \Xi & \Delta \\ \Delta^\dagger & -\Xi^T \end{pmatrix} \begin{pmatrix} c_\uparrow \\ c_\downarrow \\ c_\uparrow^\dagger \\ c_\downarrow^\dagger \end{pmatrix} \equiv \Psi^\dagger H \Psi . \tag{4.9}$$

When writing this way, we account for the possibility that the system may have several bands, in which case the blocks Ξ e Δ are multidimensional, and where c_σ denotes a column-vector of destruction operators at all sites, with spin σ. When we wish to go over to momentum space, we apply the Fourier transformation rules explained in section 2.3.

Let t_x be the first Pauli matrix operating in particle-hole space[1]:

$$t_x = \begin{pmatrix} 0 & 1 \\ 1 & 0 \end{pmatrix} ,$$

where $\mathbf{1}$ denotes the unit matrix with the appropriate dimension for the number of bands and spin projections. This is half the dimension of H. Then, the operation $t_x \Psi = (\Psi^\dagger)^T$ means placing *daggers* on the operators

---

[1]We write the Pauli matrices as $t_\mu, \sigma_\mu, \tau_\mu$ when operating in particle-hole, spin and sublattice space, respectively.

c and also $\Psi^T = \Psi^\dagger t_x$. Because $\Psi^\dagger H \Psi$ is a scalar, it is equal to its own transpose. However, when we write down this condition, a - sign must be added, because of the fermionic statistics of the operators:

$$\Psi^\dagger H \Psi = -\Psi^T H^T (\Psi^\dagger)^T = -\Psi^\dagger t_x H^T t_x \Psi \;\Rightarrow\; H = -t_x H^T t_x. \qquad (4.10)$$

This relation may be explicitly verified for the matrix H in equation (4.9). It then follows that the pairing matrix for fermions is anti-symmetric: $\Delta = -\Delta^T$.

The last equality in (4.10) shows that a BdG matrix enjoys charge conjugation (or particle-hole) symmetry in the form:

$$-t_x H^* t_x = H \;\Leftrightarrow\; -t_x H^*(-\mathbf{k}) t_x = H(\mathbf{k}) \;\Leftrightarrow\; -H^*(-\mathbf{k}) t_x = t_x H(\mathbf{k}). \qquad (4.11)$$

Let $\psi(\mathbf{k})$ be an eigenvector of the BdG matrix with energy E. Then, the state $t_x \psi^*(-\mathbf{k})$ has energy $-E$, as can be verified as follows:

$$H(\mathbf{k})\psi(\mathbf{k}) = E\ \psi(\mathbf{k})$$
$$\Rightarrow\ t_x H(\mathbf{k})\psi(\mathbf{k}) = E\ t_x \psi(\mathbf{k})$$
$$\Rightarrow\ -H^*(-\mathbf{k}) t_x \psi(\mathbf{k}) = E\ t_x \psi(\mathbf{k})$$
$$\Rightarrow\ H(\mathbf{k}) t_x \psi^*(-\mathbf{k}) = -E\ t_x \psi^*(-\mathbf{k}). \qquad (4.12)$$

Equations (4.5)-(4.6) show how the elementary excitations may be obtained from the eigenvector $\psi(\mathbf{k})$. In particular, an eigenvector $\psi(\mathbf{k})$ with positive energy allows us to write the corresponding elementary excitation operator, $\hat{\gamma}(\mathbf{k})$, in accordance with (4.5). On the other hand, the fundamental particle-hole symmetry (4.12) shows that the column vector $t_x \psi^*(-\mathbf{k})$ represents the state with symmetrical energy. And that same column vector allows us to construct $\hat{\gamma}^\dagger(-\mathbf{k})$, as shown in equation (4.6).

We may rewrite equations (4.5)-(4.6) in matrix form: to that end we only need to read the left-hand side as a column $\left(\hat{\gamma}(\mathbf{k}), \hat{\gamma}^\dagger(-\mathbf{k})\right)^T$ and the right-hand side as the product between a unitary matrix and the column of operators $\left(\hat{a},\ \hat{a}^\dagger\right)^T$. The number of entries in the column $\hat{\gamma}(\mathbf{k})$ is the number of positive energy bands in the BdG Hamiltonian matrix. The lines of that unitary matrix are, from top to bottom, $\psi^\dagger(\mathbf{k})$ and $\psi^T(-\mathbf{k})t_x$. This means that

$$\begin{pmatrix} \hat{\gamma}(\mathbf{k}) \\ \hat{\gamma}^\dagger(-\mathbf{k}) \end{pmatrix} = \begin{pmatrix} \psi^\dagger(\mathbf{k}) \\ \psi^T(-\mathbf{k})t_x \end{pmatrix} \begin{pmatrix} \hat{a} \\ \hat{a}^\dagger \end{pmatrix}. \qquad (4.13)$$

One may invert equation (4.13) and obtain the operators $\hat{a}$, $\hat{a}^\dagger$:

$$\begin{pmatrix} \hat{a} \\ \hat{a}^\dagger \end{pmatrix} = \begin{pmatrix} \psi(\mathbf{k}) & t_x \psi^*(-\mathbf{k}) \end{pmatrix} \begin{pmatrix} \hat{\gamma}(\mathbf{k}) \\ \hat{\gamma}^\dagger(-\mathbf{k}) \end{pmatrix}. \qquad (4.14)$$

We can write this in more explicit form. Denoting by j the band index (with positive energy),

$$\hat{a}_{\mathbf{k},\sigma} = \sum_j u_{j\sigma}(\mathbf{k})\hat{\gamma}_j(\mathbf{k}) + v_{j\sigma}^*(-\mathbf{k})\gamma_j^\dagger(-\mathbf{k}), \qquad (4.15)$$

$$\hat{a}_{-\mathbf{k},\sigma}^\dagger = \sum_j v_{j\sigma}(\mathbf{k})\hat{\gamma}_j(\mathbf{k}) + u_{j\sigma}^*(-\mathbf{k})\gamma_j^\dagger(-\mathbf{k}). \qquad (4.16)$$

4.1.2 *Superconducting pairing*

In this section, we adopt Einstein's summation convention, according to which a summation over repeated indices is implied.

We can write the interaction responsible for superconducting pairing as follows:

$$\hat{V} = \sum_{\mathbf{k},\mathbf{k}'} V_{\sigma_1\sigma_2\sigma_3\sigma_4}(\mathbf{k},\mathbf{k}')\hat{a}_{\mathbf{k},\sigma_1}^\dagger \hat{a}_{-\mathbf{k},\sigma_2}^\dagger \hat{a}_{-\mathbf{k}',\sigma_3}\hat{a}_{\mathbf{k}',\sigma_4}. \qquad (4.17)$$

It has the following properties:

$$V_{\sigma_1\sigma_2\sigma_3\sigma_4}(\mathbf{k},\mathbf{k}') = -V_{\sigma_2\sigma_1\sigma_3\sigma_4}(-\mathbf{k},\mathbf{k}'), \qquad (4.18)$$

$$V_{\sigma_1\sigma_2\sigma_3\sigma_4}(\mathbf{k},\mathbf{k}') = -V_{\sigma_1\sigma_2\sigma_4\sigma_3}(\mathbf{k},-\mathbf{k}'), \qquad (4.19)$$

$$V_{\sigma_1\sigma_2\sigma_3\sigma_4}(\mathbf{k},\mathbf{k}') = V_{\sigma_4\sigma_3\sigma_2\sigma_1}^*(\mathbf{k}',\mathbf{k}). \qquad (4.20)$$

We can now introduce the anomalous coupling terms in (4.17) as:

$$\hat{V}_{eff} = \sum_{\mathbf{k},\mathbf{k}'} \left[V_{\sigma_1\sigma_2\sigma_3\sigma_4}^*(\mathbf{k},\mathbf{k}')\langle\hat{a}_{\mathbf{k}',\sigma_4}^\dagger \hat{a}_{-\mathbf{k}',\sigma_3}^\dagger\rangle\hat{a}_{-\mathbf{k},\sigma_2}\hat{a}_{\mathbf{k},\sigma_1} \right.$$

$$\left. + V_{\sigma_1\sigma_2\sigma_3\sigma_4}(\mathbf{k},\mathbf{k}')\langle\hat{a}_{-\mathbf{k}',\sigma_3}\hat{a}_{\mathbf{k}',\sigma_4}\rangle\hat{a}_{\mathbf{k},\sigma_1}^\dagger \hat{a}_{-\mathbf{k},\sigma_2}^\dagger \right], \qquad (4.21)$$

whereby we define:

$$\Delta_{\sigma_1\sigma_2}(\mathbf{k}) \equiv \sum_{\mathbf{k}'} V_{\sigma_1\sigma_2\sigma_3\sigma_4}(\mathbf{k},\mathbf{k}')\langle\hat{a}_{-\mathbf{k}',\sigma_3}\hat{a}_{\mathbf{k}',\sigma_4}\rangle. \qquad (4.22)$$

It is possible to see, then, that $\Delta_{\sigma_2\sigma_1}(\mathbf{k}) = -\Delta_{\sigma_1\sigma_2}(-\mathbf{k})$, or

$$\hat{\Delta}(\mathbf{k}) = -\hat{\Delta}^T(-\mathbf{k}), \qquad (4.23)$$

where the matrix $\hat{\Delta}$ has spin indices. The equality (4.23) follows immediately from the fact that we are pairing up fermionic operators:

$$\hat{\psi}\Delta\hat{\psi} = \hat{\psi}_\alpha \Delta_{\alpha\beta}\hat{\psi}_\beta = -\hat{\psi}_\beta \Delta_{\alpha\beta}\hat{\psi}_\alpha = -\hat{\psi}\Delta^T\hat{\psi}, \qquad (4.24)$$

(where the indices denote space or spin variables) and this implies the antisymmetry property $\Delta = -\Delta^T$.

If the Cooper pair is a **singlet**, then $\hat{\Delta}$ is an even function of $\mathbf{k}$: $\Delta_{\sigma_2\sigma_1}(\mathbf{k}) = -\Delta_{\sigma_1\sigma_2}(\mathbf{k})$, therefore, it is anti-symmetric in the spin indices. We write:

$$\Delta_{\sigma_2\sigma_1}(\mathbf{k}) = \begin{pmatrix} 0 & \psi(\mathbf{k}) \\ -\psi(\mathbf{k}) & 0 \end{pmatrix}_{\uparrow,\downarrow} = \psi(\mathbf{k})i\sigma_y \ , \text{ with } \psi(\mathbf{k}) \text{ even.} \quad (4.25)$$

If the Cooper pair is a **triplet**, then $\hat{\Delta}$ in an odd function of $\mathbf{k}$ and the matrix $\hat{\Delta}$ is symmetric in the spin indices: $\Delta_{\sigma_1\sigma_2}(\mathbf{k}) = \Delta_{\sigma_2\sigma_1}(\mathbf{k})$. We write:

$$\hat{\Delta}(\mathbf{k}) = \hat{\Delta}^T(\mathbf{k}) = i\mathbf{d}(\mathbf{k}) \cdot \boldsymbol{\sigma}\sigma_y \quad (4.26)$$

$$= \begin{pmatrix} -d_x(\mathbf{k}) + id_y(\mathbf{k}) & d_z(\mathbf{k}) \\ d_z(\mathbf{k}) & d_x(\mathbf{k}) + id_y(\mathbf{k}) \end{pmatrix}_{\uparrow,\downarrow} , \text{ with } \mathbf{d}(\mathbf{k}) \text{ odd}. \quad (4.27)$$

There is a reason for writing in the form (4.26). The Hamiltonian includes $\Delta_{\alpha\beta}\hat{a}_\alpha\hat{a}_\beta$. Under the unitary transformation $\hat{a}_\alpha = U_{\alpha\mu}\hat{a}'_\mu$, we have

$$\Delta_{\alpha\beta}\hat{a}_\alpha\hat{a}_\beta = \Delta_{\alpha\beta}U_{\alpha\nu}U_{\beta\mu}\hat{a}'_\nu\hat{a}'_\mu = \Delta'_{\nu\mu}\hat{a}'_\nu\hat{a}'_\mu ,$$

with $\hat{\Delta}' = U^T\hat{\Delta}U = D\hat{\Delta}D^T$ where $D = U^T$. If we write D in the form $D = u - i\boldsymbol{v} \cdot \boldsymbol{\sigma}$, then[2] $D^T = \sigma_y D^\dagger \sigma_y$, therefore,

$$\hat{\Delta}' = D\hat{\Delta}\sigma_y D^\dagger \sigma_y \Rightarrow \hat{\Delta}'\sigma_y = D\hat{\Delta}\sigma_y D^\dagger . \quad (4.28)$$

This means that $\hat{\Delta}\sigma_y$ transforms as a vector in SU(2). More precisely, if a vector $\mathbf{d}$ undergoes a rotation $\mathbf{d}' = R(\mathbf{d}), R \in$ in SO(3), this can be expressed, in SU(2), as $\mathbf{d}' \cdot \boldsymbol{\sigma} = D\mathbf{d} \cdot \boldsymbol{\sigma}D^\dagger$. Therefore, equation (4.28) tells us that $\hat{\Delta}\sigma_y$ is a vector $\mathbf{d}$ in SU(2): $\hat{\Delta}\sigma_y = \mathbf{d}\cdot\boldsymbol{\sigma}$, hence, we can write (4.26).

If time reversal symmetry is present, then

$$i\sigma_y\hat{\Delta}^*(-\mathbf{k})(-i\sigma_y) = \hat{\Delta}(\mathbf{k}) .$$

This implies that the functions $\psi(\mathbf{k})$ and $\mathbf{d}(\mathbf{k})$ are real.

Finally, how does the introduction of a sublattice (pseudo-spin) degree of freedom modify the conditions (4.25) and (4.26)? Taking pseudo-spin into account, we write:

$$\hat{\Delta} = [\psi(\mathbf{k}) + \mathbf{d}(\mathbf{k}) \cdot \boldsymbol{\sigma}] \, i\sigma_y\tau , \quad (4.29)$$

$$\hat{\Delta}^T = \tau^T i\sigma_y^T \left[\, \psi(-\mathbf{k}) + \mathbf{d}(-\mathbf{k}) \cdot \boldsymbol{\sigma}^T \, \right]$$

$$= \left[\, -\psi(-\mathbf{k}) + \mathbf{d}(-\mathbf{k}) \cdot \boldsymbol{\sigma} \, \right] i\sigma_y\tau^T , \quad (4.30)$$

---

[2]The matrix D is unitary. Therefore, u and $\boldsymbol{v}$ are real. Also, $det D = u^2 + \boldsymbol{v} \cdot \boldsymbol{v} = 1$.

where τ is a Pauli matrix in pseudo-spin space, which commutes with the spin matrices $\boldsymbol{\sigma}$. The last equality follows from $\sigma_y^T \boldsymbol{\sigma}^T = \boldsymbol{\sigma}\sigma_y$. Therefore,

$$\Delta = -\Delta^T \implies \psi(\mathbf{k})\tau = \psi(-\mathbf{k})\tau^T \text{ and } \boldsymbol{d}(\mathbf{k})\tau = -\boldsymbol{d}(-\mathbf{k})\tau^T. \quad (4.31)$$

Hence it follows that if $\tau = \tau^T = \tau_{0,x,z}$, then $\psi(\mathbf{k})$ is even and $\boldsymbol{d}(\mathbf{k})$ odd, as in (4.25)–(4.26). But if $\tau = -\tau^T = \tau_y$, then $\psi(\mathbf{k})$ is odd and $\boldsymbol{d}(\mathbf{k})$ even. Such conditions are invariant under rotations in pseudo-spin space: indeed, for fermions $\hat{a}' = U\hat{a}$, with unitary U, the pairing term $\hat{\Delta}' = U^T \hat{\Delta} U$. The (anti-)symmetry of τ is conserved under the rotation $U^T \tau U$.

The gap equation

We now derive the gap equation from (4.22). Using expressions (4.15)-(4.16), one can calculate the expectation value,

$$\langle \hat{a}_{-\mathbf{k}',\sigma_3} \hat{a}_{\mathbf{k}',\sigma_4} \rangle = \sum_{j,j'} \langle \left[u_{j\sigma_3}(-\mathbf{k}')\hat{\gamma}_j(-\mathbf{k}') + v_{j\sigma_3}^*(\mathbf{k}')\gamma_j^\dagger(\mathbf{k}') \right]$$

$$\cdot \left[u_{j'\sigma_4}(\mathbf{k}')\hat{\gamma}_j'(\mathbf{k}') + v_{j'\sigma_4}^*(-\mathbf{k}')\gamma_j'^\dagger(-\mathbf{k}') \right] \rangle. \quad (4.32)$$

Obviously, only the terms $j = j'$ survive. The expectation value can be obtained for the ground state or at finite temperature, T. Let $f(E)$ denote the Fermi-Dirac distribution for energy E. Then,

$$\langle \hat{\gamma}_j(-\mathbf{k}')\gamma_j^\dagger(-\mathbf{k}') \rangle = 1 - f(E_j(-\mathbf{k}')), \quad (4.33)$$

$$\langle \gamma_j^\dagger(\mathbf{k}')\hat{\gamma}_j(\mathbf{k}') \rangle = f(E_j(\mathbf{k}')). \quad (4.34)$$

The Fermi-Dirac distribution has the property

$$2f(x) - 1 = -\tanh\frac{x}{2}.$$

Inserting this into (4.22) and using (4.19), we obtain

$$\Delta_{\sigma_1\sigma_2}(\mathbf{k}) = \sum_{\mathbf{k}',j} V_{\sigma_1\sigma_2\sigma_3\sigma_4}(\mathbf{k},\mathbf{k}')v_{j\sigma_3}^*(\mathbf{k}')u_{j\sigma_4}(\mathbf{k}')\left[2f(E_j(\mathbf{k}')) - 1\right] \quad (4.35)$$

$$= -\sum_{\mathbf{k}',j} V_{\sigma_1\sigma_2\sigma_3\sigma_4}(\mathbf{k},\mathbf{k}')v_{j\sigma_3}^*(\mathbf{k}')u_{j\sigma_4}(\mathbf{k}')\tanh\frac{E_j(\mathbf{k}')}{2T}. \quad (4.36)$$

Note that the dummy index j runs over the positive energy bands of the BdG Hamiltonian matrix.

4.1.3 The BCS wave function

The anticommutation relation between excitation operators in the same band,

$$[\hat{\gamma}_{\mathbf{k}}, \hat{\gamma}_{-\mathbf{k}}]_+ = 0\,,$$

implies that

$$u_\sigma^*(\mathbf{k})v_\sigma^*(-\mathbf{k}) = -u_\sigma^*(-\mathbf{k})v_\sigma^*(\mathbf{k}) \equiv u^*(\mathbf{k})v^*(-\mathbf{k})\,. \tag{4.37}$$

To put it another way, the dot product $u^*(\mathbf{k})v^*(-\mathbf{k})$ in a given band is anti-symmetric in momentum[3]. Let $|0\rangle$ denote the vacuum. The ground state of the superconductor can be written as:

$$|BCS\rangle = \Pi_{\mathbf{k}j}\hat{\gamma}_{\mathbf{k}}|0\rangle\,, \tag{4.38}$$

because any $\hat{\gamma}_{\mathbf{p}}|BCS\rangle = 0$ by virtue of $\hat{\gamma}_{\mathbf{p}}^2 = 0$. The dummy index j in the product denotes the energy band, and the product only includes positive energy bands of the BdG matrix (4.7)-(4.8). This means that the operators $\hat{\gamma}_{\mathbf{k}}$ are constructed from eigenvectors $\psi(\mathbf{k})$ with positive energy. We can expand the expression (4.38),

$$|BCS\rangle = \Pi_{\mathbf{k}j}'\hat{\gamma}_{\mathbf{k}}\hat{\gamma}_{-\mathbf{k}}|0\rangle \propto \Pi_{\mathbf{k}j}' \left(1 + \left[\frac{v_\alpha^*(\mathbf{k})v_\beta^*(-\mathbf{k})}{u^*(\mathbf{k})v^*(-\mathbf{k})}\right]_j \hat{a}_{-\mathbf{k}\alpha}^\dagger \hat{a}_{\mathbf{k}\beta}^\dagger\right)|0\rangle\,, \tag{4.39}$$

where $\Pi_{\mathbf{k}}'$ runs over half a sphere in $\mathbf{k}$ space. The square of the second term (operator) in expression (4.39) is identically zero because of the fermionic statistics of the field operators. Then, the *unnormalized* BCS ground state wave function, (4.39), may be rewritten in exponential form as:

$$|BCS\rangle \propto \Pi_{\mathbf{k}j}' e^{\left[\frac{v_\alpha^*(\mathbf{k})v_\beta^*(-\mathbf{k})}{u^*(\mathbf{k})v^*(-\mathbf{k})}\right]_j \hat{a}_{-\mathbf{k}\alpha}^\dagger \hat{a}_{\mathbf{k}\beta}^\dagger}|0\rangle\,. \tag{4.40}$$

4.1.4 Majorana fermions

If the $E = 0$ level exists and is non-degenerate, we can write, apart from a phase factor,

$$\psi(\mathbf{k}) = t_x\psi^*(-\mathbf{k}) \Rightarrow \hat{\gamma}(\mathbf{k}) = \hat{\gamma}^\dagger(-\mathbf{k})\,. \tag{4.41}$$

Then, the excitation $\hat{\gamma}$ is *Majorana fermion*: its creation and annihilation operators are equal (apart from a phase factor). This line of argumentation

---

[3]This is simply the orthogonality condition between $\psi(\mathbf{k})$ and $t_x\psi^*(-\mathbf{k})$ explained in section 4.1.4.

may be presented in real space as follows. Let $\psi(\mathbf{r})$ be an eigenvector of the BdG equations (4.9) in real space, given by

$$\psi(\mathbf{r}) = (u_\uparrow(\mathbf{r})\ u_\downarrow(\mathbf{r})\ v_\uparrow(\mathbf{r})\ v_\downarrow(\mathbf{r}))^T .$$

An excitation $\hat{\gamma}$ is then given by:

$$\hat{\gamma} = \int d\mathbf{r} \left[u_\uparrow^*(\mathbf{r})\hat{\psi}_\uparrow(\mathbf{r}) + u_\downarrow^*(\mathbf{r})\hat{\psi}_\downarrow(\mathbf{r}) + v_\uparrow^*(\mathbf{r})\hat{\psi}_\uparrow^\dagger(\mathbf{r}) + v_\downarrow^*(\mathbf{r})\hat{\psi}_\downarrow^\dagger(\mathbf{r}) \right] . \quad (4.42)$$

Particle-hole symmetry implies that if ψ has energy E, then the state $t_x\psi^*$ has energy $-E$. The entries of ψ can be used to construct the operator $\hat{\gamma}$ and the entries of $t_x\psi^*$ can be used to construct the operator $\hat{\gamma}^\dagger$. Only for $E = 0$ is it possible to have $t_x\psi^* = \psi$. In this case, one can use equation (4.42) to verify that

$$\psi(\mathbf{r}) = t_x\psi^*(\mathbf{r}) \Leftrightarrow \hat{\gamma} = \hat{\gamma}^\dagger \qquad \text{(Majorana fermion)}. \qquad (4.43)$$

The existence of zero energy states must follow from a topological property of the BdG Hamiltonian matrix. If there is a winding number in a one-dimensional system, or a Chern number in a two-dimensional one, then there will be a Majorana fermion at the boundary with a topologically trivial region.

From the point of view of the ground state wave function, the states $|BCS\rangle$ and $\gamma^\dagger|BCS\rangle$ are degenerate. They differ, however, in their *fermionic parity* because $|BCS\rangle$ is, by construction, a superposition of states with an even number of fermions in Fock space.

4.1.5 *The Nambu (or Balian-Werthammer) basis*

One often finds, in the literature, the Nambu basis $\left(c_\uparrow\ c_\downarrow\ c_\downarrow^\dagger\ -c_\uparrow^\dagger\right)$ instead of the particle-hole basis we have been using so far. The relation between the Nambu and the particle-hole basis is given by:

$$\begin{pmatrix} c_\uparrow \\ c_\downarrow \\ c_\downarrow^\dagger \\ -c_\uparrow^\dagger \end{pmatrix} = \begin{pmatrix} 1 & 0 \\ 0 & i\sigma_y \end{pmatrix} \begin{pmatrix} c_\uparrow \\ c_\downarrow \\ c_\uparrow^\dagger \\ c_\downarrow^\dagger \end{pmatrix} . \qquad (4.44)$$

Hence we obtain the relation between the Hamiltonian matrices in the particle-hole and Nambu basis as[4]:

$$H_{Nambu} = \begin{pmatrix} 1 & 0 \\ 0 & i\sigma_y \end{pmatrix} H_{ph} \begin{pmatrix} 1 & 0 \\ 0 & -i\sigma_y \end{pmatrix} = \begin{pmatrix} \hat{\Xi} & -i\hat{\Delta}\sigma_y \\ i\sigma_y\hat{\Delta}^\dagger & -\sigma_y\hat{\Xi}^T\sigma_y \end{pmatrix} . \quad (4.45)$$

---

[4]If we write $\hat{\Xi} = \boldsymbol{h} \cdot \boldsymbol{\sigma} + h_0\sigma_0$, then $\hat{\Xi}^T = \hat{\Xi}^* = \boldsymbol{h}^* \cdot \boldsymbol{\sigma}^* + h_0^*\sigma_0$. Making use of the identities $-\sigma_y\boldsymbol{\sigma}^*\sigma_y = \boldsymbol{\sigma}$ and $-\sigma_y\sigma_0\sigma_y = -\sigma_0$, we obtain $-\sigma_y\hat{\Xi}^T\sigma_y = \boldsymbol{h}^* \cdot \boldsymbol{\sigma} - h_0^*\sigma_0$.

One can write the eigenproblem in both bases as:

$$H_{ph}\begin{pmatrix} u_\uparrow \\ u_\downarrow \\ v_\uparrow \\ v_\downarrow \end{pmatrix} = E\begin{pmatrix} u_\uparrow \\ u_\downarrow \\ v_\uparrow \\ v_\downarrow \end{pmatrix} \Leftrightarrow H_{Nambu}\begin{pmatrix} u_\uparrow \\ u_\downarrow \\ v_\downarrow \\ -v_\uparrow \end{pmatrix} = E\begin{pmatrix} u_\uparrow \\ u_\downarrow \\ v_\downarrow \\ -v_\uparrow \end{pmatrix}. \tag{4.46}$$

The pairing block Δ reads as in (4.25) and (4.26) in the particle-hole basis. In the Nambu matrix it reads simply as $\psi(\boldsymbol{k})\mathbf{1}$ in the singlet case, and $\boldsymbol{d}(\boldsymbol{k}) \cdot \boldsymbol{\sigma}$ in the triplet case. Note that the block (22) in matrix (4.45) is the time-reversed version of block (11): this makes it clear that the superconducting pairing occurs between time-reversed states.

In order to see the role of fermionic statistics in the Nambu basis in a more general way, we may consider the off-diagonal matrix term in (4.45) and write:

$$\hat{\Delta}(\boldsymbol{k}) = i\hat{\Delta}_{Nambu}(\boldsymbol{k})\sigma_y, \tag{4.47}$$

$$\hat{\Delta}^T(-\boldsymbol{k}) = -i\sigma_y\hat{\Delta}_{Nambu}^T(-\boldsymbol{k}), \tag{4.48}$$

but since $\hat{\Delta}(\mathbf{k}) = -\hat{\Delta}^T(-\mathbf{k})$, then

$$\hat{\Delta}_{Nambu}(\mathbf{k}) = \sigma_y\hat{\Delta}_{Nambu}^T(-\mathbf{k})\sigma_y \quad \text{(fermionic statistics)}. \tag{4.49}$$

Particle-hole and time reversal symmetry take the form, in the Nambu basis:

$$H = -t_x H^T t_x \Leftrightarrow H_{Nambu} = -\tau_y \otimes \sigma_y H_{Nambu}^T \tau_y \otimes \sigma_y, \tag{4.50}$$

$$H = \mathbf{1} \otimes i\sigma_y H^T \mathbf{1} \otimes (-i\sigma_y) \Leftrightarrow H_{Nambu} = \mathbf{1} \otimes i\sigma_y H_{Nambu}^T \mathbf{1} \otimes (-i\sigma_y), \tag{4.51}$$

respectively.

We now consider how a unitary transformation modifies the Hamiltonian matrix in each of the two bases. A unitary matrix has the general form

$$U = \begin{pmatrix} a & b \\ -b^* & a^* \end{pmatrix}, \quad \text{with } |a|^2 + |b|^2 = 1,$$

and we have:

$$\begin{pmatrix} u' \\ v' \end{pmatrix} = U\begin{pmatrix} u \\ v \end{pmatrix} \Leftrightarrow \begin{pmatrix} v'^* \\ -u'^* \end{pmatrix} = U\begin{pmatrix} v^* \\ -u^* \end{pmatrix} = Ui\sigma_y\begin{pmatrix} u^* \\ v^* \end{pmatrix}$$

$$\Leftrightarrow \mathcal{T}\begin{pmatrix} u' \\ v' \end{pmatrix} = U\mathcal{T}\begin{pmatrix} u \\ v \end{pmatrix}. \tag{4.52}$$

This means that U and $\mathcal{T}$ commute: **for spin $\frac{1}{2}$, time reversal commutes with any unitary matrix.** This renders the Nambu basis convenient. Indeed, a unitary transformation of the BdG matrix in the Nambu basis can be simply written as:

$$H'_{Nambu} = \begin{pmatrix} U & \mathbf{0} \\ \mathbf{0} & U \end{pmatrix} H_{Nambu} \begin{pmatrix} U^\dagger & \mathbf{0} \\ \mathbf{0} & U^\dagger \end{pmatrix}, \qquad (4.53)$$

whereas such a transformation takes the form, in the particle-hole basis,

$$H'_{ph} = \begin{pmatrix} U & \mathbf{0} \\ \mathbf{0} & \sigma_y U \sigma_y \end{pmatrix} H_{ph} \begin{pmatrix} U^\dagger & \mathbf{0} \\ \mathbf{0} & \sigma_y U^\dagger \sigma_y \end{pmatrix}. \qquad (4.54)$$

The charge conjugation operation is defined as $\Psi \rightarrow (\Psi^\dagger)^T \equiv \Psi^*$. In the Nambu basis it takes the form:

$$\sigma_2 t_2 \begin{pmatrix} c_\uparrow \\ c_\downarrow \\ c_\downarrow^\dagger \\ -c_\uparrow^\dagger \end{pmatrix} = \begin{pmatrix} 0 & 0 & 0 & -1 \\ 0 & 0 & 1 & 0 \\ 0 & 1 & 0 & 0 \\ -1 & 0 & 0 & 0 \end{pmatrix} \begin{pmatrix} c_\uparrow \\ c_\downarrow \\ c_\downarrow^\dagger \\ -c_\uparrow^\dagger \end{pmatrix} = \begin{pmatrix} c_\uparrow^\dagger \\ c_\downarrow^\dagger \\ c_\downarrow \\ -c_\uparrow \end{pmatrix}. \qquad (4.55)$$

4.2   One-dimensional Kitaev model

The simplest model of a topological superconductor is the Kitaev model of a one-dimensional superconductor of spinless fermions, with pairing symmetry of p-wave type. This problem is equivalent to a triplet pairing with p wave symmetry and is described by the Hamiltonian

$$H = \sum_{j=1}^{\bar{N}} \left[-t \left(c_j^\dagger c_{j+1} + c_{j+1}^\dagger c_j \right) + \Delta \left(c_j c_{j+1} + c_{j+1}^\dagger c_j^\dagger \right) \right]$$
$$- \sum_{j=1}^{N} \mu \left(c_j^\dagger c_j - \frac{1}{2} \right), \qquad (4.56)$$

where $\bar{N} = N$ if we use periodic boundary conditions (and $N + 1 = 1$) or $\bar{N} = N - 1$ if we use open boundary conditions. Here N is the number of lattice sites, t is the hopping amplitude taken as the energy scale, Δ is the pairing amplitude and μ the chemical potential. The operator c_j destroys a spinless fermion on site j. Fluctuations do not allow the existence of superconductivity in one dimension. However, it is possible to justify this simple model considering, for instance, a semiconductor wire with a strong spin-orbit interaction, in proximity with a conventional superconductor with s-wave symmetry, and in the presence of an external magnetic field, as explained in Appendix A.

The BdG equations for the wave functions may be written as

$$H \begin{pmatrix} u_n(i) \\ v_n(i) \end{pmatrix} = \epsilon_n \begin{pmatrix} u_n(i) \\ v_n(i) \end{pmatrix} , \qquad (4.57)$$

where

$$H = \begin{pmatrix} -t(s_1 + s_{-1}) - \mu & -\Delta(s_i - s_{-1}) \\ \Delta(s_1 - s_{-1}) & t(s_1 + s_{-1}) + \mu \end{pmatrix} , \qquad (4.58)$$

where $s_{\pm 1} f(i) = f(i \pm 1)$ for each function of the lattice point i. The solution of these equations involves the diagonalizata $2N \times 2N$ matrix (using periodic boundary conditions).

In momentum space, we may write the Kitaev model as

$$\hat{H} = \frac{1}{2} \sum_k \left(c_k^\dagger, c_{-k} \right) H_k^c \begin{pmatrix} c_k \\ c_{-k}^\dagger \end{pmatrix} , \qquad (4.59)$$

where

$$H_k^c = \begin{pmatrix} \epsilon_k - \mu & -2i\Delta \sin k \\ 2i\Delta \sin k & -\epsilon_k + \mu \end{pmatrix} , \qquad (4.60)$$

with $\epsilon_k = -2t \cos k$. Here c_k is the Fourier transform of c_j.

A criterium for the existence of topology is the winding number. Let us consider, as before, an operator of chiral symmetry whose eigenvectors allow to reduce the Hamiltonian to an antidiagonal form, as in equation (2.72), with element $q(k) = \epsilon_k - \mu + 2i\Delta \sin k$. The winding number is defined as in equation (2.76) and may be written

$$W = \frac{1}{4\pi i} \int_{-\pi}^{\pi} dk \left[q^{-1} \frac{dq(k)}{dk} - (q^*)^{-1} \frac{dq^*(k)}{dk} \right] . \qquad (4.61)$$

Calculating W, we may characterize the various phases of the model. These are represented in figure 4.1. We obtain that in region I of the phase diagram $W = 1$, in region II we obtain $W = -1$ and in the trivial phases we obtain $W = 0$. We know that W counts the number of protected edge states, on each end of the chain. Additionally, its sign depends on the circulation direction around the origin of the Brillouin zone.

4.2.1 Representation of the Kitaev model by Majorana fermions

The topological nature of the model leads to the existence of edge states in the topological regimes. In order to give emphasis to their existence and

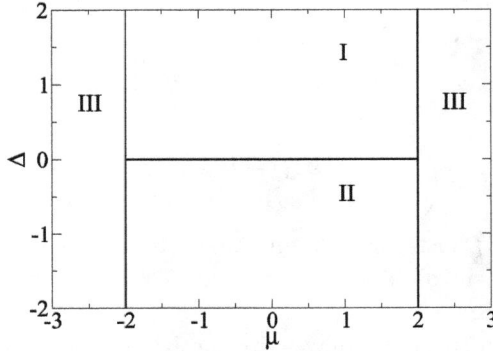

Fig. 4.1   Phases of the Kitaev model.

understand their nature, it is convenient to carry out a transformation to new operators.

The operators of spinless fermions may be written in terms of hermitian operators, called Majoranas, as

$$c_j = \frac{1}{2}\left(\gamma_{1,j} + i\gamma_{2,j}\right),$$
$$c_j^\dagger = \frac{1}{2}\left(\gamma_{1,j} - i\gamma_{2,j}\right). \tag{4.62}$$

The γ operators are hermitian and satisfy a Clifford algebra

$$\{\gamma_m, \gamma_n\} = 2\delta_{nm}. \tag{4.63}$$

In terms of these operators, the Hamiltonian may be written as

$$H = \frac{i}{2}\sum_j \left[(-t+\Delta)\,\gamma_{1,j}\gamma_{2,j+1} + (t+\Delta)\,\gamma_{2,j}\gamma_{1,j+1}\right]$$
$$- \frac{i}{2}\mu\sum_j \gamma_{1,j}\gamma_{2,j}. \tag{4.64}$$

The chemical potential term involves all the Majorana operators. Taking $\mu = 0$ and selecting the special point $t = \Delta$, the Hamiltonian reduces significantly (a similar result may be obtained for $\Delta = -t$):

$$H = it\sum_j \gamma_{2,j}\gamma_{1,j+1}. \tag{4.65}$$

It is easy to see that the operators $\gamma_{1,1}$ and $\gamma_{N,2}$ are absent from the Hamiltonian. There are, therefore, two modes of zero energy, as can be shown by the diagonalization of the Hamiltonian using open boundary conditions.

If we define two non-local operators,

$$d_j = \frac{1}{2} \left(\gamma_{2,j} + i\gamma_{1,j+1} \right) ,$$

$$d_j^\dagger = \frac{1}{2} \left(\gamma_{2,j} - i\gamma_{1,j+1} \right) , \tag{4.66}$$

we can see that these new operators are related with the other fermionic operators by

$$d_j^\dagger = \frac{1}{2} \left[-i(c_j - c_j^\dagger) - i(c_{j+1} + c_{j+1}^\dagger) \right] . \tag{4.67}$$

We may note that

$$2d_j^\dagger d_j - 1 = i\gamma_{2,j}\gamma_{1,j+1} . \tag{4.68}$$

Also,

$$2d_j d_j^\dagger = 1 + \left(c_j^\dagger - c_j \right) \left(c_{j+1}^\dagger + c_{j+1} \right) . \tag{4.69}$$

Defining $d_N = 1/2 \left(\gamma_{N,2} + i\gamma_{1,1} \right)$, we may write the Hamiltonian as

$$H = t \sum_{j=1}^{N-1} \left(2d_j^\dagger d_j - 1 \right) + \epsilon_N \left(2d_N^\dagger d_N - 1 \right) , \tag{4.70}$$

with $\epsilon_N = 0$. Therefore, the fermionic mode d_N does not appear in the Hamiltonian and the state may be occupied or empty ($d_N^\dagger d_N = 1, 0$, respectively), with no energy cost. These two states are, therefore, degenerate. The solution of the Bogoliubov-de Gennes (BdG) equations of the Kitaev model, using open boundary conditions, leads to two modes of zero energy, that at the special points in the phase diagram are perfectly localized at the edges of the chain, as peaks of the δ-Dirac type (with exponential accuracy when the system size increases).

The phase diagram of the Kitaev model has three types of phases: two topological phases with zero energy modes (gapless) if the system is finite, and two trivial phases with no edge states. In the various phases, the spectrum has a gap that results from the modes in the bulk, and at the transition lines between the phases the gap closes, allowing the possibility of topology change. The transition lines are located at $\Delta = 0$ and at $|\mu| = 2t$.

The structure of the phases may be understood in terms of the Majorana representation of the fermionic operators and the two types of phases are illustrated in figure 4.2. The structure is particularly clear at the special points ($\mu = 0, \Delta = \pm t$) but, due to the protection of topological origin of the Hamiltonian (class BDI), its nature is not changed as long as the gap does not close and, therefore, are representative of the behavior of each phase.

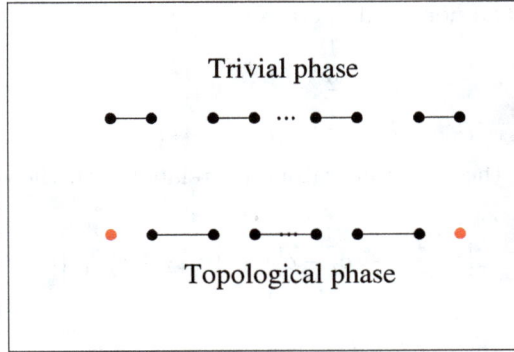

Fig. 4.2   Phases of the Kitaev model: trivial phase with $\Delta = 0, |\mu| > 2t$ and topological phase with $\mu = 0, \Delta = t$. At each lattice site, j, the two points represent the Majorana operators, $\gamma_{j,1}$ and $\gamma_{j,2}$ (real and imaginary parts of c_j). In the trivial phase the two Majoranas at each site are connected in the Hamiltonian and constitute normal fermionic modes. In the topological phase the Majoranas are connected between sites that are nearest neighbors and the first and last operators are decoupled and, therefore, have zero energy.

4.2.2   *Fermionic parity of the groundstate*

In addition to the winding number, W, and the decoupled Majorana modes, we may also characterize the groundstate by the fermionic parity (even or odd fermionic number).

Let us define the vacuum states of the operators c in the usual way

$$c_j|0\rangle_j = 0\,, \tag{4.71}$$

for each site j. We may now define two states with different fermionic parities,

$$|I\rangle = \left(\prod_{j=1}^{N-1} d_j d_j^\dagger\right)|0\rangle\,,$$

$$|II\rangle = \left(\prod_{j=1}^{N-1} d_j d_j^\dagger\right) c_N^\dagger|0\rangle\,, \tag{4.72}$$

where

$$|0\rangle = |0\rangle_N \cdots |0\rangle_1\,. \tag{4.73}$$

The states $|I\rangle$ and $|II\rangle$ do not have excitations of the operators d. That is, an operator d_j annihilates these states. Let us analyse the state $|I\rangle$. TThe action on the vacuum is given by

$$\prod_{j=1}^{N-1}\left[1+\left(c_j^\dagger-c_j\right)\left(c_{j+1}^\dagger+c_{j+1}\right)\right]|0\rangle = \prod_{j=1}^{N}\left(1+c_j^\dagger\right)|0\rangle_p, \qquad (4.74)$$

where only even powers of the operators c contribute. In a similar way, considering the state $|II\rangle$, we find

$$\prod_{j=1}^{N-1}\left[1+\left(c_j^\dagger-c_j\right)\left(c_{j+1}^\dagger+c_{j+1}\right)\right]c_N^\dagger|0\rangle = \prod_{j=1}^{N}\left(1+c_j^\dagger\right)|0\rangle_i. \qquad (4.75)$$

where only odd powers of the operators c contribute. These states are states of zero occupation of the operators d and may be characterized by their fermionic parity $|I\rangle = |\psi_p^0\rangle$ and $|II\rangle = |\psi_i^0\rangle$. We may now define two new states as linear combinations of these,

$$|\psi_0^\pm\rangle = |\psi_p^0\rangle \pm |\psi_i^0\rangle$$

$$|\psi_0^\pm\rangle = \prod_{j=1}^{N}\left(1\pm c_j^\dagger\right)|0\rangle. \qquad (4.76)$$

The state of zero occupation of the operators d is the groundstate of the system, at this special point (with $\mu = 0, t = \Delta$)

$$|GS\rangle = |n_d = 0; n_d = 0; \cdots ; n_d = 0; \cdots\rangle, \qquad (4.77)$$

written in terms of the operators d. The states $|\psi_0^\pm\rangle$ are the groundstates written in terms of the operators c.

Recall that $d_N = \frac{i}{2}(c_N^\dagger - c_N + c_1 + c_1^\dagger)$. Therefore, $d_N^\dagger + d_N = i(c_N^\dagger - c_N)$ and $d_N - d_N^\dagger = i(c_1 + c_1^\dagger)$. Also recall that

$$|\psi_0^p\rangle = \prod_{j=1}^{N}\left(1+c_j^\dagger\right)_p |0\rangle_c,$$

$$|\psi_0^i\rangle = \prod_{j=1}^{N}\left(1+c_j^\dagger\right)_i |0\rangle_c. \qquad (4.78)$$

Then,

$$\left(d_N + d_N^\dagger\right)|\psi_0^p\rangle = i|\psi_0^i\rangle,$$

$$\left(d_N + d_N^\dagger\right)|\psi_0^i\rangle = -i|\psi_0^p\rangle,$$

$$\left(d_N - d_N^\dagger\right)|\psi_0^i\rangle = i|\psi_0^p\rangle,$$

$$\left(d_N - d_N^\dagger\right)|\psi_0^p\rangle = i|\psi_0^i\rangle. \qquad (4.79)$$

This implies

$$d_N|\psi_0^i\rangle = 0. \tag{4.80}$$

Also

$$d_N^\dagger|\psi_0^i\rangle = -i|\psi_0^p\rangle. \tag{4.81}$$

We may conclude that the groundstate is non-degenerate and has odd parity. In the topological region the parity is odd and in the trivial region the parity is even (since there are no decoupled modes). Note that the parity is preserved, independently of the representation used.

4.2.3 *Extended Kitaev model*

Let us consider, now, an extension of the Kitaev model that includes terms of longer range, both in the hopping amplitude and in the pairing between the electrons. As explained in Appendix A, these terms arise, for instance, in the process of dimensional reduction from a higher dimension to an effective one-dimensional system.

The model in real space may be written as

$$H = - t \sum_i \left(c_i^\dagger c_{i+1} + c_{i+1}^\dagger c_i \right) - t' \sum_i \left(c_i^\dagger c_{i+2} + c_{i+2}^\dagger c_i \right) - \mu \sum_i c_i^\dagger c_i$$
$$+ d \sum_i \left(c_i c_{i+1} + c_{i+1}^\dagger c_i^\dagger \right) + d' \sum_i \left(c_i c_{i+2} + c_{i+2}^\dagger c_i^\dagger \right). \tag{4.82}$$

where t and t' are the hopping amplitudes between first and second neighbors, respectively, and d and d' the pairing between first and second neighbors, respectively. This Hamiltonian may be diagonalized for a finite system of size N_x using open boundary conditions, giving rise to edge modes of zero energy.

In momentum space $(k = k_x)$, the model is represented by the matrix

$$H_k = \begin{pmatrix} \epsilon_k - \mu & id \sin k + id' \sin(2k) \\ -id \sin k - id' \sin(2k) & -\epsilon_k + \mu \end{pmatrix}, \tag{4.83}$$

with $\epsilon_k = -2t \cos k - 2t' \cos(2k)$. Diagonalizing, we obtain

$$\omega_k^2 = (-\epsilon_k + \mu)^2 + [d \sin k + d' \sin(2k)]^2. \tag{4.84}$$

Transitions between different topological phases occur at the gapless points $-\epsilon_k + \mu = 0$ and $d \sin k + d' \sin(2k) = 0$. This occurs when $\cos k = -d/(2d')$, which leads to $t' = -d(\mu d' - td)/(d^2 - 2(d')^2)$. If $t' = 0, d' = 0$, there are

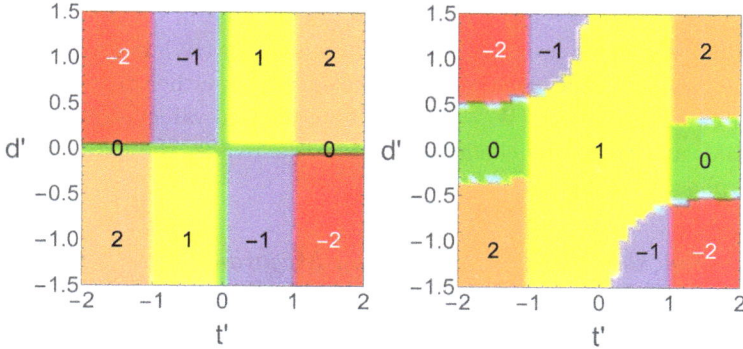

Fig. 4.3 Phase diagram for the extended Kitaev model with second neighbors both in the hopping amplitude and in the pairing, indexed by the winding number. In the left panel $t = 1, d = 0$ and in the right panel $t = 1, d = 0.6$. The chemical potential $\mu = 0$.

zeros for $k = 0$ and $\mu = -2t$ or $k = \pi$ and $\mu = 2t$. If only $d' = 0$, we obtain zeros for $k = 0$ and $2t + 2t' + \mu = 0$ or $k = \pi$ and $-2t + 2t' + \mu = 0$. If only $t' = 0$, we obtain $\mu = td/d'$.

To calculate the winding number and characterize the different topological phases, we may identify

$$q(k) = \epsilon_k - \mu - id \sin k - id' \sin(2k). \qquad (4.85)$$

The phase diagram is represented in figure 4.3, where the results are presented for $\mu = 0$ and the cases where $t = 1, d = 0$ and $t = 1, d = 0.6$ are considered, as examples.

As in the usual Kitaev model and any superconductor, the model has particle-hole symmetry. Also, has time reversal symmetry for spinless systems, which implies that, in agreement with the table 3.1, it belongs to the class BDI and, therefore, is in the class $\mathbb{Z}$ of topological systems.

Usually, it is expected that the couplings between second neighbors have smaller magnitudes than the first neighbor terms ($t'/t < 1, d'/d < 1$). We will show, however, results away from this regime. In the limits of high values of t'/t and/or d'/d, we expect changes with respect to the original Kitaev model. However, the results of figure 4.3 show that, even in the regime where $t'/t < 1, d'/d < 1$, the winding number changes with respect to the Kitaev model, and the topology and the number of edge states changes. For high values of $|t'|$ or $|d'|$, the winding number is $W = 2, -2$ giving evidence for the presence of two edge states (two at each end). In the neighborhood of the central region $W = 1, 0, -1$, as expected from the limit $t'/t << 1, d'/d << 1$. The topology is altered in some cases, even if the additional terms are small, turning the system trivial or topological.

One can also show that if $t = 0$ the system is either trivial ($W = 0$), or has two edge states ($W = 2, -2$), for different values of the chemical potential, and when the chemical potential increases, the extension of the trivial phase also increases.

4.2.4 *Shockley model expressed by Majorana fermions*

The Shockley model discussed before in equation (2.61), with the change $t_0 \to t_1^*, t_1 \to t_2$, may also be written as

$$H = \sum_{j=1}^{N} \psi^\dagger(j) \left[U\psi(j) + V\psi(j-1) + V^\dagger\psi(j+1) \right], \tag{4.86}$$

where the 2×2 matrices U and V are given by

$$U = \begin{pmatrix} 0 & t_1^* \\ t_1 & 0 \end{pmatrix}; V = \begin{pmatrix} 0 & t_2^* \\ 0 & 0 \end{pmatrix} \tag{4.87}$$

and the spinor ψ represents two orbitals that are hybridized by the matrices U and V:

$$\psi(j) = \begin{pmatrix} c_{j,A} \\ c_{j,B} \end{pmatrix}. \tag{4.88}$$

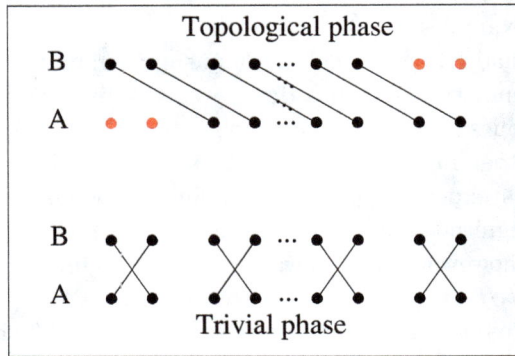

Fig. 4.4   Phases of the Shockley model: trivial phase with $t_2 = 0$ and topological phase with $t_1 = 0$. In the trivial phase the Majorana fermions are coupled to form usual fermionic modes in a given lattice site. In the topological phase, the bonds occur between sites that are nearest neighbors and there are four uncoupled Majorana fermions that give origin to usual fermionic modes of zero energy, one on each end of the chain.

The edge states may be understood using the Majorana representation as for the Kitaev model. We may define Majorana operators, γ_1, γ_2, as

$$c_{j,A} = \frac{1}{2}\left(\gamma_{j,A,1} + i\gamma_{j,A,2}\right),$$

$$c_{j,B} = \frac{1}{2}\left(\gamma_{j,B,1} + i\gamma_{j,B,2}\right). \tag{4.89}$$

Taking $t_1^* = t_1, t_2^* = t_2$, the Hamiltonian may be written as

$$
\begin{aligned}
H = {} & \frac{it_1}{2} \sum_{j=1}^{N} \left(\gamma_{j,A,1}\gamma_{j,B,2} + \gamma_{j,B,1}\gamma_{j,A,2}\right) \\
& + \frac{t_2}{4} \sum_{j=2}^{N} \left(\gamma_{j,A,1}\gamma_{j-1,B,1} + \gamma_{j,A,2}\gamma_{j-1,B,2}\right) \\
& + \frac{it_2}{4} \sum_{j=2}^{N} \left(\gamma_{j,A,1}\gamma_{j-1,B,2} - i\gamma_{j,A,2}\gamma_{j-1,B,1}\right) \\
& + \frac{t_2}{4} \sum_{j=1}^{N-1} \left(\gamma_{j,B,1}\gamma_{j+1,A,1} + \gamma_{j,B,2}\gamma_{j+1,A,2}\right) \\
& + \frac{it_2}{4} \sum_{j=1}^{N-1} \left(\gamma_{j,B,1}\gamma_{j+1,A,2} - i\gamma_{j,B,2}\gamma_{j+1,A,1}\right). \tag{4.90}
\end{aligned}
$$

Taking $t_1 = 0$, we obtain that some Majorana fermions do not contribute, $\gamma_{1,A,1}, \gamma_{1,A,2}, \gamma_{N,B,1}, \gamma_{N,B,2}$, and give origin to zero energy modes.

In figure 4.4, the structure of the Hamiltonian terms is presented for two points in the phase diagram, that correspond to the trivial and topological phase. In the topological phase there are zero energy modes that are decoupled, but that, however, have a usual fermionic nature, since the Majoranas that are decoupled are localized on the two edges of the chain, at sites A and B, respectively. The existence of an edge state on each end is confirmed by the winding number, $W = 1$, obtained previously.

4.2.5 *SSH model with triplet pairing*

This model may be seen as a dimerized Kitaev superconductor, and is a generalization of Shockley model, or of its equivalent Su-Schrieffer-Heeger (SSH) model. The dimerization is parametrized by η and the superconductivity by Δ. Due to the dimerization, the unit cell contains two atoms of types A and B. The lattice sites are indexed by j. The Hamiltonian is given by

$$H = -\mu \sum_j \left(c_{j,A}^\dagger c_{j,A} + c_{j,B}^\dagger c_{j,B} \right)$$
$$-t \sum_j \left[(1+\eta)c_{j,B}^\dagger c_{j,A} + (1+\eta)c_{j,A}^\dagger c_{j,B} \right.$$
$$+ \; (1-\eta)c_{j+1,A}^\dagger c_{j,B} + (1-\eta)c_{j,B}^\dagger c_{j+1,A} \Big]$$
$$+\Delta \sum_j \left[(1+\eta)c_{j,B}^\dagger c_{j,A}^\dagger + (1+\eta)c_{j,A}c_{j,B} \right.$$
$$+ \; (1-\eta)c_{j+1,A}^\dagger c_{j,B}^\dagger + (1-\eta)c_{j,B}c_{j+1,A} \Big] \qquad (4.91)$$

(t is the hopping amplitude, Δ the pairing amplitude and μ the chemical potential). The model with no superconductivity ($\Delta = 0$) is related with the Shockley model taking $t_1 = t(1+\eta)$ and $t_2 = t(1-\eta)$. The region $\eta > 0$ corresponds to $t_1 > t_2$ and vice-versa for $\eta < 0$. The Hamiltonian in real space mixes first neighbors and also has local terms.

In momentum space, the model is given by the matrix

$$H_k = \begin{pmatrix} -\mu & z(k) & 0 & w(k) \\ z^*(k) & -\mu & -w^*(k) & 0 \\ 0 & -w(k) & \mu & -z(k) \\ w^*(k) & 0 & -z^*(k) & \mu \end{pmatrix}, \qquad (4.92)$$

and it acts on the spinors

$$\begin{pmatrix} c_A(k) \\ c_B(k) \\ c_A^\dagger(-k) \\ c_B^\dagger(-k) \end{pmatrix}, \qquad (4.93)$$

and

$$z(k) = -t \left[(1+\eta) + (1-\eta)e^{-ik} \right],$$
$$w(k) = -\Delta \left[(1+\eta) - (1-\eta)e^{-ik} \right]. \qquad (4.94)$$

We may define hermitian Majorana fermions, $\gamma_{j,\alpha,\beta}$ (with $\beta = 1, 2$), as

$$c_{j,A} = \frac{1}{2} \left(\gamma_{j,A,1} + i\gamma_{j,A,2} \right),$$
$$c_{j,B} = \frac{1}{2} \left(\gamma_{j,B,1} + i\gamma_{j,B,2} \right). \qquad (4.95)$$

In terms of Majorana operators, the Hamiltonian is written as

$$H = -\frac{\mu}{2} \sum_{j=1}^{N} (2 + i\gamma_{j,A,1}\gamma_{j,A,2} + i\gamma_{j,B,1}\gamma_{j,B,2})$$

$$- \frac{it}{2}(1 + \eta) \sum_{j=1}^{N} (\gamma_{j,B,1}\gamma_{j,A,2} + \gamma_{j,A,1}\gamma_{j,B,2})$$

$$- \frac{it}{2}(1 - \eta) \sum_{j=1}^{N-1} (\gamma_{j+1,A,1}\gamma_{j,B,2} + \gamma_{j,B,1}\gamma_{j+1,A,2})$$

$$+ \frac{i\Delta}{2}(1 + \eta) \sum_{j=1}^{N} (\gamma_{j,A,1}\gamma_{j,B,2} + \gamma_{j,A,2}\gamma_{j,B,1})$$

$$+ \frac{i\Delta}{2}(1 - \eta) \sum_{j=1}^{N-1} (\gamma_{j,B,1}\gamma_{j+1,A,2} + \gamma_{j,B,2}\gamma_{j+1,A,1}) \,. \tag{4.96}$$

Choosing, as before, $\mu = 0$, there is a pair of special points: i) Taking $\eta = -1$ and $\Delta = 0$, we have a state similar to the SSH or Shockley models with two edge states of zero energy of the fermionic type, since the four operators $\gamma_{1,A,1}$, $\gamma_{1,A,2}$; $\gamma_{N,B,1}$, $\gamma_{N,B,2}$ are absent from the Hamiltonian. ii) $\eta = 0$ and $t = \Delta$ is a state of the Kitaev type since two Majorana operators are absent from the Hamiltonian, $\gamma_{1,A,1}$ and $\gamma_{N,B,2}$, one on each end. iii) An example of the trivial phase is the point $\eta = 1$ and $\Delta = 0$ with no zero energy edge modes. In figure 4.5, the phases with edge modes are presented for special points of the parameter space.

Considering, for instance, a regime with no superconductivity, $\Delta = 0$, and at the special point $\mu = 0, \eta = -1$ shown in the figure, the Hamiltonian reduces to

$$H = it \sum_{j=1}^{N-1} (\gamma_{j,B,2}\gamma_{j+1,A,1} - \gamma_{j,B,1}\gamma_{j+1,A,2}) \,. \tag{4.97}$$

Let us define new non-local fermionic operators,

$$d_j = \frac{1}{2}(\gamma_{j,B,2} + i\gamma_{j+1,A,1}) \,,$$

$$d_j^\dagger = \frac{1}{2}(\gamma_{j,B,2} - i\gamma_{j+1,A,1}) \,, \tag{4.98}$$

and

$$f_j = \frac{1}{2}(\gamma_{j,B,1} - i\gamma_{j+1,A,2}) \,,$$

$$f_j^\dagger = \frac{1}{2}(\gamma_{j,B,1} + i\gamma_{j+1,A,2}) \,. \tag{4.99}$$

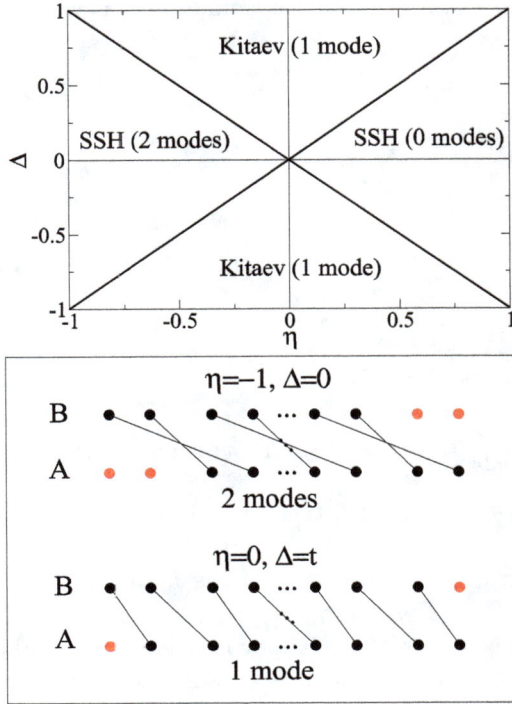

Fig. 4.5 Phases of the SSH-Kitaev model. The SSH phases (two Majoranas on each end that combine to form usual fermionic modes) and SSH with no edge modes (trivial phase) are of the type of the SSH model. The Kitaev phases (with one Majorana at each end) are of the type of the Kitaev model.

We may show that

$$i\gamma_{j,B,2}\gamma_{j+1,A,1} = 2d_j^\dagger d_j - 1\,,$$
$$-i\gamma_{j,B,1}\gamma_{j+1,A,2} = 2f_j^\dagger f_j - 1\,. \tag{4.100}$$

In terms of these operators, we may write

$$H = t\sum_{j=1}^{N-1}\left(2d_j^\dagger d_j - 1 + 2f_j^\dagger f_j - 1\right) \tag{4.101}$$

and, therefore, the problem is diagonalized. It is clear, then, that the groundstate is obtained taking at each site $d_j^\dagger d_j = 0$ and $f_j^\dagger f_j = 0$. This new Hamiltonian in terms of the operators d and f is like an Hamiltonian with no hopping amplitudes and has only a chemical potential term with $\tilde{\mu} = -2t$.

Note that the new operators may be written in terms of the original ones, using a non-local transformation as

$$d_j = \frac{i}{2} \left(c_{j,B}^\dagger - c_{j,B} + c_{j+1,A} + c_{j+1,A}^\dagger \right) ,$$

$$f_j = \frac{1}{2} \left(c_{j,B}^\dagger + c_{j,B} - c_{j+1,A} + c_{j+1,A}^\dagger \right) . \tag{4.102}$$

Also

$$c_{j,A} = \frac{1}{2} \left(-i(-d_{j-1}^\dagger + d_{j-1}) - (f_{j-1} - f_{j-1}^\dagger) \right) ,$$

$$c_{j,B} = \frac{1}{2} \left(f_j^\dagger + f_j + i(d_j + d_j^\dagger) \right) . \tag{4.103}$$

This model allows us to compare the usual fermionic modes with the Majorana modes. Since in this model the spin is not taken into account, the time reversal operator is simply the complex conjugation operation, $\mathcal{T} = K$. The model has time reversal symmetry $\mathcal{T}H(k)\mathcal{T}^{-1} = H(-k)$. In the case of $\mu = 0$, the system has sublattice symmetry. The operator that describes this symmetry is given by

$$C_1 = \tau_z = \begin{pmatrix} 1 & 0 & 0 & 0 \\ 0 & -1 & 0 & 0 \\ 0 & 0 & 1 & 0 \\ 0 & 0 & 0 & -1 \end{pmatrix} , \tag{4.104}$$

where τ_i is a Pauli matrix acting on the sublattice degrees of freedom. It may be checked that $C_1 H(k) C_1^{-1} = -H(k)$. The topological phase is class BDI since $\mathcal{T}^2 = 1$ and $C_1^2 = 1$.

If the chemical potential does not vanish, the sublattice symmetry is absent. However, the model has the particle-hole symmetry whose operator is given by $\mathcal{C} = t_x K$, where t_i is a Pauli matrix acting in the particle-hole space. It may be verified that $\mathcal{C}H(k)\mathcal{C}^{-1} = -H(-k)$. We may, therefore, define a chiral operator by the product of $\mathcal{T}$ and $\mathcal{C}$ in the form

$$C_2 = \mathcal{T}\mathcal{C} = t_x = \begin{pmatrix} 0 & 0 & 1 & 0 \\ 0 & 0 & 0 & 1 \\ 1 & 0 & 0 & 0 \\ 0 & 1 & 0 & 0 \end{pmatrix} . \tag{4.105}$$

It may be checked that $C_2 H(k) C_2^{-1} = -H(k)$ and $C_2^2 = 1$ and, therefore, in this case the system is also in class BDI. So, in both cases the topological invariant is an integer.

We can define two topological invariants associated with the two symmetries

$$N_1 = \text{Tr} \int_{-\pi}^{\pi} \frac{dk}{4\pi i} C_1 H(k) \partial_k H(k)^{-1} \, ,$$

$$N_2 = \text{Tr} \int_{-\pi}^{\pi} \frac{dk}{4\pi i} C_2 H(k) \partial_k H(k)^{-1} \, . \qquad (4.106)$$

Transforming the Hamiltonian into an antidiagonal form, defining a transformation

$$U_1 = \begin{pmatrix} 1 & 0 & 0 & 0 \\ 0 & 0 & 1 & 0 \\ 0 & 1 & 0 & 0 \\ 0 & 0 & 0 & 1 \end{pmatrix} , \qquad (4.107)$$

that leads to

$$U_1 C_1 U_1^\dagger = \tau_z, \quad U_1 H(k) U_1^\dagger = \begin{pmatrix} 0 & V_1 \\ V_1^\dagger & 0 \end{pmatrix} , \qquad (4.108)$$

with

$$V_1 = \begin{pmatrix} z & w \\ -w & -z \end{pmatrix} , \qquad (4.109)$$

from which we obtain

$$N_1 = -\text{Tr} \int_{-\pi}^{\pi} \frac{dk}{2\pi i} V_1^{-1} \partial_k V_1 = - \sum_{n=1,2} \int_{\pi}^{\pi} \frac{dk}{2\pi i} \partial_k \ln z_n(k) \, , \qquad (4.110)$$

where

$$z_1(Fk) = (t - \Delta)(1 + \eta) + (t + \Delta)(1 - \eta)e^{-ik} \, ,$$

$$z_2(k) = (t + \Delta)(1 + \eta) + (t - \Delta)(1 - \eta)e^{-ik} \, . \qquad (4.111)$$

It may be, then, shown that

$$N_1 = \Theta(\Delta - t\eta) + \Theta(-\Delta - t\eta) \, , \qquad (4.112)$$

where Θ is the Heaviside function. This expression counts the number of edge states: in the trivial phase, $N_1 = 0$; in the Kitaev phases, $N_1 = 1$; and in the SSH phase with two modes at each end of the chain, $N_1 = 2$.

In the case of the calculation of N_2, we consider a unitary transformation

$$U_2 = \frac{1}{\sqrt{2}} \begin{pmatrix} 1 & 0 & 1 & 0 \\ 0 & 1 & 0 & 1 \\ -i & 0 & i & 0 \\ 0 & -i & 0 & i \end{pmatrix} , \qquad (4.113)$$

that leads to

$$U_2 C_2 U_2^\dagger = \tau_z, \qquad U_2 H(k) U_2^\dagger = \begin{pmatrix} 0 & V_2 \\ V_2^\dagger & 0 \end{pmatrix}, \tag{4.114}$$

with

$$V_2 = \begin{pmatrix} -i\mu & i(z-w) \\ i(z^*+w^*) & -i\mu \end{pmatrix}, \tag{4.115}$$

from which we obtain

$$N_2 = -\text{Tr} \int_{-\pi}^{\pi} \frac{dk}{2\pi i} V_2^{-1} \partial_k V_2 = -\int_{\pi}^{\pi} \frac{dk}{2\pi i} \partial_k \ln Z(k), \tag{4.116}$$

where

$$Z(k) = Det V_2(k). \tag{4.117}$$

We obtain that in the SSH phases $N_2 = 0$ and in the Kitaev phases $N_2 = \pm 1$. Therefore, the invariant N_2 counts the number of Majoranas, while the invariant N_1 counts the number of zero energy modes, at each end. Combining the information of the two invariants, confirms the representation in terms of the Majorana operators with open boundary conditions and that these are coupled to form usual fermions in the SSH phases.

4.3 Bound states in Josephson junctions

We now present a Josephson junction model which serves as an example for the solution of BdG equations in a way that is analogous to a scattering problem in one dimension. It allows us to obtain expressions for Majorana fermions explicitly.

4.3.1 *Majorana states in a π junction*

Let us consider a one-dimensional superconductor with singlet pairing. The $\Delta(x)$ matrix takes the anti-symmetric form (4.25) and changes sign at $x = 0$. The BdG equations for a $E = 0$ state can be written as:

$$\begin{pmatrix} -\frac{\hbar^2}{2m}\partial^2 - \mu & \Delta(x) \\ \Delta(x) & \frac{\hbar^2}{2m}\partial^2 + \mu \end{pmatrix} \begin{pmatrix} u_\uparrow(x) \\ v_\downarrow(x) \end{pmatrix} = 0, \tag{4.118}$$

$$\begin{pmatrix} -\frac{\hbar^2}{2m}\partial^2 - \mu & -\Delta(x) \\ -\Delta(x) & \frac{\hbar^2}{2m}\partial^2 + \mu \end{pmatrix} \begin{pmatrix} u_\downarrow(x) \\ v_\uparrow(x) \end{pmatrix} = 0. \tag{4.119}$$

The solutions have, as spatial part, $e^{i\lambda x}$, and for $x > 0$ we impose $\text{Im}\lambda > 0$; for $x < 0$ we impose $\text{Im}\lambda < 0$. We always obtain the condition

$$\left(\frac{\hbar^2\lambda^2}{2m} - \mu\right)^2 + \Delta^2 = 0 \ \Rightarrow\ \lambda = \eta_1 k_F\sqrt{1 + \eta_2 i\frac{\Delta}{\mu}} \approx \eta_1 k_F\left(1 + \eta_2 i\frac{\Delta}{2\mu}\right),$$

where $\eta_{1,2}$ denote independent $\pm$ signs. The four independent solutions read:

$$\psi_1 : \begin{pmatrix} u_\uparrow(x) \\ v_\downarrow(x) \end{pmatrix} = \begin{pmatrix} 1 \\ i \end{pmatrix} e^{ik_F x - (\Delta/2\mu)k_F|x|}, \tag{4.120}$$

$$\psi_2 : \begin{pmatrix} u_\downarrow(x) \\ v_\uparrow(x) \end{pmatrix} = \begin{pmatrix} i \\ 1 \end{pmatrix} e^{ik_F x - (\Delta/2\mu)k_F|x|}, \tag{4.121}$$

$$t_x\psi_1^* : \begin{pmatrix} u_\downarrow(x) \\ v_\uparrow(x) \end{pmatrix} = \begin{pmatrix} -i \\ 1 \end{pmatrix} e^{-ik_F x - (\Delta/2\mu)k_F|x|}, \tag{4.122}$$

$$t_x\psi_2^* : \begin{pmatrix} u_\uparrow(x) \\ v_\downarrow(x) \end{pmatrix} = \begin{pmatrix} 1 \\ -i \end{pmatrix} e^{-ik_F x - (\Delta/2\mu)k_F|x|}. \tag{4.123}$$

Thus, we can explicitly write the solutions in the form of Majorana fermions,

$$\gamma_1 = \psi_1 + t_x\psi_1^* : \left[\begin{pmatrix} 1 \\ 0 \\ 0 \\ i \end{pmatrix} e^{ik_F x} + \begin{pmatrix} 0 \\ -i \\ 1 \\ 0 \end{pmatrix} e^{-ik_F x}\right] e^{-(\Delta/2\mu)k_F|x|}, \tag{4.124}$$

$$\gamma_2 = \psi_2 + t_x\psi_2^* : \left[\begin{pmatrix} 0 \\ i \\ 1 \\ 0 \end{pmatrix} e^{ik_F x} + \begin{pmatrix} 1 \\ 0 \\ 0 \\ -i \end{pmatrix} e^{-ik_F x}\right] e^{-(\Delta/2\mu)k_F|x|} = \gamma_1^*. \tag{4.125}$$

What is the role of time reversal symmetry? If we put $\mathcal{T}\psi = i1\otimes\sigma_y\psi^*$, we obtain a state orthogonal to ψ, according to Kramers theorem. The Majorana states $\mathcal{T}\gamma_1$ and $\mathcal{T}\gamma_2$ are, then, Kramers partners of γ_1 and γ_2. A time-reversal invariant perturbation cannot couple[5] ψ with $\mathcal{T}\psi$.

Examining the terms proportional to $e^{ik_F x}$, for instance, we see that they are eigenstates of spin σ_z (one might as well have chosen another spin orientation by making a different linear combination of ψ_1 and ψ_2). If the

---

[5]See section 3.6.

spin along z is conserved, then no perturbation can couple the Majorana states γ_1 and γ_2. Such a conservation law may stem from a reflection symmetry implemented by the operator $\mathcal{M}_z = i\sigma_z$ (and the inversion of the coordinate, $z \rightarrow -z$, although the latter does not appear explicitly). Indeed, one can explicitly calculate the Hamiltonian term that would couple the Majorana fermions. It would have to be proportional to $i\gamma_1\gamma_2$ at $x = 0$ (we here assume the localization length to be small). We obtain

$$i\gamma_1\gamma_2 \propto - \left[\hat{\psi}_\uparrow(0)\hat{\psi}_\downarrow(0) + \hat{\psi}_\downarrow^\dagger(0)\hat{\psi}_\uparrow^\dagger(0)\right] - \hat{\psi}_\alpha^\dagger(0)\sigma_{\alpha\beta}^x\hat{\psi}_\beta(0). \quad (4.126)$$

The second term is a transverse Zeeman field and breaks reflection symmetry, as mentioned above. The first term implies a local singlet coupling and it means that the junction would have a phase $\phi \neq \pi$. This is the case we shall study next.

4.3.2 *Andreev bound states in a ϕ junction*

Consider a one-dimensional system where $\Delta(x) = \Delta e^{-i\phi/2}$ for $x < 0$, and $\Delta(x) = \Delta e^{i\phi/2}$ for $x > 0$. In the Nambu basis, the Hamiltonian reads:

$$\hat{H} = \left(-\frac{\hbar^2}{2m}\partial_x^2 - \mu\right)\tau_z + \Delta\cos\frac{\phi}{2}\tau_x - \text{sgn}(x)\Delta\sin\frac{\phi}{2}\tau_y \quad (4.127)$$

It then decouples into two 2×2 equal matrices, one in the basis $(\hat{c}_\uparrow, \mathcal{T}\hat{c}_\uparrow = \hat{c}_\downarrow^\dagger)$, and the other in the basis $(\hat{c}_\downarrow, \mathcal{T}\hat{c}_\downarrow = -\hat{c}_\uparrow^\dagger)$. There is a spin degeneracy. Diagonalizing this matrix we have,

$$\begin{pmatrix} -\frac{\hbar^2}{2m}\partial_x^2 - \mu & \Delta(x) \\ \Delta^*(x) & \frac{\hbar^2}{2m}\partial_x^2 + \mu \end{pmatrix}\begin{pmatrix} u \\ v \end{pmatrix}e^{i\lambda x} = E\begin{pmatrix} u \\ v \end{pmatrix}e^{i\lambda x}, \quad (4.128)$$

where the column vector could either be $(u\ v) = (u_\uparrow\ v_\downarrow)$, or $(u\ v) = (u_\downarrow\ -v_\uparrow)$. For excitation energies smaller than the energy gap,

$$\frac{\hbar^2}{2m}\lambda^2 - \mu = \pm i\sqrt{\Delta^2 - E^2}. \quad (4.129)$$

The solutions for the positive semi-axis, $x > 0$, require $\text{Im}\lambda > 0$. So,

$$\lambda \approx \pm k_F + i\frac{\sqrt{\Delta^2 - E^2}}{\hbar v_F}, \quad (4.130)$$

and the corresponding solutions are:

$$\begin{pmatrix} u \\ v \end{pmatrix}e^{ik_F x - \rho x} \quad \text{where} \quad u = \frac{\Delta e^{i\phi/2}}{E - i\sqrt{\Delta^2 - E^2}}v, \quad (4.131)$$

$$\begin{pmatrix} u \\ v \end{pmatrix}e^{-ik_F x - \rho x} \quad \text{where} \quad u = \frac{\Delta e^{i\phi/2}}{E + i\sqrt{\Delta^2 - E^2}}v, \quad (4.132)$$

which imply $|u| = |v| = 1/\sqrt{2}$.

We consider now the negative semi-axis, $x < 0$. We must have $\text{Im}\lambda < 0$, so,

$$\lambda \approx \pm k_F - i\frac{\sqrt{\Delta^2 - E^2}}{\hbar v_F}, \qquad (4.133)$$

and the corresponding solutions are:

$$\binom{u}{v} e^{ik_F x + \rho x} \quad \text{where} \quad u = \frac{\Delta e^{-i\phi/2}}{E + i\sqrt{\Delta^2 - E^2}} v, \qquad (4.134)$$

$$\binom{u}{v} e^{-ik_F x + \rho x} \quad \text{where} \quad u = \frac{\Delta e^{-i\phi/2}}{E - i\sqrt{\Delta^2 - E^2}} v. \qquad (4.135)$$

For $0 < \phi < 2\pi$ we always have $\sin(\phi/2) > 0$. In order to write wave functions for the whole real axis, it is necessary to satisfy both (4.131) and (4.134), as well as (4.132) and (4.135):

$$\sqrt{\frac{\rho}{2}} \binom{1}{1} e^{-ik_F x - \rho|x|} \qquad \text{with} \qquad E = \Delta\cos(\phi/2), \qquad (4.136)$$

$$\sqrt{\frac{\rho}{2}} \binom{-1}{1} e^{ik_F x - \rho|x|} \qquad \text{with} \qquad E = -\Delta\cos(\phi/2). \qquad (4.137)$$

Here, we have introduced a normalization factor for the wave function. As mentioned above, each of these solutions is spin degenerate, because the column vector can either represent $(u_\uparrow \ v_\downarrow)$ or $(u_\downarrow \ -v_\uparrow)$.

The $\phi \neq \pi$ junction breaks time reversal symmetry because $i\sigma_y H^*(-i\sigma_y) = H(\Delta \to \Delta^*) \neq H$. By time-reversing the state $(u_\uparrow \ v_\downarrow) = (1\ 1)e^{-ik_F x - \rho|x|}$, with energy $E = \Delta\cos(\phi/2)$, we obtain the state $(u_\downarrow \ v_\uparrow) = (-1\ 1)e^{ik_F x - \rho|x|}$, with symmetrical energy. For the π junction, Δ could be chosen to be real and the two states above would then be related by time reversal.

How do these states contribute to the supercurrent? The charge current in a many-body system is given by

$$\hat{J}_Q(x) = -\frac{\delta\mathcal{H}}{\delta\boldsymbol{A}(x)},$$

where we next take $\boldsymbol{A} \to 0$. The Hamiltonian is given by (4.2) or (4.9):

$$\mathcal{H} = \frac{1}{2} \int d\boldsymbol{r}\,\hat{\psi}^\dagger(\boldsymbol{r}) H \hat{\psi}(\boldsymbol{r}). \qquad (4.138)$$

But, in the meantime,

$$\hat{\Xi} = \frac{(-i\hbar\nabla - e\boldsymbol{A})^2}{2m} \to \frac{\partial\hat{\Xi}}{\partial\boldsymbol{A}} = -\frac{e}{m}(-i\hbar\nabla - e\boldsymbol{A}),$$

$$\hat{\Xi}^T = \frac{(i\hbar\nabla - e\boldsymbol{A})^2}{2m} \to \frac{\partial\hat{\Xi}^T}{\partial\boldsymbol{A}} = -\frac{e}{m}(i\hbar\nabla - e\boldsymbol{A}).$$

So, for the one-dimensional case, we put $\boldsymbol{A} = 0$, and obtain

$$\hat{J}_Q(x) = \frac{e}{2m}\hat{\psi}^\dagger(x)\begin{pmatrix} -i\hbar\partial_x & 0 \\ 0 & -i\hbar\partial_x \end{pmatrix}\hat{\psi}(x). \tag{4.139}$$

The expectation value of the supercurrent is given by

$$J_Q = \mathrm{Re}\langle\hat{J}_Q(x)\rangle = \mathrm{Re}\frac{-ie\hbar}{2m}\sum_n \left(u_{\uparrow n}^*(x)...v_{\downarrow n}^*(x)\right)\partial_x\begin{pmatrix} u_{\uparrow n}(x) \\ ... \\ v_{\downarrow n}(x) \end{pmatrix}\langle\hat{\gamma}_n^\dagger\hat{\gamma}_n\rangle, \tag{4.140}$$

where Re denotes the real part. The sum over all eigenstates, n, of the matrix H, with positive and negative energies is weighted by the average thermal occupation. The contribution of each state to the supercurrent can be obtained from the variation of J_Q in equation (4.140) with respect to $\langle\hat{\gamma}_n^\dagger\hat{\gamma}_n\rangle$, which means calculating $\delta J_Q/\delta\langle\hat{\gamma}_n^\dagger\hat{\gamma}_n\rangle$. In particular, the contributions of the states (4.136)-(4.137) to the supercurrents are given by

$$j_Q = \frac{e}{2\hbar}\Delta\sin\frac{\phi}{2}e^{-2\rho|x|} \quad \text{in state with} \quad E = -\Delta\cos\frac{\phi}{2}, \tag{4.141}$$

$$j_Q = \frac{-e}{2\hbar}\Delta\sin\frac{\phi}{2}e^{-2\rho|x|} \quad \text{in state with} \quad E = \Delta\cos\frac{\phi}{2}. \tag{4.142}$$

We finish this section by stating the following result: a thermodynamic Josephson relation involving the free energy, F, holds as

$$j_Q = \frac{2e}{\hbar}\frac{\partial F}{\partial\phi} \quad \text{at the junction } x = 0. \tag{4.143}$$

4.4 Two-dimensional superconductors

4.4.1 *The spinless p+ip superconductor*

The spinless superconductor with pairing symmetry p_x+ip_y is often used as a *case-study* for topological Majorana fermions. The order parameter is a complex number, $\Delta(\mathbf{r}) = \Delta e^{i\phi(\mathbf{r})}$. Assuming that there is a single parabolic band, the continuum version of the Hamiltonian can be written as:

$$\hat{H} = \int d\mathbf{r}\,\left[\hat{\psi}^\dagger(\mathbf{r})\left(-\frac{\hbar^2\nabla^2}{2m} - \mu\right)\hat{\psi}(\mathbf{r}) + \frac{\Delta e^{-i\phi}}{2}\hat{\psi}(\mathbf{r})\left(\partial_x + i\partial_y\right)\hat{\psi}(\mathbf{r})\right.$$

$$+\frac{\Delta e^{i\phi}}{2}\hat{\psi}^\dagger(\mathbf{r})\left(-\partial_x + i\partial_y\right)\hat{\psi}^\dagger(\mathbf{r})\bigg] . \tag{4.144}$$

The field operator is made up of all excitations γ, as:

$$\hat{\psi}(\mathbf{r}) = \sum_n \left[\, u_n(\mathbf{r})\hat{\gamma}_n + v_n^*(\mathbf{r})\hat{\gamma}_n^\dagger \,\right],$$

where the amplitudes u, v obey BdG equations[6]:

$$E\begin{pmatrix} u(\mathbf{r}) \\ v(\mathbf{r}) \end{pmatrix} = \left[\begin{matrix} -\frac{\hbar^2\nabla^2}{2m} - \mu & \Delta e^{i\phi}\left(-\partial_x + i\partial_y\right) \\ \Delta e^{-i\phi}\left(\partial_x + i\partial_y\right) & \frac{\hbar^2\nabla^2}{2m} + \mu \end{matrix}\right]\begin{pmatrix} u(\mathbf{r}) \\ v(\mathbf{r}) \end{pmatrix}. \tag{4.145}$$

One can check that

$$\hat{\gamma} = \int d\mathbf{r} \,\left[\, u^*(\mathbf{r})\hat{\psi}(\mathbf{r}) + v^*(\mathbf{r})\hat{\psi}^\dagger(\mathbf{r}) \,\right], \tag{4.146}$$

where $\left[\hat{\gamma}, \hat{H}\right] = E\hat{\gamma}$. Indeed, we can explicitly calculate the commutators:

$$\left[\int u^*\hat{\psi}, \int \hat{\psi}^\dagger\nabla^2\hat{\psi}\right] = \int \hat{\psi}\nabla^2 u^*,$$

$$\left[\int v^*\hat{\psi}^\dagger, \int \hat{\psi}^\dagger\nabla^2\hat{\psi}\right] = -\int \hat{\psi}\nabla^2 v^*,$$

where we have integrated by parts (ignoring surface terms), so as to make the derivatives act on the functions u and v. We also have:

$$\left[\int u^*\hat{\psi}, \int \hat{\psi}^\dagger\left(-\partial_x + i\partial_y\right)\hat{\psi}^\dagger\right] = 2\int \hat{\psi}^\dagger\left(\partial_x - i\partial_y\right)u^*,$$

$$\left[\int v^*\hat{\psi}^\dagger, \int \hat{\psi}\left(\partial_x + i\partial_y\right)\hat{\psi}\right] = 2\int \hat{\psi}\left(-\partial_x - i\partial_y\right)v^*. \tag{4.147}$$

Using these commutators and the condition $\left[\hat{\gamma}, \hat{H}\right] = E\hat{\gamma}$, the BdG equations (4.145) are recovered.

In the bulk of the superconductor, the BdG equations take the form:

$$\left[\begin{matrix} \frac{\hbar^2\mathbf{k}^2}{2m} - \mu & -i\frac{\Delta}{k_F}\left(k_x - ik_y\right) \\ i\frac{\Delta}{k_F}\left(k_x + ik_y\right) & -\frac{\hbar^2\mathbf{k}^2}{2m} + \mu \end{matrix}\right]\begin{pmatrix} u \\ v \end{pmatrix} = E\begin{pmatrix} u \\ v \end{pmatrix}, \tag{4.148}$$

and the eigenenergies obey

$$E^2 = \left(\frac{\hbar^2\mathbf{k}^2}{2m} - \mu\right)^2 + \left(\frac{\Delta}{k_F}\right)^2\mathbf{k}^2. \tag{4.149}$$

---

[6]The superconducting term should contain the covariant derivative $\tilde{\partial} \equiv \partial + \frac{i}{2}\left(\partial\phi\right)$, where ϕ is the phase of the condensate. But in this case we simply have $\tilde{\partial} = \partial$ because $\hat{\psi}\hat{\psi} = 0$. Even if Δ has spatial dependence, the derivatives of Δ do not contribute to the commutators (4.147) for the same reason.

The spectrum is gapped for all $\mathbf{k}$.

We now show that the semi-infinite system, $y < 0$, has edge states. In particular, there is a solution with $E = 0$ and $k_x = 0$. We make the replacement $k_y \to -i\partial_y$ in matrix (4.148). The wave functions are proportional to $e^{\rho y}$ and

$$\left[\left(-\frac{\hbar \rho^2}{2m} - \mu \right) \sigma_z + i \frac{\Delta \rho}{k_F} \sigma_x \right] \begin{pmatrix} u \\ v \end{pmatrix} = 0 \,, \tag{4.150}$$

which, after multiplied by σ_z, can be written as:

$$\frac{\Delta \rho}{k_F} \sigma_y \begin{pmatrix} u \\ v \end{pmatrix} = - \left(\frac{\hbar \rho^2}{2m} + \mu \right) \begin{pmatrix} u \\ v \end{pmatrix} \,. \tag{4.151}$$

The eigenvectors of σ_y are ϕ_+ and ϕ_- because $\sigma_y \phi_\pm = \pm \phi_\pm$. Then, the solution can generally be written as:

$$\begin{pmatrix} u(\mathbf{r}) \\ v(\mathbf{r}) \end{pmatrix} = A e^{\rho_+ y} \phi_+ + B e^{\rho_- y} \phi_- \,.$$

The value of ρ_+ is obtained from the equation

$$\frac{\hbar \rho_+^2}{2m} + \frac{\Delta \rho_+}{k_F} + \mu = 0 \Rightarrow \rho_+ \approx -\frac{\Delta}{2\mu} k_F \pm i k_F \,,$$

and the value of ρ_- from the equation

$$\frac{\hbar \rho_-^2}{2m} - \frac{\Delta \rho_-}{k_F} + \mu = 0 \Rightarrow \rho_- \approx \frac{\Delta}{2\mu} k_F \pm i k_F \,.$$

The above approximations assume that $\Delta \ll |\mu|$. Because we need $\text{Re}[\rho] > 0$, only the solutions ρ_- are acceptable. We need a linear combination of the solutions that vanishes at the boundary $(y = 0)$:

$$\begin{pmatrix} u(\mathbf{r}) \\ v(\mathbf{r}) \end{pmatrix} = A \sin(k_F y) e^{\frac{\Delta}{2\mu} k_F y} \phi_- \,, \qquad y \le 0 \,. \tag{4.152}$$

Looking back at the operator $\hat{\psi}$ above, we see that part of it has a zero-energy excitation term,

$$\hat{\psi} \to A \sin(k_F y) e^{\frac{\Delta}{2\mu} k_F y} \left(i \hat{\gamma}_0 + \hat{\gamma}_0^\dagger \right) \,, \tag{4.153}$$

and is equal to its own adjoint (times a phase): it is a Majorana fermion. Indeed, looking at (4.146), we see that $\hat{\gamma}_0^\dagger = i \hat{\gamma}_0$. This means that the operator $e^{i\pi/4} \hat{\gamma}_0$ is rigorously of the Majorana type.

We should now try to understand why this is not possible in a conventional s-wave superconductor. If we look for a state with $E = k_x = 0$, the BdG equations read:

$$\begin{bmatrix} \frac{\hbar k_y^2}{2m} - \mu & \Delta \\ \Delta & -\frac{\hbar k_y^2}{2m} + \mu \end{bmatrix} \begin{pmatrix} u \\ v \end{pmatrix} = 0 \,,$$

or,

$$\left[\left(\frac{\hbar k_y^2}{2m} - \mu \right) \sigma_z + \Delta \sigma_x \right] \begin{pmatrix} u \\ v \end{pmatrix} = 0$$

$$\Leftrightarrow \left[\frac{\hbar k_y^2}{2m} - \mu + i\Delta\sigma_y \right] \begin{pmatrix} u \\ v \end{pmatrix} = 0 \,. \tag{4.154}$$

We look for a solution of the form:

$$\begin{pmatrix} u \\ v \end{pmatrix} = A\phi_+ + B\phi_- \,.$$

It is then easily seen that $\Delta \neq 0$ implies either $A = 0$ or $B = 0$. If, for instance, $B = 0$ and $A \neq 0$, then

$$\frac{\hbar k_y^2}{2m} - \mu + i\Delta = 0 \Rightarrow k_y \approx \pm k_F \left(1 - i\frac{\Delta}{2\mu} \right) \,.$$

It is however necessary that $\text{Im}[k_y] < 0$ in order for the wave function to decay inside the semi-infinite superconductor ($y < 0$). But then it is not possible to make the wave function vanish at the boundary $y = 0$ because the real part of k_y is also fixed.

If there is a vortex in the $p + ip$ superconductor, then a Majorana fermion appears at the vortex core. It is, in fact, an edge state between the superconducting and the normal phases at the vortex center. If there are several such vortices, the Majorana fermions obey non-abelian statistics. This is of high interest for possible applications in quantum computation. This subject will be discussed in section 8.4.

4.4.2   *Dirac cone with s-wave superconductivity*

The spinfull Dirac cone with singlet s-wave pairing is equivalent to a spinless superconductor with $p + ip$ pairing. We now prove this result. Consider the kinetic energy for the Dirac cone, $\Xi(\mathbf{k}) = \hbar v\hat{z} \cdot \boldsymbol{\sigma} \times \mathbf{k} - \mu$. This is the first block along the diagonal of the BdG matrix in the particle-hole basis,

$$\Xi(\mathbf{k}) = \begin{pmatrix} -\mu & \hbar v(k_y + ik_x) \\ \hbar v(k_y - ik_x) & -\mu \end{pmatrix} \,, \tag{4.155}$$

where v is a velocity. We write $k_y + ik_x = ke^{i\phi}$. The unitary matrix, U, that diagonalizes this kinetic term is given by

$$U(\mathbf{k}) = \frac{1}{\sqrt{2}} \begin{pmatrix} 1 & 1 \\ -e^{-i\phi} & e^{-i\phi} \end{pmatrix}. \qquad (4.156)$$

Note that the change $\mathbf{k} \to -\mathbf{k}$ implies $\phi \to \phi + \pi$. The blocks on the diagonal of the BdG matrix, in the particle-hole basis, are now diagonalized:

$$U^\dagger(\mathbf{k})\Xi(\mathbf{k})U(\mathbf{k}) = \begin{pmatrix} -\mu - \hbar vk & 0 \\ 0 & -\mu + \hbar vk \end{pmatrix} \qquad (4.157)$$

$$= U^T(-\mathbf{k})\Xi^T(-\mathbf{k})U^*(-\mathbf{k}). \qquad (4.158)$$

The spectrum has two bands with energies $E_\pm = -\mu \pm \hbar vk$. We write the pairing block for s-wave singlet pairing as [see (4.25)]:

$$\hat{\Delta} = \begin{pmatrix} 0 & \Delta \\ -\Delta & 0 \end{pmatrix}. \qquad (4.159)$$

We now write the pairing block in the basis of the operators that diagonalize the kinetic energy. A unitary transformation of the BdG matrix,

$$\begin{pmatrix} U^\dagger(\mathbf{k}) & 0 \\ 0 & U^T(-\mathbf{k}) \end{pmatrix} \begin{pmatrix} \Xi(\mathbf{k}) & \hat{\Delta} \\ -\hat{\Delta} & -\Xi^T(-\mathbf{k}) \end{pmatrix} \begin{pmatrix} U(\mathbf{k}) & 0 \\ 0 & U^*(-\mathbf{k}) \end{pmatrix}, \qquad (4.160)$$

yields the following transformation for the pairing block:

$$\hat{\Delta} \to U^\dagger(\mathbf{k})\hat{\Delta}U^*(-\mathbf{k}) = \Delta e^{i\phi} \begin{pmatrix} 1 & 0 \\ 0 & -1 \end{pmatrix}. \qquad (4.161)$$

Pairing does not couple different bands. It is intraband. The BdG matrix (4.160) decouples into two 2×2 matrices in band space. For example, the matrix for the lower band reads

$$\begin{pmatrix} -\mu - \hbar vk & \Delta e^{i\phi} \\ \Delta e^{-i\phi} & \mu + \hbar vk \end{pmatrix}, \qquad (4.162)$$

and is of the same form as that in equation (4.148).

The Dirac cone may describe surface states of a strong topological insulator. The pairing term can be induced by the proximity effect to a conventional superconductor which is deposited on the surface of the insulator.

4.4.3   *The $\mathbb{Z}_2$ superconductor*

We now include the spin degree of freedom. For triplet pairing, $d(\mathbf{k}) = (-\sin k_y, \sin k_x, 0)$, we obtain the Hamiltonian in real space:

$$\hat{H} = \int d\mathbf{r} \left[\sum_\sigma \psi_\sigma^\dagger(\mathbf{r}) \left(-\frac{\nabla^2}{2m} - \mu \right) \psi_\sigma(\mathbf{r}) \right.$$

$$+ \frac{\Delta}{2} \psi_\uparrow^\dagger(\mathbf{r}) \left[\partial_x - i\partial_y \right] \psi_\uparrow^\dagger(\mathbf{r}) + \frac{\Delta}{2} \psi_\downarrow^\dagger(\mathbf{r}) \left[\partial_x + i\partial_y \right] \psi_\downarrow^\dagger(\mathbf{r})$$

$$\left. - \frac{\Delta}{2} \psi_\uparrow(\mathbf{r}) \left[\partial_x + i\partial_y \right] \psi_\uparrow(\mathbf{r}) - \frac{\Delta}{2} \psi_\downarrow(\mathbf{r}) \left[\partial_x - i\partial_y \right] \psi_\downarrow(\mathbf{r}) \right]. \quad (4.163)$$

The BdG Hamiltonian matrix acts on the column vector $\psi(\mathbf{r}) = (u_\uparrow(\mathbf{r})\ u_\downarrow(\mathbf{r})\ v_\uparrow(\mathbf{r})\ v_\downarrow(\mathbf{r}))^T$. In order to consider a more general case, we include the local singlet pairing, Δ_s. Then, the Hamiltonian reads:

$$\begin{bmatrix} -\frac{\nabla^2}{2m} - \mu & 0 & \Delta(\partial_x - i\partial_y) & \Delta_s \\ 0 & -\frac{\nabla^2}{2m} - \mu & -\Delta_s & \Delta(\partial_x + i\partial_y) \\ \Delta(-\partial_x - i\partial_y) & -\Delta_s & \frac{\nabla^2}{2m} + \mu & 0 \\ \Delta_s & \Delta(-\partial_x + i\partial_y) & 0 & \frac{\nabla^2}{2m} + \mu \end{bmatrix} \psi(\mathbf{r}) = E\psi(\mathbf{r}).$$

$$(4.164)$$

This model enjoys particle-hole symmetry (indeed, $t_x \psi^*(\mathbf{r})$ has energy $-E$) and time reversal symmetry. Therefore, it belongs in the DIII symmetry class and has a $\mathbb{Z}_2$ topological index in one and two spatial dimensions. The BdG matrix decouples into two smaller 2×2 matrices. The latter are two copies of the $p + ip$ superconductor related to each other by time reversal. Indeed, the time reversed state, $\mathbf{1} \otimes i\sigma_y \psi^*(\mathbf{r})$, satisfies BdG equations. So, if there is a $E = 0$ Majorana fermion,

$$\hat{\gamma} = \hat{\gamma}^\dagger \Leftrightarrow \psi(\mathbf{r}) = t_x \psi^*(\mathbf{r}) = \left(u_\uparrow(\mathbf{r})\ u_\downarrow(\mathbf{r})\ u_\uparrow^*(\mathbf{r})\ u_\downarrow^*(\mathbf{r}) \right)^T,$$

then there is also its time reversed counterpart, which is orthogonal and reads

$$\left(u_\downarrow^*(\mathbf{r})\ -u_\uparrow^*(\mathbf{r})\ u_\downarrow(\mathbf{r})\ -u_\uparrow(\mathbf{r}) \right)^T.$$

4.4.4   *Inclusion of pseudo-spin*

Suppose that electrons with spin $\uparrow$ are paired with their time reversed counterparts with spin $\downarrow$. Then, the relation between their kinetic energies reads $\Xi_\downarrow(\mathbf{k}) = \Xi_\uparrow^*(-\mathbf{k})$. The BdG matrix in the particle-hole basis (4.9) takes the form:

$$H_{ph} = \begin{pmatrix} \Xi_\uparrow(\mathbf{k}) & 0 & & \hat{\Delta}(\mathbf{k}) \\ 0 & \Xi_\downarrow(\mathbf{k}) & & \\ \hat{\Delta}^\dagger(\mathbf{k}) & & -\Xi_\uparrow^*(-\mathbf{k}) & 0 \\ & & 0 & -\Xi_\downarrow^*(-\mathbf{k}) \end{pmatrix}, \quad (4.165)$$

where

$$\Xi_\uparrow(\mathbf{k}) = \mathbf{h}(\mathbf{k}) \cdot \boldsymbol{\tau} - \mu, \quad (4.166a)$$

$$\Xi_\downarrow(\mathbf{k}) = \Xi_\uparrow^*(-\mathbf{k}), \quad (4.166b)$$

and the Pauli matrices τ_i act on the pseudo-spin. The block $\hat{\Delta}(\mathbf{k})$ in the BdG matrix takes the form (4.25) or (4.26). If the pairing couples $u_\uparrow$ with $v_\downarrow$, the matrix H_{ph} will decouple into two different subspaces. We consider the submatrix that acts in the $(u_\uparrow v_\downarrow)$ subspace,

$$\begin{pmatrix} \mathbf{h}(\mathbf{k}) \cdot \boldsymbol{\tau} - \mu & \Delta(\mathbf{k})\tau_0 \\ \Delta^*(\mathbf{k})\tau_0 & -\mathbf{h}(\mathbf{k}) \cdot \boldsymbol{\tau} + \mu \end{pmatrix} \begin{pmatrix} (u_\uparrow) \\ (v_\downarrow) \end{pmatrix} = E(\mathbf{k}) \begin{pmatrix} (u_\uparrow) \\ (v_\downarrow) \end{pmatrix}. \quad (4.167)$$

Here, $(u_\uparrow)$ e $(v_\downarrow)$ are two-dimensional (the dimension of pseudo-spin or sublattice space). The function $\Delta(\mathbf{k})$ may be even, equal to $\psi(\mathbf{k})$ in the matrix (4.25), or it may be odd, equal to $d_z(\mathbf{k})$ in the matrix (4.26). The eigenvectors take the form:

$$\begin{pmatrix} (u_\uparrow) \\ (v_\downarrow) \end{pmatrix} = \begin{pmatrix} u \begin{pmatrix} \alpha \\ \beta \end{pmatrix} \\ v \begin{pmatrix} \alpha \\ \beta \end{pmatrix} \end{pmatrix}. \quad (4.168)$$

The wave function's normalization condition reads $|u|^2 + |v|^2 = |\alpha|^2 + |\beta|^2 = 1$, and the spectrum is obtained from:

$$[\mathbf{h}(\mathbf{k}) \cdot \boldsymbol{\tau} - \mu] \begin{pmatrix} \alpha \\ \beta \end{pmatrix} = \xi(\mathbf{k}) \begin{pmatrix} \alpha \\ \beta \end{pmatrix}, \quad (4.169)$$

$$\begin{cases} \xi(\mathbf{k})u + \Delta(\mathbf{k})v = E(\mathbf{k})u \\ \Delta^*(\mathbf{k})u - \xi(\mathbf{k})v = E(\mathbf{k})v \end{cases} \Leftrightarrow \mathbf{h}' \cdot \mathbf{r} \begin{pmatrix} u \\ v \end{pmatrix} = E(\mathbf{k}) \begin{pmatrix} u \\ v \end{pmatrix}. \quad (4.170)$$

We here introduced the vector $\mathbf{h}'$ such that $h'_x - ih'_y = \Delta(\mathbf{k})$ and $h'_z = \xi(\mathbf{k})$, and the Pauli matrices, $\mathbf{r}$, acting in $(u\ v)$ space.

There are two bands, $\xi(\mathbf{k}) = \pm|\mathbf{h}(\mathbf{k})| - \mu$ and the Chern number is obtained from the summation over the two negative energy BdG bands,

$$E_\pm(\mathbf{k}) = -\sqrt{\xi_\pm^2(\mathbf{k}) + |\Delta(\mathbf{k})|^2}, \quad (4.171)$$

of the system (4.170).

The Berry connection defined in (2.6) can be obtained from (4.168) as

$$\mathcal{A} = i\left((u_\uparrow)^\dagger \ (v_\downarrow)^\dagger\right) \frac{\partial}{\partial \mathbf{k}} \begin{pmatrix} (u_\uparrow) \\ (v_\downarrow) \end{pmatrix}$$

$$= i(u^* \ v^*)\frac{\partial}{\partial \mathbf{k}} \begin{pmatrix} u \\ v \end{pmatrix} + i(\alpha^* \ \beta^*)\frac{\partial}{\partial \mathbf{k}} \begin{pmatrix} \alpha \\ \beta \end{pmatrix} = \mathbf{a}' + \mathbf{a}, \qquad (4.172)$$

where $\mathbf{a}'$ and $\mathbf{a}$ denote the Berry connections corresponding to $(u \ v)$ and $(\alpha \ \beta)$, respectively. Using (2.6), the total Chern number, C, is obtained from the line integral around the Brillouin zone and summing over the two negative energy bands (4.171),

$$C = \sum \oint \mathcal{A} \cdot \delta \mathbf{k} = \sum \oint \mathbf{a}' \cdot \delta \mathbf{k} + \sum_{\pm} \oint \mathbf{a} \cdot \delta \mathbf{k}. \qquad (4.173)$$

Or, using monopole fluxes,[7],

$$C = \frac{1}{2\pi} \sum \int \mathbf{\Omega}'^- \cdot d\mathbf{S}' + \frac{1}{2\pi} \sum_{\pm} \int \mathbf{\Omega} \cdot d\mathbf{S}, \qquad (4.174)$$

where the last summation runs over the two kinetic energy bands of $\mathbf{h} \cdot \boldsymbol{\tau}$ and, therefore, adds up to zero. Hence we see that the topology of the kinetic term $\mathbf{h} \cdot \boldsymbol{\tau}$ does not contribute to C because the summation over all the kinetic energy bands cancels.

The Chern number then comes only from $\mathbf{a}'$, and is equal to the sum of the fluxes of the monopoles, $\mathbf{\Omega}'$, of the negative energy BdG bands. The vector we have to consider is, then, $\mathbf{h}' = \left(\text{Re}\Delta(\mathbf{k}), \text{Im}\Delta(\mathbf{k}), \xi_\pm\right)$. The Dirac points follow from the choice made for $\Delta(\mathbf{k})$. This leads us to the natural choice of $p + ip$ pairing, which must be triplet,

$$\Delta(\mathbf{k}) = d_z(\mathbf{k}).$$

The third component, $h'_z = \xi_-$, does not lead to topology because it is always negative: the vector $\mathbf{h}'$ lives on the southern hemisphere. However, the band ξ_+ changes sign across the Brillouin zone. So, $\mathbf{h}'$ is in the southern (northern) hemisphere when it is inside (outside) the Fermi surface.

4.4.5 *Examples of superconductors in a two-dimensional lattice*

One may use the vector $\mathbf{h}' = \left(\text{Re}\Delta(\mathbf{k}), \text{Im}\Delta(\mathbf{k}), \xi_\pm\right)$, defined in the previous section, to write down models for superconductors with a given Chern number.

---

[7]See equation (2.87).

A convenient choice to start with is a pairing term that allows the decoupling of the BdG matrix into two independent submatrices. This is possible if $\hat{\Delta}$ has either a singlet form, as in(4.25), or a triplet form with $\boldsymbol{d} = (0, 0, d_z)$ in equation (4.26). The simplest models do not need to include the pseudo-spin degree of freedom. The BdG matrix can be written, in the particle-hole basis, as

$$\begin{pmatrix} \xi_{\uparrow\uparrow}(\mathbf{k}) & 0 & 0 & \psi(\mathbf{k}) + d_z \\ 0 & \xi_{\downarrow\downarrow}(\mathbf{k}) & -\psi(\mathbf{k}) + d_z & 0 \\ 0 & -\psi^*(\mathbf{k}) + d_z^* & -\xi_{\uparrow\uparrow}(-\mathbf{k}) & 0 \\ \psi^*(\mathbf{k}) + d_z^* & 0 & 0 & -\xi_{\downarrow\downarrow}(-\mathbf{k}) \end{pmatrix} , \quad (4.175)$$

which obviously decouples into blocks of dimension 2×2. We can write these blocks in the form $h_0 r_0 + \boldsymbol{h}(\mathbf{k}) \cdot \boldsymbol{r}$.

In the following examples, we consider $\xi_{\downarrow\downarrow} = \xi_{\uparrow\uparrow} = \xi$.

Example: singlet pairing $s + id$:

$$\psi(\boldsymbol{k}) = \Delta_s(\cos k_x + \cos k_y) + i\Delta_d(\cos k_x - \cos k_y) ,$$
$$\xi(\boldsymbol{k}) = \delta + t(\sin k_x + \sin k_y) , \quad (4.176)$$

with $\delta \neq 0$, then $C = 1$.

Example: $s + id$ pairing:

$$\psi(\boldsymbol{k}) = \Delta_0 + \Delta_s(\cos k_x + \cos k_y) + i\Delta_d \sin k_x \sin k_y ,$$
$$\xi(\boldsymbol{k}) = -t(\cos k_x + \cos k_y) , \quad (4.177)$$

then $C = 2$.

Example: triplet $p + ip$ pairing:

$$\xi(\boldsymbol{k}) = -\cos(2k_x) - \cos(2k_y) - 0.3 \left[\cos(k_x) + \cos(k_y)\right] - \mu , \quad (4.178)$$
$$d_z(\boldsymbol{k}) = \Delta \left(\sin k_x - i \sin k_y\right) , \quad (4.179)$$

with $\Delta = 0.1$. If the chemical potential $\mu = -1$, the Fermi surface has four pockets around the points $(0, 0)$, $(0, \pi)$, $(\pi, 0)$, (π, π). Then $C = 0$. But if $\mu = -1.5$, the Fermi pocket around (π, π) disappears and $C = 1$.

We now study a model that includes pseudo-spin, as an application of the theorem presented in section 4.4.4.

Example: For the kinetic energy term we choose $\boldsymbol{h}$ given by

$$h_x = \sin k_y , \qquad h_y = -\sin k_x ,$$
$$h_z = 2t_1 \left(\cos k_x + \cos k_y\right) + 4t_2 \cos k_x \cos k_y ,$$
$$h_0 = -\mu - t_1 \left(\cos k_x + \cos k_y\right) . \tag{4.180}$$

It can easily be seen that h_x breaks time reversal symmetry. Space inversion symmetry, or parity, is also broken. If we use the matrix $P = \tau_x$ to implement parity, we see that the terms h_x and h_z break the symmetry. We may choose a pairing term with even or odd symmetry, according to the definition:

$$\mathcal{I} : \hat{\Delta}(\mathbf{k}) \rightarrow \tau_x \hat{\Delta}(-\mathbf{k})\tau_x = \pm\hat{\Delta}(\mathbf{k}) \begin{cases} \text{even} \\ \text{odd} \end{cases} . \tag{4.181}$$

An intraorbital triplet pairing with odd parity may be written as [see (4.26)]:

$$\hat{\Delta}(\mathbf{k}) = d_z(\mathbf{k})\sigma_x \otimes \tau_0 = d_z(\mathbf{k}) \begin{pmatrix} 0 & \tau_0 \\ \tau_0 & 0 \end{pmatrix} . \tag{4.182}$$

Changing the parity from odd to even can be achieved by making the substitution $\tau_0 \rightarrow \tau_z$ in (4.182). The BdG matrix decouples into two four-dimensional submatrices. We consider, now, only the $((u_\uparrow)(v_\downarrow))$ subspace.

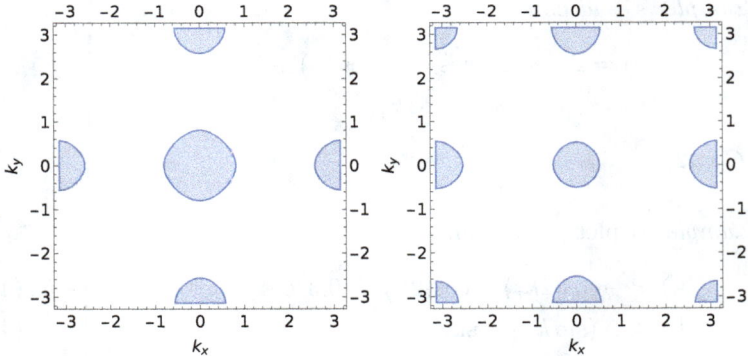

Fig. 4.6   Fermi surface for the model (4.180)-(4.182) , with $t_2 = -0.08$ (left panel) and $t_2 = 0.08$ (right panel).

The choice $\boldsymbol{d}(\mathbf{k}) = \Delta\left(0, 0, \sin k_x - i \sin k_y\right)$ corresponds to the $p + ip$ superconductor. The parameter choice $\mu = 0.6$, $t_1 = 0.07$, $t_2 = -0.08$, produces a Fermi surface with three pockets around the points $(0, 0)$, $(0, \pi)$,

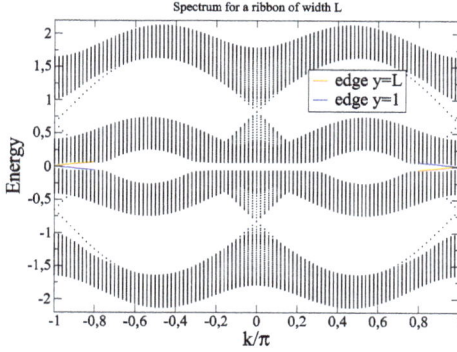

Fig. 4.7 Spectrum of model (4.180)-(4.182) for a ribbon of width L along the y axis and infinite along x. The chirality of the edge states (direction of propagation along the border) confirms $C = +1$. The Majorana fermion has $k = \pi$.

$(\pi, 0)$. See figure 4.6. Choosing, for instance, $\Delta = 0.1$, we obtain $C = 1$. For momentum (π, π), the vector $\boldsymbol{h}'$ is in the northern hemisphere.

The subsystem $((u_\downarrow)(v_\uparrow))$ has the same Chern number, $C = +1$, as the $((u_\uparrow)(v_\downarrow))$ subsystem. Although the spin $\downarrow$ electrons are the time-reversed counterparts of the spin $\uparrow$ electrons, they have the same pairing amplitude, $d_z(\mathbf{k})$. The latter determines the topology, as follows from the theorem proved in section 4.4.4. The north and south poles attained in the various Fermi surface pockets are the same for both subsystems. So, the total Chern number $C = +2$.

When we change the sign of t_2 to $t_2 = 0.08$, the Fermi surface has four pockets. The extra pocket is located around the point (π, π), as shown in figure 4.6, and we obtain $C = 0$.

In figure 4.7 the energy spectrum for a ribbon is displayed. The edge states are shown in color. At energy $E = 0$ they are Majorana fermions.

4.4.6 *Sato and Fujimoto model of a triplet superconductor*

We will, now, consider in detail a model of spinful electrons, triplet pairing symmetry and spatial symmetry of p-wave type, with spin-orbit interaction and Zeeman coupling to an external magnetic field. This model may describe non-centrosymmetric superconductors. In the basis $\left(\psi_{\boldsymbol{k}}, \psi_{-\boldsymbol{k}}^\dagger\right)^T = \left(\psi_{\boldsymbol{k}\uparrow}, \psi_{\boldsymbol{k}\downarrow}, \psi_{-\boldsymbol{k}\uparrow}^\dagger, \psi_{-\boldsymbol{k}\downarrow}^\dagger\right)^T$, the Hamiltonian matrix may be written as

$$
\begin{pmatrix}
\epsilon_{\mathbf{k}} - h_z & \alpha(sk_y + isk_x) & -d_x + id_y & d_z + \Delta_s \\
\alpha(sk_y - isk_x) & \epsilon_{\mathbf{k}} + h_z & d_z - \Delta_s & d_x + id_y \\
-d_x - id_y & d_z - \Delta_s & -\epsilon_{\mathbf{k}} + h_z & \alpha(sk_y - isk_x) \\
d_z + \Delta_s & d_x - id_y & \alpha(sk_y + isk_x) & -\epsilon_{\mathbf{k}} - h_z
\end{pmatrix} .
\tag{4.183}
$$

Here, $\epsilon_{\mathbf{k}} = -2t(\cos k_x + \cos k_y) - \varepsilon_F$ is the kinetic part, t denotes the hopping chosen as the energy scale $(t = 1)$, $\mathbf{k}$ is a wave vector in the plane xy, and we take the lattice constant as unity. $sk_x = \sin k_x$ and $sk_y = \sin k_y$. M_z is the Zeeman term responsible for the magnetization, expressed in units of t. The spin-orbit term is written as $\hat{H}_R = \mathbf{s} \cdot \boldsymbol{\sigma} = \alpha (\sin k_y \sigma_x - \sin k_x \sigma_y)$, where α is also measured in the same units and $\mathbf{s} = \alpha(\sin k_y, -\sin k_x, 0)$. The matrices $\sigma_x, \sigma_y, \sigma_z$ are Pauli matrices that act in the spin sector, and σ_0 is the identity matrix.

The pairing matrix of a superconductor with p symmetry in general satisfies $\hat{\Delta}\hat{\Delta}^\dagger = |\mathbf{d}|^2 \sigma_0 + \mathbf{q} \cdot \boldsymbol{\sigma}$, where $\mathbf{q} = i\mathbf{d} \times \mathbf{d}^*$. If the vector $\mathbf{q}$ is zero, the pairing is called unitary. Otherwise, it is called non-unitary and breaks time reversal invariance, giving origin to a spontaneous magnetization due to the pairing symmetry, as in ^{3}He. To simplify, we will consider a unitary pairing. If the spin-orbit coupling is strong, the pairing is aligned in the direction of the spin-orbit vector, $\mathbf{s}$. This case is called strong coupling. Relaxing this condition allows that the two vectors are not aligned. This case is called weak coupling. In the case of strong coupling, $\mathbf{d} = (d_x, d_y, d_z) = (d/\alpha)\mathbf{s}$, and d gives the scale of the interaction. The existence of the spin-orbit coupling breaks parity and, therefore, pairings with p ($\mathbf{d}$) and s (Δ_s) symmetry are simultaneously possible.

The eigenvalues of equation 4.183 may be written (for $\Delta_s = 0$) as

$$
\epsilon_{\mathbf{k}, \alpha_1, \alpha_2} = \alpha_1 \sqrt{z_1 + \alpha_2 2 \sqrt{z_2}},
\tag{4.184}
$$

where

$$
\begin{aligned}
z_1 &= \mathbf{d} \cdot \mathbf{d} + \mathbf{s} \cdot \mathbf{s} + \epsilon_{\mathbf{k}}^2 + h_z^2 , \\
z_2 &= (\mathbf{d} \cdot \mathbf{s})^2 + (\epsilon_{\mathbf{k}}^2 + d_z^2)(\mathbf{s} \cdot \mathbf{s} + h_z^2),
\end{aligned}
\tag{4.185}
$$

and $\alpha_1, \alpha_2 = \pm$. In the normal phase ($\mathbf{d} = 0$), the spin-orbit coupling lifts the spin degeneracy of the energy bands of the tight-binding model, except at $\mathbf{k} = (0,0)$, (π, π) and $(0, \pi)$ (and equivalent points). These degeneracies that remain are lifted including a non-zero magnetization. As can be seen from equation (4.184), the lower energy band does not have a gap at the points in which

$$\left(\mathbf{s} \cdot \mathbf{s} + h_z^2\right) + \epsilon_\mathbf{k}^2 = 2\sqrt{\left(\mathbf{s} \cdot \mathbf{s} + h_z^2\right)\epsilon_\mathbf{k}^2}. \tag{4.186}$$

In the general case ($\mathbf{d} \neq 0$), the gapless points of the lower energy band are solutions of the equation $z_1 = 2\sqrt{z_2}$, that leads to

$$\mathbf{d} \cdot \mathbf{d} + \mathbf{s} \cdot \mathbf{s} + \epsilon_\mathbf{k}^2 + h_z^2 = 2\sqrt{(\mathbf{d} \cdot \mathbf{s})^2 + (\epsilon_\mathbf{k}^2 + d_z^2)(\mathbf{s} \cdot \mathbf{s} + h_z^2)}. \tag{4.187}$$

Therefore, in the superconducting phase the spectrum has, in general, a gap. In particular, without the spin-orbit interaction the gapless points are obtained by the condition $\mathbf{d} \cdot \mathbf{d} + \epsilon_\mathbf{k}^2 = 0$, which implies particular values for the chemical potential.

In the case of strong coupling with $M_z = 0$, the system is in the $DIII$ symmetry class. In the case of s symmetry, there is only a gap in the bulk of the system and there are no gaples edge states. This is, therefore, a topologically trivial phase. In the case of p symmetry, even though the Chern number vanishes, there are zero energy edge modes. The system is in a topological phase $\mathbb{Z}_2$. Due to the time reversal symmetry, Kramers pairs of edge states, that propagate in opposite directions, give opposing contributions to the total Chern number, $C = 0$. This situation is similar to the spin-Hall effect, where there is a spin current at the edge but a vanishing charge current. If there is a mixture of the s and p components, and the p component is larger, there are edge modes and a non-trivial topological phase. Since there is coupling between spin and momentum, there is no backscattering and these states are protected, even if there are impurities at the edge, as long as they are not magnetic.

If the magnetization is finite, time reversal is broken and the symmetry class changes to D. If the magnetization is small, the system is in a trivial phase with vanishing Chern number. It is interesting to note that, even though the system is in a phase with $C = 0$, the number of edge states is 2, the same as in the $\mathbb{Z}_2$ phase when $M_z = 0$. For $M_z \neq 0$, time reversal symmetry is broken and these states are no longer topologically protected in the presence of (any type) of disorder (except if $\alpha = 0$, where the spin is conserved). In that sense the system is in a trivial phase, according to the Chern number, that vanishes. However, in the limit of no impurities, these edge states could be detected. Increasing the magnetization, a topological transition occurs to a phase with finite Chern number. This transition occurs both for p and s symmetries.

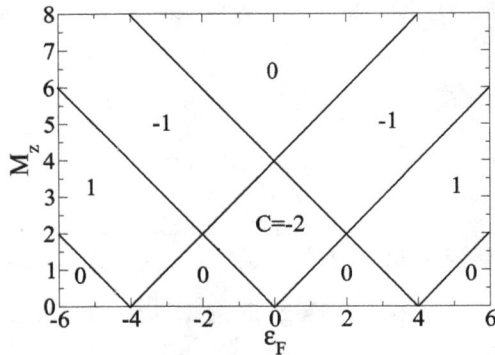

Fig. 4.8   Phase diagram of the Sato and Fujimoto model as a function of chemical potential and magnetization. Here C is the Chern number of each phase associated with the number of protected unidirectional edge modes, when the magnetization is finite.

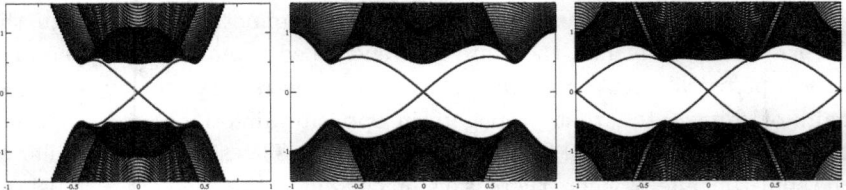

Fig. 4.9   Energy states in a ribbon geometry, for $\alpha = 0.6, \Delta_s = 0.1, d = 0.6$ and a) $M_z = 0, \epsilon_F = -3$, b) $M_z = 2, \epsilon_F = -3$, c) $M_z = 2, \epsilon_F = -1$. In the first case, the system is in class $\mathbb{Z}_2$, with two degenerate edge states inside the gap, in b) the system is in class $\mathbb{Z}$ with one edge state and in c) in class $\mathbb{Z}$ with two edge states.

The Chern number may be calculated summing the flux of the Berry curvature over plaquettes in the Brillouin zone. The results in parameter space are shown in figure 4.8, using the parameters $\alpha = 0.6$, $d = 0.6$. In the case of strong spin-orbit coupling, changing these parameters only leads to a quantitative change of the energy bands and no qualitative change of the topological properties takes place. The spin-orbit coupling does not change the topology.

Due to the bulk-edge correspondence, if the system is placed in a ribbon geometry and is in a non-trivial topological phase, there are robust edge states, where the Chern number counts the number of pairs, if time reversal symmetry is broken. In the phase $\mathbb{Z}_2$ there are also edge states that propagate in opposite directions even though the Chern number vanishes. In this phase, time reversal is preserved and the Kramers pairs give opposite contributions to the Chern number. Examples of edge states in different regimes are shown in figure 4.9.

The existence of edge states may also be shown using dimensional reduction and calculating the winding number. For $k_y = 0$ or π, the Hamiltonian $H(\boldsymbol{k})$ has chiral symmetry: $\Gamma H(\boldsymbol{k})\Gamma^\dagger = -H(\boldsymbol{k})$, (see equation (2.54)) with $\Gamma = t_x \otimes \sigma_0$, where σ_0 is the identity in spin space and t_x acts in the particle-hole space. The operator that diagonalizes Γ is $T = \sigma_0 \otimes e^{-i\frac{\pi}{4}t_y}$, and the Hamiltonian may be converted to a non-diagonal form: $TH(\boldsymbol{k})T^\dagger = q(\boldsymbol{k})\sigma^+ + q^\dagger(\boldsymbol{k})\sigma^-$, if $k_y = 0, \pi$ and $d_z = 0$. The winding number is, then, defined as in equation (2.76), $I(k_y) = \frac{1}{4\pi i} \int_{-\pi}^{\pi} dk_x Tr[q^{-1}(\boldsymbol{k})\partial_{k_x}q(\boldsymbol{k}) - (q^\dagger)^{-1}(\boldsymbol{k})\partial_{k_x}q^\dagger(\boldsymbol{k})]$. Physically, a finite value of $I(k_y)$ means that, if the system is infinite along the y axis and finite along the x axis, then there are edge states with $k_y = 0$ or π.

If the pairing term vector, $\boldsymbol{d}$, is not aligned with the spin-orbit coupling vector, the connection between the Chern number and the number of zero energy edge modes is less transparent. A careful analysis reveals that some of the zero energy modes do not have a topological nature and that the number of modes, that result from the topology, may be obtained both from the winding number and the Chern number, in the case of $C \neq 0$, in a consistent manner. Only the bands of edge states that connect the bands in the momentum/energy space of the bulk system, above and below the gap, have a topological origin.

4.5 Superconductor with impurities

The search for zero energy modes in superconductors has lead to experimental results in at least two situations: a semiconductor wire in proximity with a conventional superconductor in the presence of a magnetic field and a chain of magnetic impurities deposited on a conventional superconductor in appropriate conditions. Two examples that have been proposed are a spin configuration with orientations that form an helix or a ferromagnetic alignment of the spins in the presence of a strong spin-orbit coupling. It is, therefore, interesting to consider the effect of magnetic impurities in superconductors, both conventional and unconventional, such as, for instance, a superconductor with s symmetry and a superconductor with p symmetry.

Let us begin by considering the effect of one magnetic impurity (classical spin) in a superconductor, located at the center of a two-dimensional system. Since the impurity spin acts as a local magnetic field, the spin density of the conduction electrons tends to align with the impurity spin if the coupling is ferromagnetic. For small values of the coupling between the impurity spin and the electronic spin density, one finds a negative spin density in the vicinity of the impurity. At the impurity site the density is positive, as a result of the ferromagnetic coupling. As the coupling increases, the spin density in the vicinity of the impurity also becomes positive. If the coupling is small, the many body system screens the effect of the impurity in a way that compensates its effect, and the total magnetization of the electrons vanishes. However, if the coupling is strong, the system gets magnetized in a discontinuous way, through a quantum phase transition. An interpretation is that, if the coupling, J, is strong enough, the impurity breaks a Cooper pair and captures one of the electrons, leaving the other electron unpaired and, therefore, the spin system becomes polarized. The impurity induces a pair of boundstates inside the superconducting gap, a state with positive energy (with respect to the chemical potential) and another state with symmetric negative energy. Considering larger values of the coupling, the levels in the gap approach the Fermi level. For a critical value of the coupling the two energy levels cross in a discontinuous way and a finite magnetization arises, without the energy levels touching the Fermi energy. After the level crossing, the nature of the states changes and, when the crossing occurs, the nature of the spins also changes. The level crossing occurs between a state that describes a local uncompensated spin (when the coupling is small) and a state where the impurity spin is compensated (partially, since for the classical description to be valid the spin has to be large). The gap function decreases at the impurity site and, when the quantum phase transition takes place, it changes sign (π shift). In the case the pairing symmetry is of p type the quantum critical point occurs for higher values of the coupling and, for $\mu = 0$, the values are quite high.

Consider, now, a chain of magnetic impurities deposited on the surface of the superconductor. In the case of ferromagnetic alignment the spins are aligned perpendicularly to the surface. A case particularly interesting arises when the impurities are distributed along a chain in the center of the two-dimensional system, considered finite with dimensions $N_x \times N_y$. The classical spins are placed along a line of length $L \leq N_x$. Their contribution to the Hamiltonian may be written as

$$H_m = -\sum_j h_z(j_x, j_y)\psi^\dagger(r_j)\,(t_z \otimes \sigma_z)\,\psi(r_j)\,, \qquad (4.188)$$

where t_i are the Pauli matrices ($i = x, y, z$) acting in the particle-hole space, σ_i are Pauli matrices acting in spin space and h_z is the coupling between the magnetic impurities and the spin density of the electrons. Here r_j are, for example, the locations of the $N \leq N_x$ magnetic impurities distributed along the direction of the x axis as $r_j = x e_x$. The spinor is given by

$$\psi(r_j) = \begin{pmatrix} \psi_{j_x,j_y,\uparrow} \\ \psi_{j_x,j_y,\downarrow} \\ \psi^\dagger_{j_x,j_y,\uparrow} \\ \psi^\dagger_{j_x,j_y,\downarrow} \end{pmatrix}. \tag{4.189}$$

Here, $\psi_{j_x,j_y,\uparrow}$ destroys an electron at site j_x, j_y with $\uparrow$ spin projection.

Since the main focus is on the low energy states, we expect that these modes should be zero energy states that appear at the ends of the magnetic chain, in both cases of singlet or triplet pairing, or along the border of the two-dimensional system in the case of the triplet pairing, if we use open boundary conditions.

The Hamiltonian matrix in real space is written as

$$H = \sum_{j_x} \sum_{j_y} \begin{pmatrix} \psi^\dagger_{j_x,j_y,\uparrow} & \psi^\dagger_{j_x,j_y,\downarrow} & \psi_{j_x,j_y,\uparrow} & \psi_{j_x,j_y,\downarrow} \end{pmatrix} \hat{H}_{j_x,j_y} \begin{pmatrix} \psi_{j_x,j_y,\uparrow} \\ \psi_{j_x,j_y,\downarrow} \\ \psi^\dagger_{j_x,j_y,\uparrow} \\ \psi^\dagger_{j_x,j_y,\downarrow} \end{pmatrix}. \tag{4.190}$$

The operator $\hat{H}_{j_x,j_y}$ is given by

$$\hat{H}_{j_x,j_y} = \begin{pmatrix} A & B \\ C & D \end{pmatrix}, \tag{4.191}$$

where

$$A = \begin{pmatrix} -h_z - \mu - t\eta^x_+ - t\eta^y_+ & \frac{\alpha}{2}\eta^x_- + \frac{\alpha}{2i}\eta^y_- \\ -\frac{\alpha}{2}\eta^x_- + \frac{\alpha}{2i}\eta^y_- & h_z - \mu - t\eta^x_+ - t\eta^y_+ \end{pmatrix}, \tag{4.192}$$

$$B = \begin{pmatrix} -\frac{d}{2}\eta^x_- - \frac{d}{2i}\eta^y_- & \Delta_s \\ -\Delta_s & -\frac{d}{2}\eta^x_- + \frac{d}{2i}\eta^y_- \end{pmatrix}, \tag{4.193}$$

$$C = \begin{pmatrix} \frac{d}{2}\eta^x_- - \frac{d}{2i}\eta^y_- & -\Delta_s \\ \Delta_s & \frac{\alpha}{2}\eta^x_- + \frac{d}{2i}\eta^y_- \end{pmatrix}, \tag{4.194}$$

$$D = \begin{pmatrix} h_z + \epsilon_F + t\eta_+^x + t\eta_+^y & -\frac{\alpha}{2}\eta_-^x + \frac{\alpha}{2i}\eta_-^y \\ \frac{\alpha}{2}\eta_-^x + \frac{\alpha}{2i}\eta_-^y & -h_z + \epsilon_F + t\eta_+^x + t\eta_+^y \end{pmatrix}, \qquad (4.195)$$

where $\psi_{j_x,j_y}^\dagger \eta_\pm^x \psi_{j_x,j_y} = \psi_{j_x,j_y}^\dagger \psi_{j_x+1,j_y} \pm \psi_{j_x+1,j_y}^\dagger \psi_{j_x,j_y}$. and $\psi_{j_x,j_y}^\dagger \eta_\pm^y$ $\psi_{j_x,j_y} = \psi_{j_x,j_y}^\dagger \psi_{j_x,j_y+1} \pm \psi_{j_x,j_y+1}^\dagger \psi_{j_x,j_y}$. Here, t describes the hopping amplitude between two neighboring sites, the chemical potential is μ, α the spin-orbit coupling of Rashba type, Δ_s is the s wave pairing d is the amplitude of the p wave pairing. The exchange interaction with the magnetic moments localized at specific locations is only finite at these sites. Let us consider $h_z = J$ at these locations. σ^a are spin Pauli matrices with $a = x, y, z$ and $s, s' = \uparrow, \downarrow$.

The diagonalization of this Hamiltonian involves the solution of an eigenvalue problem of dimension $(4N_xN_y) \times (4N_xN_y)$. The energy states include, in general, states in the bulk of the system and states along the edges. Choosing appropriately $h_z(j_x, j_y)$, we may consider different distributions of magnetic impurities.

In the case of a chain of impurities, one finds different changes of the magnetization of the system, due to increased effect of the electron capture by the impurities, as the coupling increases. If the number of impurities is sufficiently large and the coupling between the impurities and the electron spins is high enough, a topological transition takes place to a phase with edge modes at the ends of the magnetic chain. This transition may be signalled by a topological invariant.

4.5.1 *Magnetic chain on a singlet superconductor*

To characterize the zero energy modes we may calculate the Majorana number M, that shows the presence ($M = -1$) or absence ($M = 1$) of these modes at the edges of the chain. For a system translationally invariant, the Majorana number is given by

$$M = \mathrm{Sgn}\Big[\mathrm{Pf}(A_p)\mathrm{Pf}(A_a)\Big], \qquad (4.196)$$

where $\mathrm{Pf}(A)$ denotes the Pfaffian of matrix A, and A_p (A_a) is the Hamiltonian with periodic boundary conditions (antiperiodic) rewritten in the Majorana basis.

Let us use a simplified notation in which $j = (j_x, j_y)$ and consider, as example, the case with s symmetry. Introducing once again Majorana fermions $a_{2j-1,s} = c_{j,s} + c_{j,s}^\dagger$, $a_{2j,s} = i(c_{j,s} - c_{j,s}^\dagger)$, the Hamiltonian becomes

$$H = -i/4 \sum_j \Big\{ \mu \sum_s a_{2j-1,s} a_{2j,s}$$
$$+ t \sum_{s, \delta = \delta x, \delta y} \big[a_{2j-1,s} a_{2(j+\delta),s} + a_{2(j+\delta)-1,s} a_{2j,s} \big] -$$
$$- \alpha \big[a_{2(j+\delta x)-1,\uparrow} a_{2j,\downarrow} + a_{2j-1,\downarrow} a_{2(j+\delta x),\uparrow} - a_{2(j+\delta x)-1,\downarrow} a_{2j,\uparrow}$$
$$- a_{2j-1,\uparrow} a_{2(j+\delta x),\downarrow} - a_{2(j+\delta y)-1,\uparrow} a_{2j-1,\downarrow} - a_{2(j+\delta y)-1,\downarrow} a_{2j-1,\uparrow}$$
$$- a_{2(j+\delta y),\uparrow} a_{2j,\downarrow} - a_{2(j+\delta y),\downarrow} a_{2j,\uparrow} \big] +$$
$$+ \mathrm{Re}(\Delta_j) \big[a_{2j-1,\uparrow} a_{2j,\downarrow} - a_{2j-1,\downarrow} a_{2j,\uparrow} \big]$$
$$+ \mathrm{Im}(\Delta_j) \big[a_{2j,\uparrow} a_{2j,\downarrow} - a_{2j-1,\uparrow} a_{2j-1,\downarrow} \big] +$$
$$+ J_j \big[a_{2j-1,\downarrow} a_{2j,\downarrow} - a_{2j-1,\uparrow} a_{2j,\uparrow} \big] \Big\}. \tag{4.197}$$

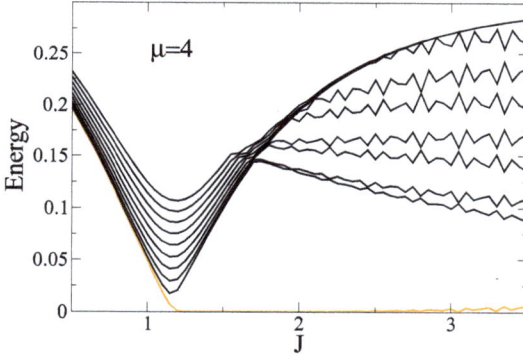

Fig. 4.10 The 10 lowest energy states as a function of the exchange coupling J, for $\mu = 4$. The parameters are $L = N_x = 90$, $N_y = 11$, $\alpha = 0.3$ and $\Delta = 0.3$. For $M = -1$ zero energy modes appear at the edges of the magnetic chain.

Taking Δ homogeneous and using the discrete Fourier transform,

$$a_{2j-1,s} = 1/\sqrt{N} \sum_k \mathrm{Exp}[-ikj] a_{k,1,s},$$
$$a_{2j,s} = 1/\sqrt{N} \sum_k \mathrm{Exp}[-ikj] a_{k,2,s}, \tag{4.198}$$

the Hamiltonian (4.197) may be rewritten in the basis $a_k = (a_{k,1,\uparrow}, a_{k,1,\downarrow}, a_{-k,2,\uparrow}, a_{-k,2,\downarrow})^T$ as

$$H = \frac{i}{4} \sum_k a_k^\dagger A(k) a_k, \tag{4.199}$$

where the matrix $A(k)$ has the following non-vanishing elements

$$A_{1,3}(k) = \varepsilon_k + J, \qquad A_{1,4}(k) = \Delta + 2i\alpha \sin(k),$$
$$A_{2,3}(k) = -\Delta - 2i\alpha \sin(k), \qquad A_{2,4}(k) = \varepsilon_k - J, \qquad (4.200)$$

with $\varepsilon_k = -\mu - 2t \cos(k)$. The Majorana number may be calculated using equation (4.196), where $A_p = A(0)$ and $A_a = A(\pi)$, therefore we find $M = -1$ for

$$\sqrt{\Delta^2 + (\mu - 2t)^2} < J < \sqrt{\Delta^2 + (\mu + 2t)^2}. \qquad (4.201)$$

In figure 4.10 we show the lowest energy states as a function of the model parameters. Note the transitions of the lower energies from finite to vanishing values when the gap closes and reopens. When $L = N_x$, the appearance of the zero energy modes coincides with a negative Majorana number. However, in some cases the Majorana number may be negative, even though the gap is closed. In these cases, the states of lower energy are not Majorana modes.

4.5.2 *Magnetic chain on a triplet superconductor*

While in the case of a conventional superconductor magnetic impurities induce topological modes, in the case of a triplet superconductor the topology may appear even in the absence of impurities. Are examples, the edge modes that arise in the Sato and Fujimoto model, even if there is time reversal symmetry; the topological modes of a superconductor of $p + ip$ type by time reversal symmetry breaking; or those that appear if we consider some type of non-unitary pairing. It is, therefore, interesting to study the coexistence/competition between topological modes of the superconductor with p symmetry and, therefore, intrinsic modes, and those that are the result of the local fields induced by the magnetic impurities.

Consider a magnetic chain that extends on the surface of a triplet superconductor described, in the absence of impurities, by the Sato and Fujimoto model. In figure 4.11 we apply open boundary conditions along both spatial directions and show the density of the eigenstate of lower energy, as a function of x and y, for two values of the chemical potential $\mu = 0, -3$, and consider that each magnetic impurity may be seen as a local magnetic field with amplitude $h_z = 2$. We also choose $\alpha = 0, d = 0.6, \Delta_s = 0$.

Let us consider first $\mu = 0$. The groundstate (lowest energy in the single-particle spectrum) is four times degenerate, with an energy of the order of 10^{-7} or 10^{-6}, and the next set of energies have values of the order of 10^{-2}.

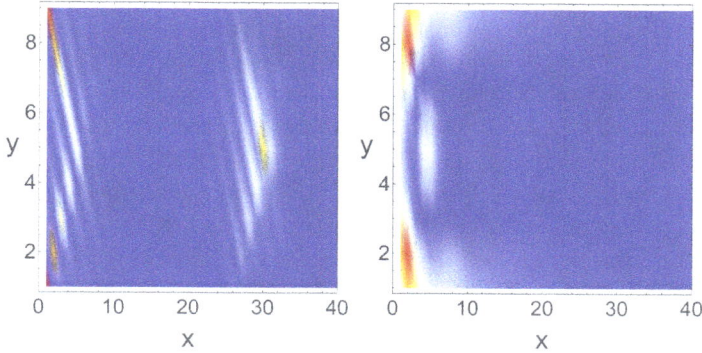

Fig. 4.11   Wave functions of the lowest energy states with open boundary conditions along x and y for $\mu = 0$ (left column) and $\mu = -3$ (right column) for $h_z = 2$, in a system of size 210×9 and with a magnetic chain of length 150 sites.

In the left column of figure 4.11, we show one of the degenerate states. In this case, the two types of edge states are clearly seen. The state along the border of the two-dimensional system has peaks on the sides along y, and the state associated with the edge of the magnetic chain has a peak at this end and decays in the direction of the sides, along y. Both states are very localized along the x direction. If the chemical potential is $\mu = -3$, the border state dominates and any peak at the end of the magnetic chain is difficult to detect if the magnetic field is small. Its height increases with the magnetic field, and becomes perfectly visible for $h_z = 5$ (not shown in the figure). However, in this case the two peaks overlap and in the region between them the wave function is clearly finite and appreciable. The lowest energy level is non-degenerate and with a very small energy, of the order of 10^{-15}. We may note that the two peaks, in a general way, are overlapped on the border of the system along the y direction. The increase of the size along y does not lead to a decrease of the spatial extension along y.

The existence of topological modes is also associated, once again, to a topological invariant. In a two-dimensional system with translation invariance we may use the Chern number, as seen previously. However, in the presence of impurities translational invariance is broken and a description in terms of momentum space is not useful. We need, therefore, a real space description. This point is discussed next.

4.5.3 *Chern number in real space*

A possible choice for a topological invariant that characterizes each topological phase is obtained calculating the Chern number associated with the integration over twisted boundary conditions in the two-dimensional system.

As we saw in section 2.1, if a quantum system is described by an Hamiltonian that depends on a given parameter, $\boldsymbol{R}$, and if it is periodic in this vector, then the integral of the z component of the Berry curvature over the surface that the vector traverses, S, is an integer, called Chern number:

$$C_n = \frac{1}{2\pi} \int_S dS \Omega_z^{(n)}(\boldsymbol{R}). \tag{4.202}$$

The index n respects to the state n of the adiabatic energy levels (resulting from the transformation along the vector trajectory)

$$H(\boldsymbol{R})|u_n(\boldsymbol{R})\rangle = E_n(\boldsymbol{R})|u_n(\boldsymbol{R})\rangle. \tag{4.203}$$

The component z of the Berry curvature is defined as (here $(1,2) = (x,y)$)

$$\Omega_z^{(n)} = \frac{\partial}{\partial R_1} \mathcal{A}_2^{(n)}(\boldsymbol{R}) - \frac{\partial}{\partial R_2} \mathcal{A}_1^{(n)}(\boldsymbol{R}), \tag{4.204}$$

where the Berry connection is defined as in equation (2.7),

$$\mathcal{A}^{(n)}(\boldsymbol{R}) = \langle u_n(\boldsymbol{R})|i\nabla_{\boldsymbol{R}}|u_n(\boldsymbol{R})\rangle. \tag{4.205}$$

In a translational invariant system, the parameter $\boldsymbol{R}$ may be chosen as the momentum, and the integration is carried out along the Brillouin zone. In a problem with no translational invariance, the procedure may be chosen in terms of twisted boundary conditions in the two spatial directions.

The integration of the Berry curvature over the Brillouin zone has, in general, numerical problems if we come close to a transition point (for which the gap between bands vanishes) and is more convenient to calculate the Chern number on a lattice by the plaquette method discussed in section 2.7.

The twisted boundary conditions are chosen as

$$u_n^\theta(\boldsymbol{r} + N_i \boldsymbol{a}_i) = e^{i\theta_i} u_n^\theta(\boldsymbol{r}), \tag{4.206}$$

where $\theta = (\theta_1, \theta_2)$ and $H(\theta)|u_n^\theta\rangle = E_n^\theta|u_n^\theta\rangle$, and $\boldsymbol{a}_i$ are the lattice basis vectors, along the direction i. Here $u_n^\theta(\boldsymbol{r}) = \langle \boldsymbol{r}|u_n^\theta\rangle$. The problem is solved for a finite system with size $N_1 \times N_2$, typically chosen large enough. In general, degeneracies in the energy spectrum occur that render difficult the calculation of the Chern number and, therefore, a lattice method is more efficient. We consider the Slater determinant of the one-particle eigenfunctions of the Hamiltonian $H(\theta)$. The groundstate may, then, be represented by a matrix, in terms of the expansion coefficients

$$|u_n^\theta\rangle = \sum_{r_i} \phi_{r_i}^{n,\theta}|r_i\rangle \,, \tag{4.207}$$

as

$$\Phi_\theta = \begin{pmatrix} \phi_{r_1}^{1,\theta} & \phi_{r_1}^{2,\theta} & \cdots & \phi_{r_1}^{M,\theta} \\ \phi_{r_2}^{1,\theta} & \phi_{r_2}^{2,\theta} & \cdots & \phi_{r_2}^{M,\theta} \\ \cdots & \cdots & \cdots & \cdots \\ \phi_{r_N}^{1,\theta} & \phi_{r_N}^{2,\theta} & \cdots & \phi_{r_N}^{M,\theta} \end{pmatrix} \,, \tag{4.208}$$

where $N = N_1 N_2$ and M is the number of occupied states. The Chern number may be obtained in a similar way,

$$C = \frac{1}{2\pi} \sum_{l=1}^{L_1 L_2} arg\left(\langle \Psi^{\theta_l}|\Psi^{\theta_l+1}\rangle\langle\Psi^{\theta_l+1}|\Psi^{\theta_l+1+2}\rangle\langle\Psi^{\theta_l+1+2}|\Psi^{\theta_l+2}\rangle\langle\Psi^{\theta_l+2}|\Psi^{\theta_l}\rangle\right) \,, \tag{4.209}$$

where the vectors **1** and **2** denote the two spatial directions that correspond to the two types of boundary conditions, and L_1 and L_2 are the number of times the twist angles are divided. Since each state Ψ^{θ_l} is a many-body state, described by a Slater determinant, the superposition between two states with different boundary conditions is given by

$$\langle \Psi^\theta|\Psi^{\theta'}\rangle = det\left(\Phi_\theta^\dagger \Phi_{\theta'}\right) \,, \tag{4.210}$$

and the Chern number is obtained by

$$C = \frac{1}{2\pi} \sum_{l=1}^{L_1 L_2} arg\lambda_p \,, \tag{4.211}$$

where λ_p are the eigenvalues of the matrix product

$$\prod_{l=1}^{L_1 L_2} \Phi_{\theta_l}^\dagger \Phi_{\theta_l+1}\Phi_{\theta_l+1}^\dagger \Phi_{\theta_l+1+2}\Phi_{\theta_l+1+2}^\dagger \Phi_{\theta_l+2}\Phi_{\theta_l+2}^\dagger \Phi_{\theta_l} \,. \tag{4.212}$$

However, this method is very time consuming. It was shown that it is enough to calculate the Chern number using only periodic boundary conditions, and not twisted boundary conditions. The Chern number is obtained by a similar expression, but the eigenvalues λ_p are the eigenvalues of the matrix of dimension $M \times M$,

$$F = C_{q_0 q_1} C_{q_1 q_2} C_{q_2 q_3} C_{q_3 q_0} \,, \tag{4.213}$$

where the momenta are defined by

$$\boldsymbol{q}_0 = (0,0); \; \boldsymbol{q}_1 = \left(\frac{2\pi}{N_1}, 0\right); \; \boldsymbol{q}_2 = \left(\frac{2\pi}{N_1}, \frac{2\pi}{N_2}\right); \; \boldsymbol{q}_3 = \left(0, \frac{2\pi}{N_2}\right). \quad (4.214)$$

The matrices are defined by their matrix elements, mn, as

$$C_{\boldsymbol{q},\boldsymbol{q}'}^{mn} = \sum_{\boldsymbol{r}_i} \left(\phi_{\boldsymbol{r}_i}^{m,\theta=0}\right)^* e^{i(\boldsymbol{q}-\boldsymbol{q}')\cdot \boldsymbol{r}_i} \phi_{\boldsymbol{r}_i}^{n,\theta=0}. \quad (4.215)$$

Applying this method to the magnetic chain problem, we obtain a zero Chern number, unless the magnetic field is large, since the magnetic impurity concentration is small. We expect, therefore, that in the case of $\mu = -3$ the Chern number vanishes. In the case of $\mu = 0$, the Chern number may be finite, since in a uniform magnetic field any small magnetic field changes the Chern number from $C = 0$ to $C = -2$. These results may be confirmed by the calculation of Chern numbers, that describe equally well the case of uniform magnetization (considered in the Sato and Fujimoto model) as the case of variable impurity concentrations.

Further reading:

A detailed discussion of the Bogoliubov-de Gennes equations can be found in:

- P. G. de Gennes, *Superconductivity of Metals and Alloys*, Reading, MA: Addison-Wesley (1989).

The symmetry classes for Bogoliubov-de Gennes matrices in the particle-hole basis are discussed in:

- Schnyder et. al., *Physical Review* B **78**, 195125 (2008).

and a study of reflection symmetry in superconductors can be found in:

- C.-K. Chiu, H. Yao, and S. Ryu, *Physical Review* B **88** , 075142 (2013).

On Bogoliubov-de Gennes matrices in the Nambu basis and Majorana fermions in a Josephson junction:

- L. Fu, and C. L. Kane, *Physical Review Letters* **100**, 096407 (2008).
- R. M. Lutchyn, J. D. Sau, and S. Das Sarma, *Physical Review Letters* **105**, 077001 (2010).

The Kitaev model is presented in:

- A. Y. Kitaev, *Phys. Usp.* **44**, 131 (2001).

On the properties of states and parity in the Kitaev model, and in its extended version:

- M. Greiter, V. Schnells and R. Thomale, *Annals of Physics* **351**, 1026 (2014).
- P. D. Sacramento, *Journal of Physics: Condensed Matter* **27**, 445702 (2015).

On the Su-Schrieffer-Heeger model and its extension to include triplet pairing:

- W. P. Su, J. R. Schrieffer, and A. J. Heeger, *Physical Review Letters* **42**, 1698 (1979).
- W. P. Su, J. R. Schrieffer, and A. J. Heeger, *Physical Review* B **22**, 2099 (1980).
- A. J. Heeger, S. Kivelson, J. R. Schrieffer, and W. P. Su, *Reviews of Modern Physics* **60**, 781 (1988).
- R. Wakatsuki, M. Ezawa, Y. Tanaka, and N. Nagaosa, *Physical Review* B **90**, 014505 (2014).

A detailed discussion on Majorana fermions is presented in:

- J. Alicea, *Reports Progress Physics* **75**, 076501 (2012).
- M. Leijnse and K. Flensberg, *Semiconductor Science Technology* **27**, 124003 (2012).
- C. W. J. Beenakker, *Annual Reviews Condensed Matter Physics* **4**, 113 (2013).
- R. Aguado, *Rivista Nuovo Cimento* **40**, 523 (2017).

The equivalence between the Dirac cone with s-wave superconductivity and the spinless $p + ip$ superconductor, as well as the realization of topological superconductors by the proximity effect, can be found in:

- A. C. Potter, and P. A. Lee, *Physical Review* B **83**, 184520 (2011).

The model by Sato and Fujimoto is presented in

- M. Sato, and S. Fujimoto, *Physical Review* B **79**, 094504 (2009).

On proposals for the realization and observation of Majorana fermions:

- Roman M. Lutchyn, Jay D. Sau, and S. Das Sarma, *Physical Review Letters* **105**, 077001 (2010).
- Yuval Oreg, Gil Refael, and Felix von Oppen, *Physical Review Letters* **105**, 177002, (2010).
- V. Mourik, K. Zuo, S. M. Frolov, S. R. Plissard, E. P. A. M. Bakkers, and L. P. Kouwenhoven, *Science* **336**, 1003 (2012).
- Falko Pientka, Leonid I. Glazman, and Felix von Oppen, *Physical Review* B **88**, 155420 (2013).
- S. Nadj-Perge, I. K. Drozdov, J. Li, H. Chen, S. Jeon, J. Seo, A. H. MacDonald, B. A. Bernevig, and A. Yazdani, *Science* **346**, 602 (2014).
- P. M. R. Brydon, S. Das Sarma, Hoi-Yin Hui, and Jay D. Sau, *Physical Review* B **91**, 064505 (2015).
- M. T. Deng, S. Vaitiekenas, E. B. Hansen, J. Danon, M. Leijnse, K. Flensberg, J. Nygard, P. Krogstrup, and C. M. Marcus, *Science* **354**, 1557 (2016).
- H. Zhang, C. X. Liu, S. Gazibegovic, D. Xu, J. A. Logan, G. Wang, N. Van Loo, J. D. Bommer, M. W. De Moor, D. Car, R. L. Op Het Veld, P. J. Van Veldhoven, S. Koelling, M. A. Verheijen, M. Pendharkar, D. J. Pennachio, B. Shojaei, J. S. Lee, C. J. Palmström, E. P. Bakkers, S. D. Sarma, and L. P. Kouwenhoven, *Nature* (London) **556**, 74 (2018).

The effect of magnetic impurities on conventional and triplet superconductors, as well as their effect on topological properties, is discussed, for instance, in:

- L. Yu, *Acta Physica Sinica* **21**, 75 (1965).
- H. Shiba, *Progress Theoretical Physics* **40**, 435 (1968).
- A. Rusinov, *Soviet Physics* JETP **29**, 1101 (1969).
- A. Sakurai, *Progress Theoretical Physics* **44**, 1472 (1970).
- M. I. Salkola, A. V. Balatsky, and J. R. Schrieffer, *Physical Review* B **55**, 12648 (1997).
- A. V. Balatsky, I. Vekhter, and J-X. Zhu, *Reviews Modern Physics* **78**, 373 (2006).
- T. Cadez and P. D. Sacramento, *Journal of Physics: Condensed Matter* **28**, 495703 (2016).

- P. D. Sacramento, *Journal of Physics: Condensed Matter* **27**, 445702 (2015).

The calculation of the Chern number and its application in real space is discussed in:

- T. Fukui, Y. Hatsugai, H. Suzuki, *Journal Physical Society Japan* **74**(6), 1674, (2005).
- Y.-F. Zhang et al., *Chinese Physics* B, **22**(11), 117312 (2013).

Chapter 5

Topological semimetals

5.1 Definition and symmetries

From the previous chapters we acquired the notion that an electronic topological system is a state of matter that is insulating in the bulk and metallic on the surface. The energy bands are gapped, therefore, the infinite system would be a band insulator. As soon as a surface is created, a continuum of metallic states appears on it.

There are, however, three-dimensional systems where the valence and conduction bands touch at discrete points, or on a line, in momentum space. Such systems are *topological semimetals*. The points or lines are described by Dirac (wave function with four components) of Weyl (wave function with two components) Hamiltonians.

The concept of *codimension* plays a role here: it is the difference between the dimensionality of the system and that of the Fermi surface. If two bands touch at nodal points, then the Fermi surface has zero dimensions. The codimension is then 3=3-0. If there is a nodal line, the codimension is 2=3-1.

Close to a Weyl point, K, in the Brillouin zone, the Hamiltonian can be linearized as:

$$\hat{H} = \mathbf{h} \cdot \boldsymbol{\tau} + h_0 \tau_0 \equiv h_\mu \tau_\mu \,, \tag{5.1}$$

$$h_\mu(\boldsymbol{K} + \mathbf{q}) \approx \hbar \boldsymbol{v}_\mu \cdot \mathbf{q} \,, \tag{5.2}$$

where the momenta are three-dimensional. In what follows, we consider $\hbar = 1$. The energy spectrum is given by

$$E(\mathbf{q}) = \boldsymbol{v}_0 \cdot \mathbf{q} \pm \sqrt{\sum_{i=1}^{3} (\boldsymbol{v}_i \cdot \mathbf{q})^2} \,. \tag{5.3}$$

121

The chiral charge or chirality of a Weyl point is defined as $c = sgn\,(\boldsymbol{v}_1 \cdot \boldsymbol{v}_2 \times \boldsymbol{v}_3)$.

Because they obey a two-component Weyl equation[1], Weyl fermions do not have mass, so, they have no energy gap. The spectrum (5.3) has no mass gap. This is in contrast to Dirac fermions, with four components, because the Dirac equation may include a mass (producing an energy gap).

Let us consider the role of space inversion symmetry or parity, implemented by the operator $\mathcal{P} : H(\mathbf{p}) \to H(-\mathbf{p})$. We have, for a system having this symmetry,

$$H(\boldsymbol{K}+\mathbf{q}) = \boldsymbol{v}_\mu \cdot \mathbf{q}\,\tau_\mu = H(-\boldsymbol{K}-\mathbf{q}) \;\Rightarrow\; H(-\boldsymbol{K}+\mathbf{q}) = -\boldsymbol{v}_\mu \cdot \mathbf{q}\,\tau_\mu\,, \quad (5.4)$$

hence we conclude that the points $\pm\boldsymbol{K}$ have opposite chirality and cannot be equivalent points in the Brillouin zone. On the other hand, time reversal symmetry, $\mathcal{T}$, applied to pseudo-spin, implies that $H = H^*$, therefore,

$$H(\boldsymbol{K}+\mathbf{q}) = \boldsymbol{v}_\mu \cdot \mathbf{q}\,\tau_\mu = H^*(-\boldsymbol{K}-\mathbf{q}) \;\Rightarrow\; H(-\boldsymbol{K}+\mathbf{q}) = -\boldsymbol{v}_\mu \cdot \mathbf{q}\,\tau_\mu^*\,, \quad (5.5)$$

so, the points $\pm\boldsymbol{K}$ have the same chirality. This result still holds if the Pauli matrices represent the physical spin ($\tau \to \sigma$):

$$H(-\boldsymbol{K}+\mathbf{q}) = -\boldsymbol{v}_\mu \cdot \mathbf{q}\,\sigma_y\sigma_\mu^*\sigma_y\,.$$

We then conclude that the presence of both symmetries, $\mathcal{P}, \mathcal{T}$ would lead to $\boldsymbol{v}_2 = 0 \Rightarrow c = 0$, so, the Weyl point would not exist. The existence of Weyl points necessarily implies breaking one of those symmetries.

Nielsen-Ninomiya theorem guarantees that there exist an even number of Weyl points in the Brillouin zone, and that the sum of their chiral charges is zero.

Example: The Hamiltonian $\hat{H}(\mathbf{k}) = \boldsymbol{h} \cdot \boldsymbol{\tau}$, with

$$h_x = \sqrt{2}t_1\,(\cos k_x + \cos k_y)\,, \quad h_y = \sqrt{2}t_1\,(\cos k_x - \cos k_y)\,, \quad h_z = -\cos k_z\,,$$

has eight Weyl points at $k_z = \pm\pi/2$ and $k_{x/y} = \pm\pi/2$.

5.2 Type I Weyl points

Consider the dispersion relation (5.3). If the second term is larger than the first for all directions of $\mathbf{q}$, then the Weyl point is said to be of type I. Obviously, if $\boldsymbol{v}_0 = 0$, the Weyl point is of type I.

---

[1] $\sigma_\mu \partial_\mu \psi = 0$.

If the term v_0 is finite and dominates over the second term in (5.3) along one direction of $\mathbf{q}$, then the Weyl point is of type II. A simple example of a type II point is:

$$\hat{H}(\mathbf{q}) = aq_z + \mathbf{q} \cdot \boldsymbol{\tau}, \qquad |a| > 1, \tag{5.6}$$

$$E(\mathbf{q}) = aq_z \pm |\mathbf{q}| . \tag{5.7}$$

We now focus on type I points, and set $v_0 = 0$. Type II points will be treated in section 5.6.

Note that the Hamiltonian term $v_0\tau_0$ has no influence on the chiral charge.

Weyl points exist in pairs, as follows from Nielsen-Ninomiya theorem. It is possible that one of those points is of type I and the other of type II.

5.2.1 *Sources and drains of Berry curvature*

A point with chiral charge $+1$ behaves as a "source" of curvature in momentum space, $\mathbf{q}$, and a point with charge -1 as a "drain". Indeed, the curvature in the lowest band is

$$\boldsymbol{\Omega} = \frac{\boldsymbol{h}}{2h^3} = \frac{(\boldsymbol{v}_1 \cdot \mathbf{q}, \boldsymbol{v}_2 \cdot \mathbf{q}, \boldsymbol{v}_3 \cdot \mathbf{q})}{2h^3} . \tag{5.8}$$

Using Gauss theorem to calculate the flux of curvature in $\boldsymbol{h}$ space, we have

$$\int_{\boldsymbol{h}} \boldsymbol{\Omega} \cdot d\boldsymbol{S} = \int_{\boldsymbol{h}} \nabla_{\boldsymbol{h}} \cdot \boldsymbol{\Omega} \, dV_{\boldsymbol{h}} . \tag{5.9}$$

But because Berry curvature acts like a monopole in $\boldsymbol{h}$ space, we have

$$\nabla_{\boldsymbol{h}} \cdot \boldsymbol{\Omega} = 2\pi \, \delta(\boldsymbol{h}) = 2\pi \, \delta(\boldsymbol{v}_1 \cdot \mathbf{q})\delta(\boldsymbol{v}_2 \cdot \mathbf{q})\delta(\boldsymbol{v}_3 \cdot \mathbf{q}) . \tag{5.10}$$

We change to coordinates $\mathbf{q}$ in the volume integral (5.9). Then, we have to make use of the determinant of the Jacobian for that coordinate transformation and

$$\mathrm{Det}\left(\frac{\partial \boldsymbol{h}}{\partial \mathbf{q}}\right) = \boldsymbol{v}_1 \cdot \boldsymbol{v}_2 \times \boldsymbol{v}_3 . \tag{5.11}$$

Therefore, $dV_{\boldsymbol{h}} = \boldsymbol{v}_1 \cdot \boldsymbol{v}_2 \times \boldsymbol{v}_3 \, d^3q$ and the integral becomes

$$\int_{\boldsymbol{h}} \nabla_{\boldsymbol{h}} \cdot \boldsymbol{\Omega} \, dV_{\boldsymbol{h}} = 2\pi \int_{\mathbf{q}} \delta\left(\boldsymbol{h}\right) \boldsymbol{v}_1 \cdot \boldsymbol{v}_2 \times \boldsymbol{v}_3 \, d^3q . \tag{5.12}$$

The integral can be performed on a sphere surrounding the Weyl point. In that case, the Dirac delta function will contribute. The sign of the integral is given by the chiral charge of the Weyl point.

Exercise: Consider $H(\mathbf{k}) = \boldsymbol{h} \cdot \boldsymbol{\tau}$ for spinless fermions, with

$$\boldsymbol{h} = (\cos k_x + \cos k_y - \mu, \sin k_x, \cos k_z) .$$

Which symmetry, $\mathcal{T}$ or $\mathcal{P}$, is broken? Identify the Weyl points. Which ones have the same chirality for symmetry reasons?

5.2.2 *Density of states*

The density of states for the Hamiltonian (5.2) with $\boldsymbol{v}_0 = 0$ can be calculated as follows:

$$\rho(E) = \int \frac{d^3q}{(2\pi)^3} \delta \left(\sqrt{\sum_{i=1}^{3} (\boldsymbol{v}_i \cdot \mathbf{q})^2} - E \right) . \tag{5.13}$$

Making a change of variables to $\xi_i = \boldsymbol{v}_i \cdot \mathbf{q}$ requires the determinant of the Jacobian, $\partial q / \partial \xi$. The determinant of the inverse matrix,

$$\mathrm{Det} \left(\frac{\partial \xi}{\partial q} \right) = \boldsymbol{v}_1 \cdot \boldsymbol{v}_2 \times \boldsymbol{v}_3 , \tag{5.14}$$

therefore,

$$\rho(E) = \frac{1}{|\boldsymbol{v}_1 \cdot \boldsymbol{v}_2 \times \boldsymbol{v}_3|} \int \frac{d^3\xi}{(2\pi)^3} \delta \left(\sqrt{\sum_{i=1}^{3} \xi_i^2} - E \right) , \tag{5.15}$$

and going over to spherical coordinates where $r^2 = \sum_{i=1}^{3} \xi_i^2$ and $d^3\xi = 4\pi r^2 dr$, we obtain the result

$$\rho(E) = \frac{E^2}{2\pi^2 |\boldsymbol{v}_1 \cdot \boldsymbol{v}_2 \times \boldsymbol{v}_3| \hbar^3} , \tag{5.16}$$

where the reduced Planck constant has been restored. The density of states vanishes quadratically at type I Weyl points.

5.3   Surface states with "Fermi arcs"

Suppose that there are two Weyl points with opposite chiral charges in the three-dimensional Brillouin zone. And that when a surface is created, the projections of both points on the plane of the surface do not coincide. Then, there will be zero energy surface states along a line (on the surface Brillouin zone) that connects the projections of the Weyl points.

We can study this effect in a model with two Weyl points:

$$H(\mathbf{k}) = \sin k_x \tau_x + \sin k_y \tau_y + \frac{\cos k_z - \cos Q}{\sin Q} \tau_z$$
$$+ (\cos k_x + \cos k_y - 2)\,\tau_z\,, \tag{5.17}$$

where the Weyl points are $(0, 0, \pm Q)$. Suppose $0 < Q < \pi$. The last term has been included in order to eliminate other points that would otherwise appear at $(0, \pi)$, $(\pi, 0)$, (π, π). Linearizing $H(\mathbf{k})$ close to the Weyl points, we obtain:

$$H(\mathbf{k}) \approx k_x \tau_x + k_y \tau_y \mp (k_z - Q)\tau_z\,, \qquad \text{for } k_z \approx \pm Q\,. \tag{5.18}$$

If we fix the value of k_z between $-Q$ and Q, the Hamiltonian in two dimensions, $H_{k_z}(k_x, k_y)$, has Chern number $C = 1$. If $|k_z| > Q$ then $C = 0$. See figure 5.1 (right panel). Hence we conclude that the surface states will appear on the lateral surfaces as long as $|k_z| < Q$, but we have not yet determined their dispersion relation, as function of k_x or k_y, according to whether the surface is perpendicular to y or x. The surface states' dispersion must cross the zero energy value somewhere because it connects a lower to a higher band of bulk states. The dispersion has a zero that depends on k_z. A line of zeroes connecting the points $k_z = \pm Q$ must exist.

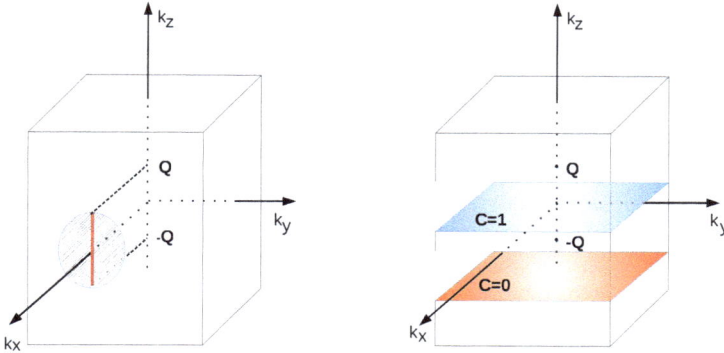

Fig. 5.1 Left: region with surface states in the plane (k_y, k_z). The segment $(k_y = 0, |k_z| < Q)$ is the Fermi arc. Right: planes with constant k_z have $C = 1$ if $|k_z| < Q$ and produce surface states.

We shall now see this in more detail by considering a surface perpendicular to x and using the method presented in section 2.13 for the calculation of edge states. Since $H(\mathbf{k}) = \boldsymbol{h}(\mathbf{k}) \cdot \boldsymbol{\tau}$, we write

$$\boldsymbol{h}(\mathbf{k}) = 2\boldsymbol{b}_i \sin k_x + 2\boldsymbol{b}_r \cos k_x + \boldsymbol{b}_0 \,, \tag{5.19}$$

$$2\boldsymbol{b}_i = (1,0,0), \quad 2\boldsymbol{b}_r = (0,0,1) \,, \tag{5.20}$$

$$\boldsymbol{b}_0 = \left(0, \sin k_y, -2 + \cos k_y + \frac{\cos k_z - \cos Q}{\sin Q}\right) = \boldsymbol{b}_{\parallel} + \boldsymbol{b}_{\perp} \,, \tag{5.21}$$

$$\boldsymbol{b}_{\perp} = (0, \sin k_y, 0) \,, \tag{5.22}$$

$$\boldsymbol{b}_{\parallel} = \left(0, 0, -2 + \cos k_y + \frac{\cos k_z - \cos Q}{\sin Q}\right) \,. \tag{5.23}$$

The condition for the existence of edge states says that $2\boldsymbol{b}_i \sin k_x + 2\boldsymbol{b}_r \cos k_x + \boldsymbol{b}_{\parallel}$ is an ellipse that includes the origin. Therefore,

$$\left| -2 + \cos k_y + \frac{\cos k_z - \cos Q}{\sin Q} \right| < 1 \Leftrightarrow \exists \text{ edge states} \,, \tag{5.24}$$

and the dispersion is given by $\pm |\boldsymbol{b}_{\perp}|$, therefore,

$$E_s(k_y, k_z) = \pm |\sin k_y| \,. \tag{5.25}$$

Figure 5.1 shows the region (shadowed) defined by equation (5.24). The full line is defined by $E(k_y, k_z) = 0 \Leftrightarrow k_y = 0$ and connects the Weyl points' projections on the surface plane.

5.4   Chiral anomaly

The chiral anomaly arises when an electric and a magnetic field are simultaneously applied, parallel to each other. The Weyl points $\mathbf{K}$ and $-\mathbf{K}$ then have different electronic occupations, with observable effects in charge transport. This effect is the Condensed Matter realization of the Adler-Bell-Jackiw chiral anomaly that was first proposed in the context of lattice field theory.

Consider the Weyl point

$$\hat{H} = v_x p_x \tau_x + v_y p_y \tau_y + v_z p_z \tau_z \,, \tag{5.26}$$

with chiral charge $c = sgn(v_x v_y v_z)$. The momentum $\mathbf{p} = -i\hbar \nabla$. A magnetic field $\boldsymbol{B} = (0,0,B)$ may be described by the vector potential $\boldsymbol{A} = B(0, x, 0)$. After making the substitution $\mathbf{p} \to \mathbf{p} - e\boldsymbol{A}$, the Hamiltonian matrix can be written as

$$H = \begin{pmatrix} v_z p_z & \mathcal{O} \\ \mathcal{O}^{\dagger} & -v_z p_z \end{pmatrix} \,, \tag{5.27}$$

where $\mathcal{O} = v_x p_x - i v_y (p_y - eBx)$. If $B > 0$, the commutator $[\mathcal{O}, \mathcal{O}^\dagger] = -2\hbar v_x v_y eB > 0$. Hence one obtains an algebra similar to that in section 3.4. We may introduce states $|\phi_n\rangle$ with $n = 0, 1, ...,$ which obey

$$\mathcal{O}|\phi_0\rangle = 0, \tag{5.28}$$

$$|\phi_n\rangle = (\mathcal{O}^\dagger)^n |\phi_0\rangle, \tag{5.29}$$

$$\mathcal{O}|\phi_n\rangle = n(-2\hbar v_x v_y eB)|\phi_{n-1}\rangle, \tag{5.30}$$

$$\langle\phi_n|\phi_n\rangle = n!(-2\hbar v_x v_y eB)^n. \tag{5.31}$$

The energy E eigenstates of the Hamiltonian (5.27) take the form, in real space,

$$\psi_n(\mathbf{r}) = e^{i(k_y y + k_z z)} \begin{pmatrix} (E + \hbar v_z k_z)\phi_{n-1}(x) \\ \phi_n(x) \end{pmatrix}, \qquad n \geq 1 \tag{5.32}$$

$$E = \pm\sqrt{(\hbar v_z k_z)^2 + n(-2\hbar v_x v_y eB)}, \tag{5.33}$$

and there is also the zero Landau level:

$$\psi_0(\mathbf{r}) = e^{i(k_y y + k_z z)} \begin{pmatrix} 0 \\ \phi_0(x) \end{pmatrix}, \tag{5.34}$$

with energy $E_0 = -\hbar v_z k_z$. Such a dispersion leads to a group velocity $-v_z$ along the Oz axis. This calculation applies to a Weyl point with chiral charge $\mathrm{sgn}(v_x v_y v_z)$.

Space inversion symmetry implies the existence of another point $(-\mathbf{K})$ with opposite chirality. If we consider inversion symmetry as in equation (5.4), the Weyl point at $-\mathbf{K}$ has symmetric velocities v_μ. The expressions for the eigenstates are the same as calculated above, but the dispersion of the zero Landau level (and the corresponding group velocity) is symmetrical. If the product $v_x v_y < 0$ at point $-\mathbf{K}$, the reader may then repeat the calculation of the wave functions and obtain the dispersion $E_0 = \hbar v_z k_z$. So, the group velocity has a sign opposite to that at point $\mathbf{K}$.

What are the physical consequences? In a system with zero chemical potential, the zero level states at $\mathbf{K}$ are occupied for $k_z > 0$. And at $-\mathbf{K}$ the states are occupied for $k_z < 0$. The application of the magnetic field made the system become metallic along the Oz axis.

The chiral anomaly manifests itself when there exist parallel electric and magnetic fields. It corresponds to a different fermionic occupation at points $\mathbf{K}$ and $-\mathbf{K}$. If we consider an electric field, $\mathbf{E} = (0, 0, E)$, semiclassical dynamics, $\hbar\dot{\mathbf{k}} = \mathbf{E}$, implies that the electronic system goes "down" the branch $E_0 = -\hbar v_z k_z$ at $\mathbf{K}$, and goes "up" the branch $E_0 = \hbar v_z k_z$ at $-\mathbf{K}$. To put it another way, the effect of a battery is to create different populations moving to the right and to the left. This then leads to different occupations. The only process that can restore equal occupations in the two branches is electron scattering, by impurities, between the Weyl points. However, the mean collision time is long in a clean system, and the Landau level degeneracy is high and proportional to the magnetic field. The application of parallel E and B then leads to a negative magnetoresistance[2] . The latter is high when compared to that in conventional metals or semiconductors, which is positive and weak.

5.5   Perturbation of a Dirac point

Weyl points result from the perturbation of a Dirac point by space inversion and/or time reversal symmetry breaking. The most simple Dirac metal is described by the Hamiltonian

$$H(\mathbf{k}) = -v_F \tau_z \mathbf{k} \cdot \boldsymbol{\sigma} \,, \tag{5.35}$$

which enjoys both time reversal symmetry, $H(\mathbf{k}) = \sigma_y H^*(-\mathbf{k})\sigma_y$, and space inversion, $H(\mathbf{k}) = \tau_x H(-\mathbf{k})\tau_x$.

Indeed, one may obtain (5.35) from the Dirac equation,

$$(i\hbar\gamma^\mu \partial_\mu - mc)\psi = 0 \Leftrightarrow (i\hbar\partial_0 + i\hbar\gamma^0\boldsymbol{\gamma} \cdot \nabla - mc\gamma^0)\psi = 0 \,. \tag{5.36}$$

If we now use the Weyl representation for the γ^μ matrices,

$$\gamma^0 = \tau_x \sigma_0, \qquad \gamma^j = i\tau_y\sigma_j \qquad (j = x, y, z) \,, \tag{5.37}$$

we can then rewrite (5.36) as

$$i\hbar\partial_0\psi = (-\tau_z\boldsymbol{\sigma} \cdot \hat{\boldsymbol{p}} + mc\tau_x\sigma_0) \, \psi \,, \tag{5.38}$$

and we only have to remember that, in Condensed Matter, the Fermi velocity, v_F, replaces the light velocity, c. Equation (5.35) holds for the massless case.

Perturbing (5.35) with a Zeeman field, $\boldsymbol{b}$, we obtain the Hamiltonian

$$H(\mathbf{k}) = -v_F\tau_z\mathbf{k} \cdot \boldsymbol{\sigma} + \tau_0\boldsymbol{b} \cdot \boldsymbol{\sigma} \,, \tag{5.39}$$

where both inversion and time reversal symmetries are broken. If, for example, $\boldsymbol{b} = b\hat{x}$, then the four bands in the spectrum are given by

$$E(\mathbf{k}) = \pm\sqrt{(v_F k_x \pm b)^2 + v_F^2 k_y^2 + v_F^2 k_z^2} \,, \tag{5.40}$$

where the signs $\pm$ are independent. There appear the points $\boldsymbol{K} = \pm b/v_F$, where two bands touch at $E = 0$. They are Weyl points.

---

[2]Magnetoresistance: MR=$[\rho(\boldsymbol{B}) - \rho(0)] \, /\rho(0)$, where ρ denotes the resistivity.

One may think of other choices for the perturbation terms. For instance, the choice

$$H(\mathbf{k}) = -v_F \tau_z \mathbf{k} \cdot \boldsymbol{\sigma} + \tau_x \boldsymbol{b} \cdot \boldsymbol{\sigma}, \tag{5.41}$$

also breaks time reversal but preserves space inversion symmetry. If we choose

$$H(\mathbf{k}) = -v_F \tau_z \mathbf{k} \cdot \boldsymbol{\sigma} + \tau_y \boldsymbol{b} \cdot \boldsymbol{\sigma}, \tag{5.42}$$

then space inversion symmetry is broken but time reversal is preserved. For $\boldsymbol{b} = b\hat{x}$, for instance, one obtains the four-band spectrum of (5.41) or (5.42),

$$E(\mathbf{k}) = \pm \sqrt{v_F^2 k_x^2 + \left(v_F \sqrt{k_y^2 + k_z^2} \pm b\right)^2}, \tag{5.43}$$

where the signs $\pm$ are independent. The case $-b$ in brackets leads to a *nodal line* defined by $\sqrt{k_y^2 + k_z^2} = b \wedge k_x = 0$. It describes a semimetal with two bands that touch on the nodal line at energy $E = 0$. We shall study nodal lines further ahead.

5.6 Type II Weyl points

The most simple model for a type II Weyl point reads

$$\hat{H}(\mathbf{k}) = ak_z + \mathbf{k} \cdot \boldsymbol{\tau}, \qquad |a| > 1, \tag{5.44}$$

$$E_\pm(\mathbf{k}) = ak_z \pm |\mathbf{k}|. \tag{5.45}$$

In order to understand this dispersion, we first consider $\mathbf{k} = (0, 0, k_z)$. The lower and upper branches, $E_\pm$, take the form

$$E_+(k_z) = \begin{cases} (1+a)k_z, & k_z > 0, \\ (-1+a)q_z, & k_z < 0. \end{cases} \tag{5.46}$$

$$E_-(k_z) = \begin{cases} (-1+a)k_z, & k_z > 0, \\ (1+a)k_z, & k_z < 0. \end{cases} \tag{5.47}$$

The effect of $a > 1$ is to incline the Dirac cone as shown in figure 5.2 (left). Note that the part $k_z < 0$ of the upper cone lies below zero energy. It is, therefore, occupied by electrons. In a similar way, the part $k_z > 0$ of the lower cone lies above zero energy, so, it is occupied by holes. The colored areas in the figure depict the band filling.

The other momentum components appear in the combination $k_\perp = \sqrt{k_x^2 + k_y^2}$. The part of the upper band that is occupied by electrons is given by

$$E_+(\mathbf{k}) = \sqrt{k_z^2 + k_\perp^2} + ak_z \leq 0, \tag{5.48}$$

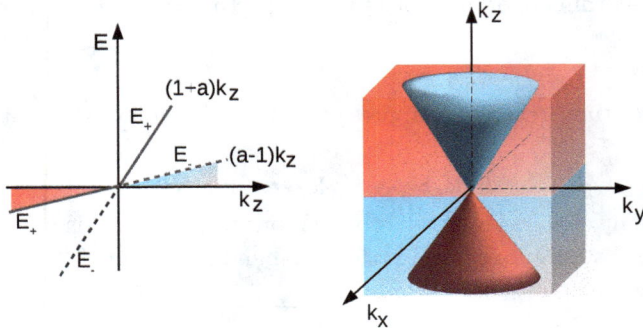

Fig. 5.2 Left: $E_+(k_z, k_\perp = 0)$ in full line and $E_-(k_z, k_\perp = 0)$ in dashed line. The colored regions identify which part of the dispersion E_+ is occupied by electrons (red), and the part of the E_- band occupied by holes (blue). Right: cone shaped Fermi surface at a type II Weyl point. States occupied by electrons (red) and holes (blue).

and is limited by the electronic Fermi surface $k_\perp = \sqrt{a^2 - 1}|k_z|$, where $k_z < 0$. This Fermi surface is cone-shaped around the $k_z < 0$ semi-axis. Inside the cone there are electrons (red color in figure 5.2), and outside it there are holes (colored blue in the figure).

The part of the lower band occupied with holes is given by

$$E_-(\mathbf{k}) = -\sqrt{k_z^2 + k_\perp^2} + ak_z \geq 0, \tag{5.49}$$

and is limited by the hole Fermi surface $k_\perp = \sqrt{a^2 - 1}k_z$, with $k_z > 0$. The Fermi surface is cone-shaped around the $k_z > 0$ semi-axis. There are holes inside the cone, and electrons outside. So, two cones touch at the origin. One with holes ($k_z > 0$) and the other with electrons ($k_z < 0$), as shown in figure 5.2.

The calculation of the density of states is more cumbersome than for the type I case. The result is qualitatively different:

$$\rho(E) = const. - \frac{E^2(a^2 + 1)}{a(a^2 - 1)^2}. \tag{5.50}$$

It is, therefore, finite at $E = 0$, and decays quadratically with energy.

Because the Fermi surface is open, type II Weyl points are neither sources nor drains of Berry curvature.

5.7 Nodal rings

In a topological three-dimensional semimetal, the contact between conduction and valence bands may occur on a line instead of on a set of discrete points. A nodal line may be either open (when it crosses the Brillouin zone from one end to another) or closed. In the latter case it is also called *Dirac ring* or *Weyl ring*.

A minimalist version of a Weyl ring Hamiltonian can be written as:

$$H(\mathbf{k}) = \hbar v_1 \left(k_\| - k_0\right) \tau_x + \hbar v_2 k_\perp \tau_y \,, \tag{5.51}$$

where $k_\|$ denotes the wave vector component parallel to the ring plane, and $k_\perp$ the perpendicular component. The velocities v_1 e v_2 are often omitted (renormalized to unity) as they do not have an influence on topological properties.

The Hamiltonian (5.51) is assumed to be valid when $k_\| \approx k_0$ and $k_\perp \to 0$. It actually is the linearized version of another Hamiltonian valid in a wider region of the Brillouin zone. The first term in (5.51), for instance, might be $\hbar^2(k_\|^2 - k_0^2)/(2m)\tau_x$, of which the linearized version is (5.51).

5.7.1 *Topological invariant for nodal lines*

Let $H(\mathbf{k}) = \left(k_0^2 - k_\perp^2\right) \tau_z + vk_z\tau_y$, where $\mathbf{k}$ is three-dimensional and $k_\perp^2 = k_x^2 + k_y^2$. The symmetries $\mathcal{P}, \mathcal{T}$ are both present, with parity implemented by $P = \tau_z$. The existence of these two symmetries implies that one component of $\mathbf{h}(\mathbf{k})$ vanishes. Then, the condition $\mathbf{h}(\mathbf{k}) = \mathbf{0}$ yields a nodal line in three-dimensional momentum space. In this model, the nodal line is a circle, $(k_\perp = k_0, k_z = 0)$, in the xy plane. One may adopt the following Bloch wave functions for the lower band:

$$\psi(\mathbf{k}) = \begin{pmatrix} \alpha \\ \beta \end{pmatrix} \,, \quad \alpha = \sqrt{\frac{1}{2}\left(1 - \frac{h_z}{h}\right)} \,, \quad \beta = -\frac{i}{h_y}\left(h + h_z\right)\alpha \,, \tag{5.52}$$

where $h^2 = h_y^2 + h_z^2$. We always have $h + h_z \geq 0$, and $h + h_z = 0$ only when $k_z = 0$. The Berry connection,

$$\mathcal{A}(\mathbf{k}) = -|\alpha|^2 \nabla arg(\alpha) - |\beta|^2 \nabla arg(\beta) \,. \tag{5.53}$$

We see that the Berry connection $\mathcal{A} = 0$ except inside a circle limited by the nodal line $(k_\perp < k_0; k_z = 0)$. Indeed, the phase of $\beta = \pi/2$ for $k_z < 0$ and it is equal to $-\pi/2$ for $k_z > 0$. Therefore, the gradient of the phase is a Dirac delta function, but it only contributes to $\mathcal{A}$ when $|\beta| \neq 0$. This occurs only at a point inside the circle where $|\beta| = 1$. We obtain, in cylindrical coordinates,

$$\mathcal{A}(\mathbf{k}) = \pi\delta(k_z)\Theta(k_0 - k_\perp)\hat{k}_z \,, \tag{5.54}$$

$$\nabla \times \mathcal{A}(\mathbf{k}) = \mathbf{\Omega} = -\pi\delta(k_z)\delta(k_0 - k_\perp)\mathbf{u}_\phi \,. \tag{5.55}$$

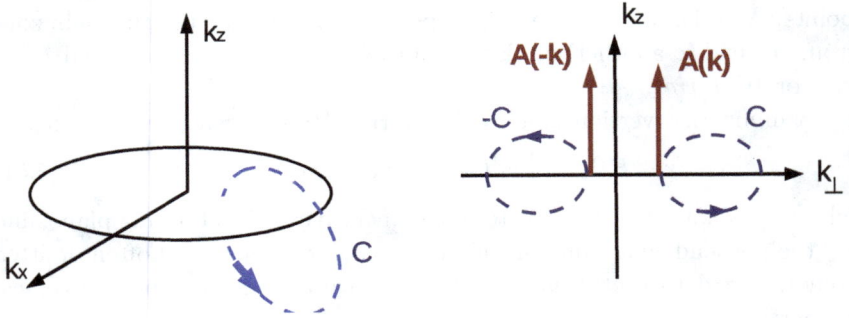

Fig. 5.3   Berry phase from the circulation (blue) around the nodal line.

The Berry curvature lives on the nodal line. It is analogous to a vortex ring. Let $\mathcal{C}$ denote a closed path around the nodal line (see figure 5.3). For the path $\mathcal{C}$ we get $\gamma(\mathcal{C}) = -\pi$. For the symmetric path, $\gamma(-\mathcal{C}) = \pi = -\gamma(\mathcal{C})$, because the vector $\mathcal{A}$ is the same and the circulation on the path $-\mathcal{C}$ is performed in the same direction as in the path $\mathcal{C}$. We get, in this case, the Berry phase exponential as $e^{i\gamma} = -1$. It is said that the nodal line has a topological charge $\mathbb{Z}_2$. If the nodal line contracts to a point, $k_0 \to 0$, the same discontinuity of $\arg(\beta)$ would occur at the origin (which would be the nodal point), where the phase change would be from π to $|\beta|^2 = 1/2$.

Not all nodal lines with $\mathcal{P}, \mathcal{T}$ symmetry have a $\mathbb{Z}_2$ charge: consider, now, the example $h_x = 0$, $h_y = vk_y$, $h_z = k_x^2$, where the k_z is the nodal line. Consider the closed path in the xy plane, around the nodal axis. We can choose the same α, β as above, with $\arg(\alpha) = 0$, and $\arg(\beta) = \pi \left[\frac{1}{2} - \Theta(h_y) \right]$. Hence we get $\mathcal{A} = \pi \delta(k_y)\hat{k}_y$ which has vanishing circulation, therefore, $\gamma = 0$.

Finally, we note that a metal's Fermi surface is, in general, a nodal line/surface for excitations. The latter are poles of the single-particle Green's function. It is then possible to define a winding number from a circulation analogous to that depicted in figure 5.3. A detailed discussion on this point is presented in Appendix 9.3.2.

Consider the Hamiltonian

$$H(\boldsymbol{p}) = (p_\parallel - p_0)\tau_x + p_z\tau_y, \qquad \text{where} \quad p_\parallel = \sqrt{p_x^2 + p_y^2}\,.$$

Exercise 5.1

Show that there is reflection symmetry in the plane $p_z = 0$ and identify the corresponding operator, $\mathcal{M}_z$.

Exercise 5.2

Using the coordinate ϕ in momentum space, $\phi = arg\left(p_\parallel - p_0 + ip_z\right)$, write down Hamiltonian eigenstates that are also eigenvectors of $\mathcal{M}_z$ with eigenvalues $\pm i$.

Exercise 5.3

Consider the lower band. Show that the states in the plane $p_z = 0$ inside the nodal line have eigenvalues of $\mathcal{M}_z$ equal to $+i$, and that the states outside have eigenvalue $-i$.

5.7.2 *Drumhead edge states*

A nodal line limits a region in momentum space (inside or outside the line). Such a region may produce zero energy surface states in a finite or semi-infinite sample. The surface states are characterized by a momentum vector parallel to the surface.

One may understand the appearance of such surface states by analyzing an example. Consider the Hamiltonian

$$H(\mathbf{k}) = \left[t_1 \left(\cos k_x + \cos k_y - b\right) + t_2 \left(\cos k_z - 1\right)\right] \tau_x + \sin k_z \tau_y. \quad (5.56)$$

This model has a nodal line given by $\cos k_x + \cos k_y - b = 0 \wedge k_z = 0$. For this line to be unique, it is necessary that $|b| < 2 \wedge |b + 2t_2/t_1| > 2$.

Consider a finite sample along the z axis. When looking for edge states we resort to the concepts of chirality and winding number. We fix k_x, k_y and consider the Hamiltonian as describing a one-dimensional system, being a function of $k_z \in] -\pi, \pi]$. It then has off-diagonal form, where the matrix element

$$H_{12}(k_z) = t_1 \left(\cos k_x + \cos k_y - b\right) + t_2 \left(\cos k_z - 1\right) - i \sin k_z. \quad (5.57)$$

The Hamiltonian enjoys chiral symmetry. As a function of k_z, H_{12} describes an ellipse in the complex plane, with semi-axes $a = t_2$ on the real axis, and $b = 1$ on the imaginary axis. The center of the ellipse is the point $(x_0, y_0) \equiv [\ t_1 (\cos k_x + \cos k_y - b) - t_2, 0\ ]$. The equation for the ellipse is $(x - x_0)^2/a^2 + y^2/b^2 = 1$. In order to have the winding number $W \neq 0$, it is necessary that the origin stays inside the ellipse: $x_0^2 < a^2$. This implies

$$(\cos k_x + \cos k_y - b)\left[\cos k_x + \cos k_y - b - 2\frac{t_2}{t_1}\right] < 0. \quad (5.58)$$

In the case $b + 2t_2/t_1 > 2$, the second factor is negative. So, the region $\cos k_x + \cos k_y > b$ will have zero energy surface states. In the case $b +

$2t_2/t_1 < -2$ the second factor is positive, therefore, the region $\cos k_x + \cos k_y < b$ has zero energy surface states.

The surface Brillouin zone $(-\pi < k_x \leq \pi; -\pi < k_y \leq \pi)$ is not complete because surface states exist only in the region limited by the projection of the nodal line on the $k_z = const$ plane.

If there are two nodal lines, only the region (k_x, k_y) between them will have edge states. Indeed, if we put $|b| < 2 \wedge |b + 2t_2/t_1| < 2$, we obtain a second nodal line $k_z = \pi \wedge \cos k_x + \cos k_y = b + 2t_2/t_1$. The condition (5.58) is fulfilled in one of the cases:

$$-2 < b < \cos k_x + \cos k_y < b + 2\frac{t_2}{t_1} < 2 \ \text{if} \ \frac{t_2}{t_1} > 0,$$

$$2 > b > \cos k_x + \cos k_y > b + 2\frac{t_2}{t_1} > -2 \ \text{if} \ \frac{t_2}{t_1} < 0.$$

5.8 $\mathbb{Z}_2$ nodal rings

In order to obtain a $\mathbb{Z}_2$ nodal ring, we start by considering the Weyl point

$$H_W(\mathbf{k}) = \mathbf{k} \cdot \boldsymbol{\tau}, \tag{5.59}$$

with chiral charge $c = +1$. In similarity to the Kane-Mele topological insulator, we introduce an additional pseudo-spin degree of freedom, σ_i with $i \in \{x, y, z\}$, multiplying one of the τ matrices, for instance,

$$H_W'(\mathbf{k}) = k_x \sigma_i \tau_x + k_y \tau_y + k_z \tau_z. \tag{5.60}$$

The particles with eigenvalue of $\sigma_i = \pm 1$ see a chiral charge ± 1. So, the states are divided in two sectors. In order to create a nodal ring we need to introduce a term $\sigma_\nu \tau_\mu$. One may choose such a term to commute with the k_x, k_y terms, and anticommute with the k_z term. Then, $(\nu, \mu) = (j, y)$, where $j \neq i$:

$$H(\mathbf{k}) = k_x \sigma_i \tau_x + k_y \tau_y + k_z \tau_z + k_0 \sigma_j \tau_y, \qquad i \neq j. \tag{5.61}$$

The spectrum of (5.61) may be written down by inspection, taking into account that the τ_z term anticommutes with all the others,

$$E^2(\mathbf{k}) = \left(\sqrt{k_x^2 + k_y^2} \pm k_0\right)^2 + k_z^2, \tag{5.62}$$

and the nodal ring is given by $\sqrt{k_x^2 + k_y^2} = k_0 \wedge k_z = 0$.

If k_0 changes continuously from negative to positive, the nodal line shrinks to a point and then opens up again: *it cannot be destroyed by varying* k_0. This is an important difference to the nodal ring (5.51). The latter is destroyed as k_0 goes from positive to negative.

Because there can only be a even number of Weyl points in the Brillouin zone, there must be an even number of $\mathbb{Z}_2$ rings. Such a restriction to an even number does not exist for rings of the type (5.51).

The justification for the existence of a $\mathbb{Z}_2$ index in this model follows from the content of section 1.6. An equivalent version of model (5.61) is:

$$H(\mathbf{k}) = k_x \tau_x + k_y \sigma_y \tau_y + k_z \tau_z + k_0 \sigma_x \tau_x , \qquad (5.63)$$

where the Hamiltonian matrix is real in momentum space. This is because of a PT symmetry with $(\mathcal{PT})^2 = 1$: see equation (2.56) with $A = 1$. Because $H(\mathbf{k})$ is real, its eigenvectors are real and possible basis changes in the occupied (or unoccupied) subspace can be achieved using orthogonal matrices, from the group $O(2)$. Consider a sphere, in $\mathbf{k}$ space, that contains the nodal line. The $\mathbb{Z}_2$ character of the model means that it is not possible to write a wave function that is single-valued all over the sphere. Suppose we write two wave functions, for the occupied subspace, in a gauge such that the functions are single-valued in the southern hemisphere. And that we write other two that are single-valued in the northern hemisphere. Then, the orthogonal $O(2)$ matrix that changes from one basis to the other has itself a topological invariant in group $\pi_1[O(2)]$, which corresponds to an odd winding number. The exercise 5.6 below helps clarify this point.

Exercise 5.4

Check the PT symmetry in the Hamiltonian (5.61) for $(i, j) = (2, 3)$, by identifying the matrix A in definition (2.56).

Exercise 5.5

Using a similar reasoning to that that led to (5.61), write down a model where the nodal ring is in the xz plane.

Exercise 5.6

Consider the Hamiltonian (5.63) for $k_0 = 0$. Consider $|\mathbf{k}| = const.$ and parametrize $\mathbf{k}$ through its spherical coordinates, θ, ϕ.

(1) Write down a wave function for the lower band that is an eigenstate of σ_y with eigenvalue 1, by taking equation (2.78) as example.
(2) Because the Hamiltonian matrix is real, the real and imaginary parts of the solutions obtained in (1) are the two lower band occupied states.

Check that these functions are single-valued in the south pole but not in the north pole. Denote the functions as $|u_1^S\rangle$, $|u_2^S\rangle$.

(3) How can one modify (1) so as to obtain a new basis of occupied states that is single-valued in the north pole (N) but not in the south pole (S)? Denote such functions as $|u_1^N\rangle$, $|u_2^N\rangle$.

(4) Write down an orthogonal matrix of dimension 2, $M(\phi)$, that changes from one basis (S) to the other (N).

(5) Consider the function $M_{11} + iM_{22}$ and check that its winding number around the origin in the complex plane is equal to 1, when ϕ varies from 0 to 2π. The value 1 so obtained means that it is impossible to choose a basis for the occupied states that is single-valued all over the sphere $|\mathbf{k}| = const.$, and therefore, in both poles[3].

Further reading:

On topological semimetals and Fermi arcs:

- X. Wan et al., *Physical Review* B **83**, 205101 (2011).
- A. A. Burkov, M. D. Hook, and L. Balents, *Physical Review* B **84**, 235126 (2011).
- T. Ojanen, *Physical Review* B **87**, 245112 (2013).

On the topological semimetals' symmetry properties:

- S. Matsuura et al., *New Journal of Physics* **15**, 065001 (2013).
- C.-K Chiu, and A. P. Schnyder, *Physical Review* B **90**, 205136 (2014).
- K. Shiozaki, and M. Sato, *Physical Review* B **90**, 165114 (2014).

On Weyl points in a magnetic field or under electromagnetic radiation, and the chiral anomaly:

- H.-Z. Lu, S.-B. Zhang, and S.-Q. Shen, *Physical Review* B **92**, 045203 (2015).
- X. Huang et al., *Physical Review* X **5**, 031023 (2015).
- C.-K. Chan et al., *Physical Review Letters* **116**, 026805 (2016).

---

[3] A more complete discussion can be found in the paper C. Fang et al., *Physical Review* B **92**, 081201 (2015).

A review on superconductivity in Weyl points:

- A. P. Schnyder and P. M. R. Brydon, *J. Physics: Condensed Matter* **27**, 243201 (2015).

On topological insulators obtained from nodal rings:

- L. Li, and M. A. N. Araújo, *Physical Review* B **94**, 165117 (2016).
- L. Li et al., *Physical Review* B **95**, 121107(R) (2017).

Some papers on superconductivity and spontaneous symmetry breaking in nodal rings:

- R. Handkishore, *Physical Review* B **93**, 020506 (2016).
- S. Sur and R. Nandkishore, *New Journal of Physics* **18**, 115006 (2016).
- Y. Wang, and R. M. Nandkishore, *Physical Review* B **95**, 060506 (2017).
- B. Roy, *Physical Review* B **96**, 041113 (2017).
- M. A. N. Araújo, and L. Li, *Physical Review* B **98**, 155114 (2018).

The $\mathbb{Z}_2$ nodal ring has been introduced in:

- C. Fang et al., *Physical Review* B **92**, 081201(R) (2015)

and has been included in a review on nodal line semimetals:

- C. Fang et al., *Chinese Physics* B **25**, 117106 (2016).

Chapter 6

Spin systems with topological properties

6.1 Representations of spin systems

A magnetic insulator may be described by an ordered set of local magnetic moments (or by local spin operators). The spin operators have commutation relations that are more complex than the commutation (anticommutation) relations of the bosonic (fermionic) operators, since the commutator of two spin operators is another spin operator, instead of a constant. In some cases, it is convenient to adopt a description of the spin operators in terms of bosons or fermions, but in general this description requires an enlargement of the Hilbert space. The spin-1/2 case is a bit special, since one may expect that the two states (up or down spins), may be related with the two states of one fermion (state occupations zero or one). Considering more than one spin operator, while the fermionic operators satisfy anticommutation relations between different space points, two spin operators on different space locations commute, since two local operators are distinguishable.

Considering the interest and importance of spin systems, a possible connection between a spin problem and fermionic (or bosonic) models establishes a relation between the topics of previous chapters and spin systems.

Among the various representations of spin operators in terms of fermionic operators, two are particularly useful. One is a representation in terms of Majorana fermions (hermitian fermionic operators) and the other the Jordan-Wigner representation.

Majorana operators allow a representation of spin-1/2 operators. One needs three Majorana operators to represent a local spin operator (there are other ways to establish the representation but we limit ourselves to this representation)

$$S_x = -(i/2)\gamma_2\gamma_3; \qquad S_y = -(i/2)\gamma_3\gamma_1; \qquad S_z = -(i/2)\gamma_1\gamma_2. \quad (6.1)$$

We may explicitly check that the commutation relations of the spin operators are satisfied, considering that the Majorana operators anticommute, $\{\gamma_i, \gamma_j\} = 2\delta_{ij}$, with $i = x, y, z$, and that the norm of the spin operator is also satisfied.

An interesting question that arises as we represent fermions and spins is if there is an enlargement of the number of states. In case one represents non-hermitian fermionic operators in terms of Majorana fermions, there is no enlargement of the number of states. Starting from the fermionic operators

$$c = \frac{1}{2}\left(\gamma_1 + i\gamma_2\right),$$

$$c^\dagger = \frac{1}{2}\left(\gamma_1 - i\gamma_2\right), \quad (6.2)$$

we may define Majorana operators

$$\gamma_1 = c^\dagger + c,$$

$$\gamma_2 = i\left(c^\dagger - c\right). \quad (6.3)$$

Defining the operator $\mathscr{S}_3 = -i\gamma_1\gamma_2 = (2c^\dagger c - 1)$, this anticommutes with both γ_1 and γ_2, but is not independent, and there is no enlargement of the number of states, that remains 2. However, when one considers a bilinear representation of the spin operators, there is an enlargement of the number of states, as a result of the symmetry $\mathbb{Z}_2$ associated with the bilinear representation, since the spin operators remain invariant if we consider $\gamma_i \to -\gamma_i$.

In the case of one spin operator, the number of states doubles, with four states instead of two. The simplest way to understand this result is to introduce an additional Majorana operator to pair with one of the three Majorana operators, used in the representation of the spin operators, which allows to obtain a second pair of non-hermitian creation and destruction fermionic operators. Therefore, we may, for instance, define a fermionic operator as $f = \frac{1}{2}\left(\gamma_x + i\gamma_y\right)$ and a second one as $g = \frac{1}{2}\left(\gamma_z + i\gamma_a\right)$, where γ_a is the additional Majorana operator. Applying this procedure to a set of N spins $\frac{1}{2}$ we have then 2^{2N} states instead of 2^N. In general, if we have N spins $\frac{1}{2}$, the number of states is 2^N. Introducing three operators for each spin, we get $3N$ Majorana operators. If the number of lattice sites N is even, we may associate the operators in pairs obtaining $\frac{3N}{2}$ pairs of fermionic operators, and the number of states is, therefore, $2^{\frac{3N}{2}}$. If the number of states is odd, we may introduce an additional operator and the number of states is $2^{\frac{3N+1}{2}}$. We should note, however, that the space enlargement appears as a simple multiplication factor in the partition function of the spin system.

In the case of spin-1/2 chains, an alternative representation is the Jordan-Wigner representation (it is also possible to consider higher dimensional systems, but this generalization will not be considered here). The Jordan-Wigner transformation of spin operators to spinless fermionic operators, is defined as

$$S_j^+ = c_j^\dagger e^{i\pi \sum_{n=1}^{j-1} c_n^\dagger c_n},$$
$$S_j^- = e^{-i\pi \sum_{n=1}^{j-1} c_n^\dagger c_n} c_j,$$
$$S_j^z = c_j^\dagger c_j - \frac{1}{2}. \tag{6.4}$$

Let us consider a one-dimensional system of spin-1/2 operators that has an anisotropy in the XY plane and that is described by the Hamiltonian

$$H_{XY} = -\sum_{j=1}^{N} \left(J_x S_j^x S_{j+1}^x + J_y S_j^y S_{j+1}^y \right). \tag{6.5}$$

Defining $S^x = (S^+ + S^-)/2; S^y = (S^+ - S^-)/(2i)$, we may write

$$H_{XY} = -\frac{1}{4} \sum_j \left[(J_x - J_y) \left(S_j^+ S_{j+1}^+ + S_j^- S_{j+1}^- \right) \right]$$
$$-\frac{1}{4} \sum_j \left[(J_x + J_y) \left(S_j^+ S_{j+1}^- + S_j^- S_{j+1}^+ \right) \right]. \tag{6.6}$$

Using the Jordan-Wigner transformation, leads to

$$H_{XY} = -\frac{1}{4} \sum_j \left[(J_x - J_y) \left(c_j^\dagger c_{j+1}^\dagger + c_{j+1} c_j \right) \right]$$
$$-\frac{1}{4} \sum_j \left[(J_x + J_y) \left(c_j^\dagger c_{j+1} + c_{j+1}^\dagger c_j \right) \right]. \tag{6.7}$$

This model is, therefore, related to the Kitaev model (with zero chemical potential) choosing $t = (J_x + J_y)/4; \Delta = (J_x - J_y)/4$.

In general, the transformation between the spin operators and the fermionic operators involves a factor, denoted string, that counts the number of fermions to the left of a given lattice site. This factor is, in general, complicated to deal with, but is absent at the level of the Hamiltonian, if only nearest-neighbors are considered. In this case, the duality between the two problems allows the diagonalization of the spin problem, due to its simplicity in terms of the fermionic operators. In general, the relation between correlation functions in the spin problem and in the fermionic problem is not trivial. In the same way, including terms that couple second neighbors or further away neighbors, or including spin terms with components along the z axis, the Hamiltonian, expressed in terms of the fermionic operators, becomes interactive and the solution complicates.

In the case of spin systems with $S > 1/2$, the relation between the spin problems and a correspondent bosonic or fermionic problem is more complex. In the next section we will consider a method for general spin-S that reveals the topological nature of one-dimensional spin systems.

6.2   Spin chains

The groundstate of one-dimensional antiferromagnetic systems, with half-integer spins, is described by the Lieb, Shultz and Mattis theorem, as being either a spin singlet state with a gapless spectrum, or to be doubly degenerate with a gap to the first excited state, corresponding to a breaking of translation symmetry, as, for example, a spontaneous dimerization of the system. The exact solution of the model by the Bethe ansatz confirms this result for a spin-1/2 system with translation symmetry, showing that the spectrum is gapless.

The Lieb, Shultz, Mattis theorem does not give information when the spins have integer values. Also, the Bethe ansatz does not provide information on this case, since, in general, the problem does not have an exact solution (there are, however, some spin models that are soluble by the Bethe ansatz and that may be solved for arbitrary spin values, half-integer and integer; these exactly soluble models are gapless since they are critical models that respect to transition points between various phases of the parameter space). Using a representation of the spin operators in terms of coherent states, Haldane concluded that antiferromagnetic spin chains of integer and half-integer spins are divided in two classes: as mentioned above, systems with half-integer spins have gapless spectra if they are not dimerized, but the integer spin systems have a gap in the spectrum. Actually, using the representation of the spin problem using coherent states, the problem may be solved approximately. The dominant term for the partition function corresponds to the limit when the spin tends to infinity (classical limit) and, in this regime, the one-dimensional system has a gapped spectrum.

Expanding around the classical limit, and considering finite spin values, this result holds. However, in the construction of the path integral for the spin system, in addition to a term that is associated with the Hamiltonian of the system, the Lagrangian involves a term that has a topological nature and that distinguishes between integer and half-integer spin values. If the spin is integer, this term does not give a contribution. But when the spin is half-integer this term contributes with a topological term, whose sign depends on the winding number of the spin configuration. This winding number measures the number of times that the spin configuration covers the Bloch sphere as one changes the spatial coordinate, along the spin chain, and the imaginary time of the construction of the path integral for the partition function. Taking into account that the Bethe ansatz leads to a gapless spectrum, Haldane conjectured that this additional topological term is responsible for a different behavior between the half-integer and integer spin chains. Indeed, this conjecture has been confirmed extensively, both with numerical calculations and experimentally, considering either integer spin systems or spin-3/2 systems, or of higher values.

6.2.1   *AKLT projection*

The conjecture of the existence of a gap in the spectrum of integer spin chains (usually referred as being in the Haldane phase), finds an explicit realization in the so-called gapped phases of AKLT (Affleck, Kennedy, Lieb, Tasaki).

A spin-1 chain with exact solution may be constructed explicitly as a system of links between nearest-neighbors, such that each link is projected in a subspace with total spin $S = 2$. As we couple two neighboring $S = 1$ spins, the total spin $\mathbf{S} = \mathbf{S}_i + \mathbf{S}_{i+1}$ may take the values $S = 0, 1, 2$. The Hamiltonian is chosen as a sum of projectors of spin operators in states of total spin of each link as $S = 2$ and may be written as

$$H = \sum_i P_2 \left(\mathbf{S}_i + \mathbf{S}_{i+1} \right), \tag{6.8}$$

where the projector P_2 is such that the spin that is the result of the sum of $\mathbf{S}_i$ and $\mathbf{S}_{i+1}$ is in the subspace of $S = 2$. One can easily see that if we define

$$P_2 \left(\mathbf{S}_i + \mathbf{S}_{i+1} \right) = \frac{1}{2} \mathbf{S}_i \cdot \mathbf{S}_{i+1} + \frac{1}{6} \left(\mathbf{S}_i \cdot \mathbf{S}_{i+1} \right)^2 + \frac{1}{3} \tag{6.9}$$

and use $\mathbf{S}_i \cdot \mathbf{S}_{i+1} = \mathbf{S}^2/2 - 2$, that leads to

$$\mathbf{S}_i \cdot \mathbf{S}_{i+1}|S = 2\rangle = |S = 2\rangle$$
$$\mathbf{S}_i \cdot \mathbf{S}_{i+1}|S = 1\rangle = -|S = 1\rangle$$
$$\mathbf{S}_i \cdot \mathbf{S}_{i+1}|S = 0\rangle = -2|S = 0\rangle, \tag{6.10}$$

and we obtain

$$P_2((\mathbf{S}_i + \mathbf{S}_{i+1})|S = 2\rangle = |S = 2\rangle$$
$$P_2((\mathbf{S}_i + \mathbf{S}_{i+1})|S = 1\rangle = 0$$
$$P_2((\mathbf{S}_i + \mathbf{S}_{i+1})|S = 0\rangle = 0, \tag{6.11}$$

and, therefore, the action of the projector is to reproduce the state of total spin $S = 2$ and to eliminate the total spin components $S = 0, 1$. As we can see, the energy of each link is zero or positive and, therefore, if it is possible to construct a state with zero energy, this will be the groundstate (or is part of the degenerate subspace of zero energy).

The idea to construct a state of zero energy is simple. If it is possible to define a state of the spin-1 system such that each link has no projection on the $S = 2$ state then, by construction, the action of the Hamiltonian on that state guarantees that this has zero energy and is, therefore, a state of minimal energy. The state, designated by VBS[1], satisfies this condition. It may be represented as

$$\bigcirc\ \bigcirc \longleftrightarrow \bigcirc\ \bigcirc \longleftrightarrow \bigcirc\ \bigcirc \longleftrightarrow \bigcirc\ \bigcirc \longleftrightarrow \bigcirc\ \bigcirc \longleftrightarrow \bigcirc\ \bigcirc \longleftrightarrow \bigcirc\ \bigcirc$$
$$\qquad\qquad j-1 \qquad\quad j \qquad\quad j+1$$

In this state on each lattice site, i, there is a localized spin $S = 1$ operator. Each spin $S = 1$ is decomposed in two spins $1/2$, in a symmetric representation. The lines represent the formation of a singlet state ($S = 0$) between two spins $1/2$ localized on neighboring sites.

As a starting point let us call ψ_α the states of a spin-$1/2$. The spin-1 states may be constructed by the symmetric combination of two spin-$1/2$ operators as

$$\psi_{\alpha\beta} = \frac{1}{\sqrt{2}}\left(\psi_\alpha \otimes \psi_\beta + \psi_\beta \otimes \psi_\alpha\right), \tag{6.12}$$

with the normalization (dot product of two states)

$$(\psi_{\alpha\beta}, \psi_{\gamma\delta}) = \delta_{\alpha\gamma}\delta_{\beta\delta} + \delta_{\alpha\delta}\delta_{\beta\gamma}. \tag{6.13}$$

---

[1]Valence bond solid.

Therefore, each spin $S = 1$ operator is the sum of two spin-1/2 operators. The states with $S = 1$ correspond to the triplet state of these two spin-1/2 operators. Next, one considers that two of the spin-1/2 operators are in a state of $S = 0$, a singlet state. Considering a model with antiferromagnetic interactions, the groundstate should be the result of the formation of spin singlets of the spin-1/2 operators, on different sites. The singlet state between two spin-1/2 operators, one on each neighboring site, may be constructed considering the combination $\psi_\alpha \otimes \psi_\beta \epsilon_{\alpha\beta}$, where $\epsilon_{\alpha\beta}$ is an antisymmetric tensor and repeated indices are summed over. The VBS state for L sites may, then, be constructed as

$$\Omega_{\alpha\beta} = \psi_{\alpha_1\beta_1} \otimes \psi_{\alpha_2\beta_2} \otimes \cdots \otimes \psi_{\alpha_L\beta} \epsilon_{\beta_1\alpha_2} \, \epsilon_{\beta_2\alpha_3} \cdots \epsilon_{\beta_{L-1}\alpha_l}, \qquad (6.14)$$

guaranteeing that at each local site, i, the symmetrical combination of two spins $1/2$ is taken to define a spin $S = 1$ at that site, and that one considers that the spins $1/2$ give rise to singlets with another spin $S = 1/2$ on the neighboring site. In this way the state with $S = 1$ that may result of the coupling between these two spin $1/2$ operators in two neighboring spins is eliminated. Two spin $1/2$ operators are left, one on site i and another on the site $i + 1$, for example. These two spins may now couple to form a spin operator that may take the values $S = 0, 1$. This spin now coupled to the spin $S = 0$, that was the result of the coupling between the other spins $S = 1/2$, on the same sites i and $i+1$, can only lead to a total spin operator of the two sites that takes the values $S = 0, 1$ and therefore the total spin $S = 2$ is not achievable. Therefore, the action of the Hamiltonian on this VBS state gives zero, as wanted.

This state does not break the translation symmetry, if the system is infinite. If the system is finite, two spin-1/2 operators are left, one on each end of the chain, that are not coupled to another spin $S = 1/2$ operator. These two spins give origin to four states at the edge, and, therefore, to a degeneracy of 4. In general, if we have spins-S we consider a partition in $2S$ spins $1/2$. If the spin is $1/2$, the establishment of links with nearest neighbors leads to a dimerization of the system and consequent breaking of the translation symmetry, since the link may be established either to the right or to the left.

One can show that the spectrum has a gap and that the correlation functions decay exponentially with distance. This result is in contrast with the result for spin-1/2 chains where the spectrum is gapless. In this case, the correlation functions decay with a power law in the distance between lattice points.

The AKLT Hamiltonian for spin-1 has the generic form

$$H = \sum_i \left(\mathbf{S}_i \cdot \mathbf{S}_{i+1} - \beta \left(\mathbf{S}_i \cdot \mathbf{S}_{i+1} \right)^2 \right), \tag{6.15}$$

with $\beta = -1/3$. It turns out that in the case of $\beta = 1$ the model is integrable. It can be shown that the spectrum is gapless and the correlation functions decay as a power law. This integrable model is called the Takhtajian-Babujian model and may be generalized to an arbitrary spin value S.

It is interesting that the difference between half-integer and integer spin chains was proposed as the result of a topological term that only appears if the spin is integer. Therefore, half-integer spin chains are not topological. As the Haldane conjecture shows, the spin $S = 1$ chain is topological. This result may be revealed either by the presence of edge states (and consequent degeneracy, or near degeneracy, of the groundstate in the case of a finite system with open boundary conditions), or through the calculation of the Berry phase, as we will see in the next section. Additionally, it can be shown that integer spin chains with odd spin have topological protection while integer spins with even spin are not topologically protected.

6.2.2 *Berry phase*

The assertiveness that a topological phase may exist on a spin $S = 1$ chain may be achieved, alternatively, calculating the Berry phase.

Let us consider that the Hamiltonian of the system depends on a cyclic variable. For instance, an angular variable, ϕ. The Berry phase may be defined as in equation(2.6), minus a sign, as

$$\gamma = \int_0^{2\pi} d\phi A(\phi), \tag{6.16}$$

where

$$A(\phi) = -i\langle GS(\phi) | \partial_\phi | GS(\phi) \rangle, \tag{6.17}$$

with $|GS(\phi)\rangle$ the groundstate, obtained for a given value of the cyclic parameter.

In general, the Berry phase is not invariant under a gauge transformation performed on the Berry connection. It is, however, possible to make a gauge choice and, if the Hamiltonian is invariant under the action of an antiunitary operator, the Berry phase is well defined, and is actually quantized.

The quantization of the Berry phase may be understood in the following way[2]. Let us represent the parameters on which the Hamiltonian depends in a simplified notation x. Then, let us consider that the parameters change sufficiently slowly so that one may define the Hamiltonian spectrum, for each set of values of x, as $H(x)|\psi(x)\rangle = E(x)|\psi(x)\rangle$. The Berry phase along a given contour C may be defined, minus a sign, as $\gamma = \int_C A$, where the Berry connection is given by $A = -i\langle\psi|d\psi\rangle$. We may perform a phase change in the state $|\psi(x)\rangle = |\psi'(x)\rangle e^{i\Omega(x)}$. The Berry connection transforms as $A = A' + d\Omega(x)$. Therefore, the Berry phase is changed. Let us consider now the projector to the state $|\psi\rangle$ and note that it is invariant under the gauge transformation $P = |\psi\rangle\langle\psi| = |\psi'\rangle\langle\psi'|$. We may then introduce an arbitrary reference state $|\Phi\rangle$ and consider its projection on the groundstate

$$|\psi_\Phi\rangle = \frac{1}{\sqrt{N_\Phi}}P|\Phi\rangle = \frac{1}{\sqrt{N_\Phi}}|\psi\rangle\langle\psi|\Phi\rangle = \frac{1}{\sqrt{N_\Phi}}\eta_\Phi|\psi\rangle. \qquad (6.18)$$

This state is normalized if we choose $N_\Phi = |\eta_\Phi|^2$. Let us call $|\Phi\rangle$ a regular state if $\eta_\Phi \neq 0$, whatever the value of x. Let us now consider another regular state $|\Phi'\rangle$. Then,

$$|\psi_\Phi\rangle = |\psi_{\Phi'}\rangle e^{i\Omega}, \qquad (6.19)$$

where $\Omega = \arg(\eta_\Phi - \eta_{\Phi'})$. As a consequence,

$$\gamma(A_{\psi_\Phi}) = \gamma(A_{\psi_{\Phi'}}) + \Delta, \qquad (6.20)$$

where $\Delta = \int_C d\Omega$. Imposing that the reference state, $|\Phi\rangle$, is well defined in the space of parameters, x, the phase difference, Δ, is also well defined and, therefore, $\Delta = 2\pi M$, where M is an integer. Therefore, the Berry phase is gauge invariant, modulo a multiple of 2π:

$$\gamma = \int_c A, \quad \mod (2\pi M). \qquad (6.21)$$

Let us consider, now, that the Hamiltonian remains invariant under the action os some antiunitary operator, $\Theta = KU_\theta$, such as, for example, time reversal. Here K denotes complex conjugation and U_θ an unitary operator. Therefore, if $[H, \Theta] = 0$, and if a given state of the Hamiltonian, $|\psi\rangle$, is unique, we conclude that $|\psi\rangle$ and $\Theta|\psi\rangle = |\psi^\theta\rangle$ only differ by a phase. Therefore, the Berry phase satisfies

$$\gamma(A_\psi) = \gamma(A_{\psi^\theta}), \quad \mod (2\pi M). \qquad (6.22)$$

---

[2]As discussed in *Journal of the Physical Society of Japan*, **75**, 123601 (2006).

We may expand the state $|\psi(x)\rangle$ in a complete basis as $|\psi(x)\rangle = \sum_J c_J(x)|J\rangle$. Then $|\psi^\theta(x)\rangle = \sum_J c_J^*(x)|J^\theta\rangle$. We obtain, therefore, that the Berry connection under the action of the operator Θ satisfies

$$A_{\psi^\theta} = \sum_J c_J dc_J^* = -\sum_J c_j^* dc_j = -A_\psi, \qquad (6.23)$$

since $\sum_J |c_J|^2 = 1$. Therefore,

$$\gamma(A_\psi) = -\gamma(A_{\psi^\theta}), \qquad (6.24)$$

which implies that the Berry phase is quantized and given by

$$\gamma(A_\psi) = 0, \pi M, \quad \mod (2\pi M), \qquad (6.25)$$

since it also satisfies the equation(6.22).

In the case of the spin problem, the operator may be chosen as time reversal. Under the action of this operator, the spin operators change sign $\mathbf{S}_i \to \Theta \mathbf{S}_i \Theta^{-1} = -\mathbf{S}_i$, and, therefore, the Hamiltonian remains invariant. In a phase that is not topological, the Berry phase vanishes and in a topological phase has the value π.

A possible way to introduce a dependence on a cyclic variable consists in considering a twist on a link between nearest neighbors (this twist may be interpreted as a change on the boundary conditions) as

$$S_i^+ S_j^- + S_i^- S_j^+ \to e^{i\phi} S_i^+ S_j^- + e^{-i\phi} S_i^- S_j^+. \qquad (6.26)$$

It is convenient to calculate the Berry phase using a lattice description. Introducing

$$\gamma_N = -\sum_{n=1}^{N} \arg A_N(\phi_n), \qquad (6.27)$$

where N is the number of points in the interval from zero to 2π, $\phi_n = (2\pi/N)n$ and

$$A_N(\phi_n) = \langle GS(\phi_n)|GS(\phi_{n+1})\rangle. \qquad (6.28)$$

We may use periodic boundary conditions $\phi_{N+1} = \phi_1$. The Berry phase may be then obtained taking the infinite size limit

$$\gamma = \lim_{N\to\infty} \gamma_N. \qquad (6.29)$$

The spin-1 Heisenberg chain does not have an exact solution, but it is possible to calculate the Berry phase numerically. This calculation has been performed using the Lanczos method, considering the subspace of

$\sum_i S_i^z = 0$. Indeed, one may consider a more general system with the addition of dimerization

$$H = \sum_{i=1}^{N/2} \left(J_1 \mathbf{S}_{2i} \cdot \mathbf{S}_{2i+1} + J_2 \mathbf{S}_{2i+1} \cdot \mathbf{S}_{2i+2} \right), \qquad (6.30)$$

and consider a system with periodic boundary conditions. We may parametrize the interactions as $J_1 = \sin\theta$ and $J_2 = \cos\theta$ with $0 < \theta < \pi/2$. In the groundstate there are $N/2$ singlet pairs for $\theta \to 0$ and $\theta \to \pi/2$ and the system is completely dimerized, with pairs either on the right or on the left of a given lattice site. The case of an antiferromagnetic chain without dimerization is obtained taking $\theta = \pi/4$.

Around the region of $\pi/4$ of an undimerized chain the Berry phase is π, indicative of a topological phase. This phase extends on a finite region from $\theta_c = 0.531237$ to a point symmetric around $\theta = \pi/4$, or until $\theta = \pi/2 - \theta_c$. Outside this region, the Berry phase vanishes and the system is in a trivial phase, that is, for values $0 \leq \theta < \theta_c$ and $\theta_c < \theta \leq \pi/2$. In the topological phase, the spins-1/2 are connected to their neighbors, maintaining translation invariance (recall the image that the spins 1 may be seen as the symmetric addition of two spins 1/2). In the trivial regimes, the state is dimerized and the local spins 1/2 form singlet states with the spins to the right or to the left, as mentioned previously.

In the models with dimerization of Shockley or SSH, considered previously, the behavior is different. The point with no dimerization is the critical point that separates the trivial phase from the topological phase. The topological phase extends for all negative values of the dimerization, η, while the model is in the trivial phase for all positive values of the dimerization.

If one considers a spin chain with $S = 2$, the behavior of the Berry phase is qualitatively different. In this case, around the region with no dimerization the Berry phase vanishes and, therefore, a system with integer spin $S = 2$ (even) is not topological. The introduction of the dimerization leads to regions of finite extension in which the Berry phase is π.

6.3 Topological defects

Various systems may have excitations (defects) of a topological nature that may be characterized by the existence of topological numbers. We will consider in this section some examples such as magnetic systems and superfluids.

Fig. 6.1   Spin configuration of a texture called hedgehog, in which the spin vectors point radially from a given origin.

6.3.1   *Hedgehogs and skyrmions*

The magnetic order parameter is a vector whose orientation may be parametrized by the angles of the spherical coordinates, θ_{op}, ϕ_{op}, considering that its amplitude is fixed. Let us define a spherical surface of radius r_s and parametrized by the angles θ_s, ϕ_s, and consider the distribution of the order parameter along the surface.

A configuration of the order parameter, along the surface, which is topologically non-trivial, may be obtained easily. Specifically, placing the origin of the order parameter vector on the points of the surface S and taking $\theta_{op} = \theta_s, \phi_{op} = \phi_s$, we may define a topological invariant as

$$N = \frac{1}{4\pi} \int_0^{2\pi} d\phi_s \int_0^{\pi} d\theta_s \sin\theta_{op} \frac{\partial(\theta_{op}, \phi_{op})}{\partial(\theta_s, \phi_s)}. \tag{6.31}$$

The Jacobian of the transformation between the two sets of angles is unity. Therefore, the invariant is $N = 1$. This configuration is called hedgehog. In figure 6.1 we represent a configuration of this type in which the spins point radially from a given origin. Choosing a spherical surface centered in the origin, along the surface the spins point outward, as the spikes of a hedgehog (justifying the origin of the name of this configuration).

Another possible choice for the angle dependence of the order parameter is $\theta_{op} = f(r_s/L), \phi_{op} = \phi_s$, where r_s is the radial coordinate of the surface and L gives the spatial scale of the spin configuration. Defining the topological invariant as

$$N = \frac{1}{4\pi} \int_0^{2\pi} d\phi_s \int_0^{\infty} dr_s \sin\theta_{op} \frac{\partial(\theta_{op}, \phi_{op})}{\partial(r_s, \phi_s)}, \tag{6.32}$$

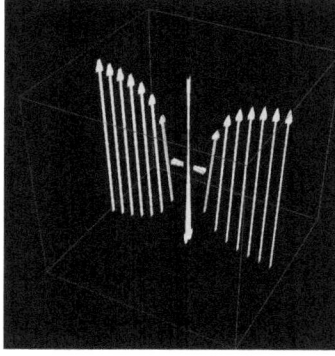

Fig. 6.2   Spin configuration in the form of a skyrmion. In a given point of space, the spin points in a given direction (for instance, downwards) and in the infinite the spins point upward. It can be considered as an example of a dependence with distance to the origin that has the form $f(r/L) = \pi/(r/L + 1)$. In the figure we represent a cut along a given direction.

and using

$$\frac{\partial \theta_{op}}{\partial r_s} = \frac{1}{L} f' \left(\frac{r_s}{L} \right), \qquad \frac{\partial \phi_{op}}{\partial r_s} = 0, \qquad \frac{\partial \theta_{op}}{\partial \phi_s} = 0, \qquad \frac{\partial \phi_{op}}{\partial \phi_s} = 1, \quad (6.33)$$

we obtain

$$N = \frac{1}{4\pi} \int_0^{2\pi} d\phi_s \int_0^{\infty} dr_s \sin \left[f \left(\frac{r_s}{L} \right) \right] \left(\frac{1}{L} f' \left(\frac{r_s}{L} \right) \right). \qquad (6.34)$$

The choice of boundary conditions $f(0) = \pi$ and $f(\infty) = 0$ leads to $N = -1$. This type of configuration is called skyrmion and is represented in figure 6.2.

6.3.2   *Vortices and Kosterlitz-Thouless transition*

Another type of topological defect is a vortex. This type of defect may occur in different systems such as spin systems, superfluids, etc.

Let us begin by considering a superfluid system, such as a set of ^{4}He atoms. Neglecting the interactions between the atoms, the system, composed of bosons, constitutes a Bose-Einstein condensate in a coherent state of zero linear momentum. In the presence of weak interactions, the system has properties similar to the free system, with the important difference that the excitation spectrum changes its nature and shows a linear dispersion relation, instead of the quadratic dispersion of the free boson system. As a consequence, the system becomes a superfluid, such that one particle that may move inside the system, with sufficiently low velocity, does not suffer friction.

The boson condensate may be represented by a macroscopic wave function, $\psi(\mathbf{r})$, that gives the probability amplitude to find one boson in position $\mathbf{r}$. The boson concentration is, therefore, $n_p(\mathbf{r}) = \psi^*(\mathbf{r})\psi(\mathbf{r})$. We may represent the wave function as

$$\psi(\mathbf{r}) = \sqrt{n_p}\,e^{i\theta(\mathbf{r})} \tag{6.35}$$

and approximate the boson concentration by a constant. The velocity of one particle is given by

$$\mathbf{v} = \frac{1}{m}\mathbf{p} = \frac{1}{m}\frac{\hbar}{i}\nabla. \tag{6.36}$$

This result implies that

$$\psi^* \mathbf{v}\psi = \frac{n_p}{m}\hbar\nabla\theta(\mathbf{r}). \tag{6.37}$$

Let us then consider the possibility that the superfluid may rotate around singular points, where the velocity circulation satisfies

$$\int_C \mathbf{v}_s \cdot d\mathbf{l} = 2\pi\chi\,,$$

$$2\pi r v_s = 2\pi\chi\,,$$

$$v_s = \frac{\chi}{r}\,. \tag{6.38}$$

Using $\mathbf{v}_s = \hbar/m\nabla\phi$ (identifying $\theta = \phi$), we obtain

$$\int_C \mathbf{v}_s \cdot d\mathbf{l} = \frac{\hbar}{m}\Delta\phi = \frac{\hbar}{m}n2\pi \tag{6.39}$$

and, for $n = 1$, $\chi = \hbar/m$. The circulation gets then quantized in a vortex. These excitations cost energy, and may be created as a result of a rotation of the superfluid around some axis. Vortices also occur in a superconductor, for example, as a result of applying a magnetic field, allowing the gradual penetration of the magnetic field in certain superconductors, designated type-*II*.

Vortices may also be defined in spin systems, as in the two-dimensional *XY* model. In this system, the spins are vectors defined in a plane and vortex-like excitations may be considered, taking the orientations of the spins as rotating around a central point, with relative orientations that integrate to a multiple of 2π.

Let us consider the classical two-dimensional *XY* model given by

$$H = -J\sum_{\langle i,j \rangle} \mathbf{S}_i \cdot \mathbf{S}_j = -J\sum_{\langle i,j \rangle} \cos(\phi_i - \phi_j), \tag{6.40}$$

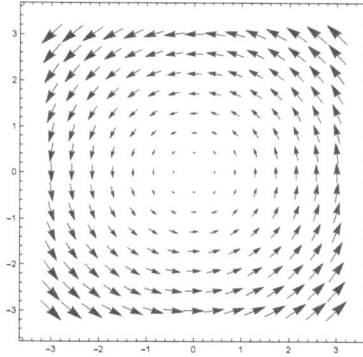

Fig. 6.3   Vortex in a spin system constituted by a rotation of the spins around a given space point, that defines the location of the vortex.

where ϕ_i and ϕ_j are the angles of the spin vectors, in relation to a given direction, and let us call a the distance between nearest neighbor spins. Considering a ferromagnetic interaction $J > 0$, in the groundstate the spins are all aligned. Let us consider deviations from this perfectly ordered state, allowing the angles between neighboring spins to be small, that is, $\phi_i - \phi_j$ is small. In this limit we obtain

$$H - E_0 \propto \frac{1}{2} J \sum_{\langle i,j \rangle} (\phi_i - \phi_j)^2 = J \sum_r (\Delta\phi(r))^2 . \qquad (6.41)$$

Let us consider a circle around a given point, using polar coordinates. The angle with respect to a given direction is given by $\varphi = l/\rho$, where l is the distance along the arc and ρ the radius of the circle. Consider a vortex defined by the orientation of the many spins that rotate around the central point and let us define φ as the angle of each spin, in relation to the same reference direction. Then,

$$\oint \nabla\varphi \cdot d\mathbf{l} = 2\pi, \qquad (6.42)$$

if the spins rotate once around the origin. As the two vectors are parallel, and considering a uniform change of the angle, we obtain that $\nabla\varphi = 1/\rho$. We may, now, identify ϕ with φ. As a consequence, considering a vortex of size R, we obtain that the excitation energy is given by

$$H - E_0 \propto J \int_a^R \rho d\rho \int_0^{2\pi} d\varphi (\nabla \varphi)^2$$

$$= J \int_a^R \rho d\rho \int_0^{2\pi} d\varphi \frac{1}{\rho^2}$$

$$= 2\pi J \log\left(\frac{R}{a}\right). \tag{6.43}$$

Therefore, when $R \to \infty$ and/or $a \to 0$ it costs infinite energy. Let us then take these quantities R, a as finite. The number of possible positions to place the vortex in the system is given by $(R/a)^2$. The entropy is, therefore, given by

$$S = k_B \log \Omega = 2k_B \log\left(\frac{R}{a}\right), \tag{6.44}$$

where Ω is the number of configurations. Evaluating the free energy, we obtain

$$F = U - TS$$

$$= 2\pi J \log\left(\frac{R}{a}\right) - T2k_B \log\left(\frac{R}{a}\right)$$

$$= 2(\pi J - k_B T) \log\left(\frac{R}{a}\right). \tag{6.45}$$

As a consequence, $F = 0$ separates a regime in which the energy contribution U is dominant and, therefore, the presence of vortices is unfavorable, since they cost energy, from a regime in which the entropy contribution is dominant and, therefore, it is favorable to have vortices, since it minimizes the free energy. The point $F = 0$ defines a temperature

$$k_B T_c = \pi J. \tag{6.46}$$

For $T < T_c$ it is unfavorable to have vortices while for $T > T_c$ the existence of vortices is favorable. The temperature T_c defines the temperature for which the Kosterlitz-Thouless transition, between the regime with no vortices and the regime in which vortices abound, takes place.

Let us consider the action of the XY model defined by $Z = \int D\theta e^{-S}$ where Z is the partition function. The action is given by

$$S = \beta J \sum_{\langle n,m \rangle} \cos(\theta_n - \theta_m)$$

$$= \frac{1}{2}\beta J \sum_{\langle n,m \rangle} \left[e^{i(\theta_n - \theta_m)} + e^{-i(\theta_n - \theta_m)} \right]. \tag{6.47}$$

Let us consider the correlation function

$$\langle e^{i\theta_0} e^{-i\theta_n} \rangle = \frac{1}{Z} \int \prod_m d\theta_m e^{i(\theta_0 - \theta_n)} e^{-\frac{J}{k_B T} \sum_{\langle n,m \rangle} \cos(\theta_n - \theta_m)}, \qquad (6.48)$$

and the case of high temperatures. We may then expand the Boltzmann factor. Note that

$$\int_0^{2\pi} d\theta_m = 2\pi, \int_0^{2\pi} d\theta_m e^{i\theta_m} = 0.$$

The first non-vanishing term of the expansion is of the order of

$$\langle e^{i\theta_0} e^{-i\theta_n} \rangle \propto \left(\frac{J}{k_B T} \right)^{|n|} = e^{-|n| \log \left(\frac{k_B T}{J} \right)}, \qquad (6.49)$$

and, therefore, decreases exponentially with the distance $|n|$.

Let us now take the opposite limit of low temperatures. In this case, we expect that θ_n changes slowly. Therefore, we may expand $\cos(\theta_n - \theta_m) = 1/2 (\theta_n - \theta_m)^2$. This result leads to an action of the form $S \propto J/2 \sum_{n,i} |\Delta_i \theta(n)|^2$. The correlation function is, then, written as

$$\langle e^{i\theta_0} e^{-i\theta_n} \rangle \propto \frac{1}{Z} \int \prod_m d\theta_m e^{i(\theta_n - \theta_m)} e^{-\frac{J}{2k_B T} \sum (\Delta \theta)^2}. \qquad (6.50)$$

Therefore, in this approximation, we get a gaussian integral. The correlation function is given by

$$\langle e^{i\theta_0} e^{-i\theta_n} \rangle \propto e^{\frac{k_B T}{J} \Delta(n)}, \qquad (6.51)$$

where for high values of $|n|$,

$$\Delta(n) \propto -\frac{1}{2\pi} \log|n|. \qquad (6.52)$$

This result leads to a power law for the decay of the phase-phase correlation function

$$\langle e^{i\theta_0} e^{-i\theta_n} \rangle \propto \left(\frac{1}{|n|} \right)^{\frac{k_B T}{2\pi J}}. \qquad (6.53)$$

As a conclusion, we find two qualitatively different regimes. A regime at high temperatures, in which the phase-phase correlation function decays exponentially with the distance between two spins, and another regime at low temperatures, in which the same correlation function decays as a power law, with an exponent that depends on the temperature and on the coupling constant between the spins. As a consequence, the low temperature phase is different from the high temperature phase that, as is well known, is a disordered phase. One still has to determine how the system evolves between the two types of behavior of the correlation function. We can determine that the transition occurs at a given temperature, as suggested by the free energy analysis. Clearly, the two regimes are distinguished by the presence of few vortices in the regime of low temperatures and by the proliferation of vortices in the high temperature regime, due to the entropy gain, and that tends to completely disorder the spins. Therefore, the presence or absence of the topological defects separates the two regimes. A complete description of the Kosterlitz-Thouless transition requires an analysis by the renormalization group method, that is, however, beyond the scope of this book and may be consulted in the appropriate literature.

Note, however, that the Mermin-Wagner theorem is satisfied. This theorem establishes that no long-range order should occur in a classical two-dimensional system. As shown above, the correlation functions vanish when $|n| \to \infty$ and there is no order parameter with finite value, since $\langle e^{i\theta_0} \rangle = \langle e^{-i\theta_n} \rangle = 0$.

6.4 Duality and topology

6.4.1 *Inverse Jordan-Wigner transformation*

As discussed in section 6.1, it is possible to relate a system of spins $1/2$ with a fermionic system. The inverse Jordan-Wigner transformation is defined by

$$c_j = e^{i\pi \sum_{l=1}^{j-1} S_l^+ S_l^-} S_j^-,$$
$$c_j^\dagger = S_j^+ e^{-i\pi \sum_{l=1}^{j-1} S_l^+ S_l^-}, \tag{6.54}$$

where $S_j^\pm$ are the raising and lowering spin operators on site j. Using this inverse transformation, the Kitaev model is equivalent to the spin-$1/2$ anisotropic XY model considered before.

As discussed extensively in chapter 4, the Kitaev model in equation (4.56) has topological regimes, characterized by a topological invariant or by edge states. On the other hand, we also discussed that the spin model in which Kitaev's model is transformed is not topological. Recalling the relation between the parameters of the two models, the gapless regime for $J_X = J_Y$ of equation (6.5) (or $\Delta = 0$ in the fermionic language) separates two phases. In the spin problem, the groundstates on the two sides of the transition are characterized, when the coupling is ferromagnetic, by some order along X or Y. As we saw, these states are topologically trivial. In the Kitaev model the corresponding regimes are topological. The topologically trivial regimes of the Kitaev model when $|\mu| > 2t$ correspond to the paramagnetic phase in the spin problem. The chemical potential term of the Kitaev model fixes the average fermion number and, therefore, multiplies the fermionic number operator. This term in the Hamiltonian corresponds to a term of spin density along the Z axis and the regime $|\mu| > 2t$ corresponds to a magnetic field that is high enough that the system is in a paramagnetic phase, with Ising symmetry. One concludes, therefore, that the correspondence between the models apparently transforms a fermionic system with topological properties, in certain regimes, in a spin problem that is not topological. Therefore, there is an apparent change of topology inspite of the duality between the two problems. Actually, something similar occurs using different representations of a given fermionic problem.

6.4.2 *Fermionic representations of the one-dimensional Kitaev model*

Let us recall the transformation in equation (4.67)

$$c_j = \frac{i}{2} \left[d^\dagger_{j-1} - d_{j-1} + d_j + d^\dagger_j \right]$$
$$c^\dagger_j = -\frac{i}{2} \left[d_{j-1} - d^\dagger_{j-1} + d_j + d^\dagger_j \right]. \tag{6.55}$$

The operators d_j are particularly useful if $\mu = 0, \Delta = t$, since they allow the diagonalization of the Hamiltonian. We want, now, to express the Hamiltonian in terms of these operators. The Hamiltonian expressed in terms of these operators has a more complex form

$$H = \frac{1}{2} \sum_j (-t + \Delta) \left(-d^\dagger_{j+1} d_{j-1} - d^\dagger_{j-1} d_{j+1} + d_{j-1} d_{j+1} + d^\dagger_{j+1} d^\dagger_{j-1} \right)$$
$$+ \frac{1}{2} \sum_j (t + \Delta) \left(2 d^\dagger_j d_j - 1 \right)$$
$$- \frac{\mu}{2} \sum_j \left[-d^\dagger_{j-1} d_j - d^\dagger_j d_{j-1} + d_{j-1} d_j + d^\dagger_j d^\dagger_{j-1} \right]. \tag{6.56}$$

The term multiplied by the chemical potential has now nearest neighbor terms, both normal and anomalous. There are also terms, both hopping-like and pairing-like, that involve second neighbors. Note that we are considering an infinite system. If the system has a finite extension, the transformation in equation (4.67) has to be applied with care near the edges.

In momentum space we may write the Hamiltonian matrix, H_k^d, as

$$\begin{pmatrix} \mu\cos k + t + \Delta + (t-\Delta)\cos(2k) & i\mu\sin k + i(t-\Delta)\sin(2k) \\ -i\mu\sin k + i(-t+\Delta)\sin(2k) & -\mu\cos k - t - \Delta - (t-\Delta)\cos(2k) \end{pmatrix}.$$

$$(6.57)$$

This expression is of the type of the Kitaev model with second neighbor terms, as seen before.

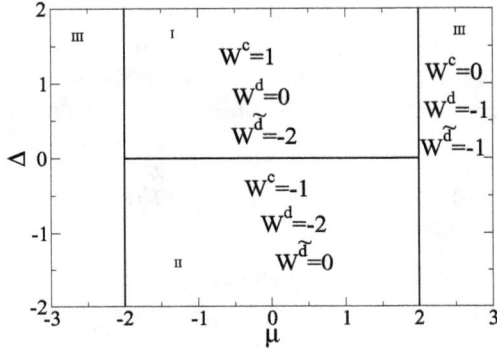

Fig. 6.4   Winding numbers W^c, W^d and $W^{\tilde{d}}$ for the various phases as a function of chemical potential and pairing amplitude in units of t.

The matrix H_k^d allows a chiral transformation that leads to

$$q^d(k) = -(-t+\Delta)\cos(2k) + (t+\Delta) + \mu\cos k + i(-t+\Delta)\sin(2k) - i\mu\sin k,$$

$$(6.58)$$

as in equation(2.72). Let us consider, for example, the point $\mu = 0, \Delta = t$. The winding number may be obtained as in equation(2.76)

$$W^d = W^d(q^d(k)) = 0. \tag{6.59}$$

Therefore, in terms of the operators d the winding number at this point indicates a trivial phase. On the other hand, on $\Delta = t, \mu > 2t$ we obtain $W^d = 1$. We may also perform a chiral transformation on the original Hamiltonian defined in terms of the fermionic operators, c_j, giving origin to a matrix with an antidiagonal element q^c and may define the associated winding number, W^c. Therefore, from the point of view of W^d versus W^c, an apparent change of topology takes place.

Since in terms of the operators d the Kitaev Hamiltonian has contributions from first and second neighbors, it is indeed an effective Kitaev model with first and second neighbors. The extended Kitaev model has a richer phase diagram with winding numbers $W = 0, \pm1, \pm2$, as discussed previously in section 4.2.3. The model is in class BDI and is characterized by an invariant $\mathbb{Z}$, as shown in Table 3.1. For example, considering $\mu = 0$, the model is equivalent to the Kitaev effective model with parameters $(\mu, t, t', \Delta, \Delta')$ where $\mu = 0$ implies that $t = \Delta = 0$ (absence of nearest neighbor terms both kinetic and pairing). As explained in section 4.2.3, this model has two Majorana modes at each edge. In the absence of nearest neighbor coupling, the model is equivalent to two decoupled chains and, therefore, the number of edge states doubles.

At the point $\mu = 0, \Delta = t$ the Kitaev Hamiltonian was diagonalized in real space, using Majorana operators and introducing new fermionic operators. We may follow the same procedure to diagonalize the problem at the point $\mu = 0, \Delta = -t$. At this point, the fermionic operator that diagonalizes the Hamiltonian may be defined as $\tilde{d}_j = \frac{-i}{2}(\gamma_{1,j} - i\gamma_{2,j+1})$. This new operator allows to write the Hamiltonian as

$$H = t \sum_j \left(2\tilde{d}_j^\dagger \tilde{d}_j - 1\right). \tag{6.60}$$

The groundstate is obtained taking at each site $\tilde{d}_j^\dagger \tilde{d}_j = 0$. We obtain the relation

$$\tilde{d}_j = \frac{i}{2}\left[c_j + c_j^\dagger + c_{j+1}^\dagger - c_{j+1}\right]. \tag{6.61}$$

Replacing the operators $c_j, c_j^\dagger$ by the operators $\tilde{d}_j, \tilde{d}_j^\dagger$ we obtain an expression similar to equation(6.58) substituting $\Delta \rightarrow -\Delta$. We obtain that in the topological phase II of the Kitaev model expressed in terms of the fermionic operators c, for which W^c is different from zero and the operators $\tilde{d}$ diagonalize the Hamiltonian, the winding number, $W^{\tilde{d}}$, also vanishes. In figure 6.4 we compare the winding number using the various representations.

6.4.3 *Berry phase and change of representation*

Let us calculate the Berry phase of the eigenstates of the Kitaev Hamiltonian using different representations and, therefore, different basis states. Even though we will concentrate on this model, the procedure is general, and it may be applied to other problems. Performing a Fourier transform of the real space relation between the operators c and d, equation(4.67), we may show that

$$\begin{pmatrix} d_k \\ d_{-k}^\dagger \end{pmatrix} = U_k^\dagger \begin{pmatrix} c_k \\ c_{-k}^\dagger \end{pmatrix},$$

$$(6.62)$$

with

$$U_k = -e^{-\frac{ik}{2}} \begin{pmatrix} \sin\frac{k}{2} & i\cos\frac{k}{2} \\ -i\cos\frac{k}{2} & -\sin\frac{k}{2} \end{pmatrix},$$

$$(6.63)$$

which implies that $H_k^d = U_k^\dagger H_k^c U_k$. Performing the transformation of c_k to d_k, corresponds to diagonalizing the problem at $\mu = 0, \Delta = t$ and carrying out the transformation from c_k to $\tilde{d}_k$ corresponds to diagonalizing the problem at $\mu = 0, \Delta = -t$. At the points $\mu = 0, \Delta = \pm t$ the eigenvalues of the Hamiltonian are ± 2, as we saw previously, and the eigenvectors are

$$\psi_+ = sgn\left[\cos\frac{k}{2}\right] \begin{pmatrix} -i\frac{\Delta}{t}\sin\frac{k}{2} \\ \cos\frac{k}{2} \end{pmatrix}$$

$$\psi_- = sgn\left[\cos\frac{k}{2}\right] \begin{pmatrix} \cos\frac{k}{2} \\ -i\frac{\Delta}{t} - \sin\frac{k}{2} \end{pmatrix}.$$

$$(6.64)$$

Taking now $\mu = 0$ but any value of Δ, the eigenvalues are

$$\lambda_\pm = \pm 2\sqrt{(t\cos k)^2 + (\Delta\sin k)^2}.$$

$$(6.65)$$

The eigenvectors are

$$\psi_+^\Delta = \begin{pmatrix} \frac{-i2\Delta\sin k}{\sqrt{2\lambda_+(\lambda_+ + 2t\cos k)}} \\ \sqrt{\frac{\lambda_+ + 2t\cos k}{2\lambda_+}} \end{pmatrix}$$

$$(6.66)$$

$$\psi_-^\Delta = \begin{pmatrix} \sqrt{\frac{\lambda_- - 2t\cos k}{2\lambda_-}} \\ \frac{-i2\Delta\sin k}{\sqrt{2\lambda_-(\lambda_- - 2t\cos k)}} \end{pmatrix}.$$

$$(6.67)$$

Let us consider now the Berry phase in momentum space (Zak phase). For an eigenstate n it is given by

$$\gamma_n = \frac{i}{\pi} \int dk \langle \psi_n(k) | \frac{\partial}{\partial k} | \psi_n(k) \rangle.$$

$$(6.68)$$

As we change variables (or the description in terms of different operators), the states also change. In the case of a transformation of the type $|\psi_n(k)\rangle \to$

$e^{i\xi(k)}|\psi_n(k)\rangle$ where the function $\xi(k)$ is periodic, Zak's phase is invariant. In our case, the state is a vector and, in general, one has a transformation of $|\psi_n\rangle$ to $|\tilde{\psi}_n\rangle$:

$$|\tilde{\psi}_n\rangle = U_k^\dagger |\psi_n\rangle$$
$$\langle\tilde{\psi}_n| = \langle\psi_n|U_k \tag{6.69}$$

and

$$\tilde{H} = U_k^\dagger H U_k. \tag{6.70}$$

Then, we define

$$\gamma_n = \frac{i}{\pi}\int dk\langle\psi_n|\partial_k|\psi_n\rangle\,,$$
$$\tilde{\gamma}_n = \frac{i}{\pi}\int dk\langle\tilde{\psi}_n|\partial_k|\tilde{\psi}_n\rangle$$
$$= \gamma_n + \delta\gamma_n, \tag{6.71}$$

where

$$\delta\gamma_n = \frac{i}{\pi}\int dk\langle\psi_n|U_k\left(\partial_k U_k^\dagger\right)|\psi_n\rangle. \tag{6.72}$$

In general, $\delta\gamma_n \neq 0$. We may see that

$$\gamma_n = \tilde{\gamma}_n + \frac{i}{\pi}\int dk\langle\tilde{\psi}_n|U_k^\dagger\left(\partial_k U_k\right)|\tilde{\psi}_n\rangle$$
$$\tilde{\gamma}_n = \gamma_n + \frac{i}{\pi}\int dk\langle\psi_n|U_k\left(\partial_k U_k^\dagger\right)|\psi_n\rangle. \tag{6.73}$$

Defining a new phase as

$$\tilde{\Gamma}_n = \frac{i}{\pi}\int dk\langle\tilde{\psi}_n|D_k|\tilde{\psi}_n\rangle, \tag{6.74}$$

with

$$D_k = \partial_k - \left(\partial_k U_k^\dagger\right)U_k, \tag{6.75}$$

this new phase is invariant in the sense that $\Gamma_n = \tilde{\Gamma}_n$. The differential operator D_k is similar to a covariant derivative and similar to a non-abelian transformation (that is, non-commutative due to the matrix nature of the transformation), that in general is necessary, for example, if the states are degenerate, or if there is a vectorial structure of the states, as in the case considered here.

Let us now calculate the Zak phase of the state $|\psi_-\rangle_c$, the lowest energy eigenstate of Kitaev Hamiltonian expressed in terms of the original fermionic operators, c_k, given in equation(6.64). This is given by equation(2.69),

$$
\begin{aligned}
\gamma_{-1} &= i \int_0^{2\pi} dk \langle \psi_- | \partial_k | \psi_- \rangle_c \\
&= i \int_0^{2\pi} dk \frac{d}{dk} \left(\ln sgn \left[\cos \frac{k}{2} \right] \right) \\
&= -\pi = \gamma_c.
\end{aligned}
\tag{6.76}
$$

Only the singular part of the function contributes, associated to the discontinuity of the $sgn \cos(k/2)$.

Let us now evaluate the Zak phase of the state correspondent to the lowest eigenvalue of $H_k^d(\mu = 0, \Delta = t)$. This state is simply

$$
|\psi_-\rangle_d = \begin{pmatrix} 0 \\ 1 \end{pmatrix},
\tag{6.77}
$$

and $\gamma_{-1} = \gamma_d = 0$. Using $\gamma_d = \gamma_c + \delta\gamma$, and calculating $\delta\gamma$ we obtain $\delta\gamma = \pi$, showing that the change of topology is hidden in the transformation, as expected. If the transformation is singular, $\delta\gamma \neq 0$ and if is non-singular, $\delta\gamma = 0$. The contribution to $\delta\gamma$ results from the factor $e^{-ik/2}$ in the operator U_k, being ill-defined at $k = 0$.

Exercise 6.1

Consider an operator, $\mathcal{U}$, that transforms the Hamiltonian at a given point in parameter space on the Hamiltonian of a different point in parameter space. Show that:

(1) If the two points are in the same topological phase and the operator is non-singular, the contribution $\delta\gamma$ vanishes;
(2) If one of the two points is in a topological phase and the other in a trivial phase, then $\delta\gamma \neq 0$ and the operator $\mathcal{U}$ is singular.
    Similar results can be obtained in other models, such as the Shockley model.

6.4.4 *Topology of the spin model in the fermionic representation*

Let us return to the question of the change of topology that is the result of the transformation of the spin system to the dual fermionic problem.

We may determine the topological properties of a system with or without interactions, considering that the Hamiltonian depends on a variable, that may be taken as cyclic. A possible way to introduce the dependence on the cyclic variable is to consider, again, a twist on a given link between spatial points. The Berry phase can be defined integrating over the values of the cyclic variable.

Let us consider again the spin problem with $S = 1/2$

$$H = -\sum_{j} \left(J_x S_j^x S_{j+1}^x + J_y S_j^y S_{j+1}^y \right). \tag{6.78}$$

The diagonalization of this Hamiltonian for a finite system shows that using periodic boundary conditions (PBC) or open (OBC) the groundstate is in general non-degenerate. For the model X with $J_y = 0$, the groundstate is doubly degenerate, for the two sets of boundary conditions. Imposing twisted boundary conditions, as in equation (6.26), we may calculate the Berry phase and this vanishes, confirming that the system is topologically trivial.

Let us consider, now, the traditional Jordan-Wigner transformation, that converts the spin problem to a spinless fermionic problem. This transformation carries out straightforwardly using OBC. However, the choice of PBC for the spins, $S_{N+1} = S_1$, leads, in the fermionic problem, to

$$c_{N+1} = e^{i\pi \hat{N}_F} c_1 = c_1 e^{i\pi(\hat{N}_F - \hat{n}_1)}, \tag{6.79}$$

where $\hat{N}_F$ is the operator that counts the total number of fermions, c, (see equation(6.4)). Diagonalizing the many-body fermionic problem that results from the Jordan-Wigner transformation and considering OBC, we obtain the same energy levels of the original spin problem, as expected. The PBC, in addition to the relation imposed by the Jordan-Wigner transformation, lead to the same spectrum of the spin problem, with the same boundary conditions. These results may be obtained diagonalizing numerically the spin problem and the corresponding fermionic problem for a finite system.

Consider, now, the Kitaev model with no reference to its possible origin in a dual spin system. The choice of periodic boundary conditions, $c_{N+1} = c_1$, leads to a spectrum that is different from the one of the spin problem. In particular, the groundstate energy is different (at least for a finite system). If we now calculate the Berry phase, this is not zero neither $\pi \pmod{2\pi}$. The Berry phase is only well defined if there is a gap that remains when the twist angle is varied, while in this problem there are various degeneracies and the gap closes. However, it can be shown that the model is topological using a single-particle description. Therefore, the apparent change of topology, when we establish a relation between the spin problem and the Kitaev model, is the result of different boundary conditions (when OBC are not used) that lead to a difference in the Berry phase. As shown above, the diagonalization of the Kitaev model requires a non-local transformation of the fermionic operators and implies that the effects of the edges have to be taken with care.

Further reading:

A discussion on spin operators representations can be found in:

- A. Auerbach, *Interacting Electrons and Quantum Magnetism*, Springer-Verlag, New York (1998).

The method of the representation of spin-1/2 operators by Majorana operators is discussed in:

- F. A. Berezin and M. S. Marinov, *Annals of Physics* **104** 336 (1977); *JETP Letters* **21**, 320 (1975).
- V. R. Vieira, *Physical Review* B **23**, 6043 (1981).
- P. D. Sacramento and V. R. Vieira, *Annals of Physics* **391**, 216 (2018).

The Jordan-Wigner representation and its generalization to two dimensions is presented in:

- P. Jordan and E. Wigner, *Zeitschrift Physik* **47**, 631 (1928).
- Y. R. Wang, *Physical Review* B **46**, 151 (1992).

Differences between half-integer and integer spins are discussed in:

- E. Lieb, T. Schultz and D. Mattis, *Annals of Physics* **16**, 407 (1961).
- F. D. M. Haldane, *Physics Letters* **93A**, 464 (1983).
- F. D. M. Haldane, *Physical Review Letters* **50**, 1153 (1983).
- H. Bethe, *Zeitschrift Physik* **71**, 205 (1931).
- Ph.D. thesis of Kong-Ju-Bock Lee, Temple University (1989).
- E. Fradkin, *Field theories of condensed matter systems*, Benjamin (1991).

Examples of Haldane's phase and its properties are discussed in:

- I. Affleck, T. Kennedy, E. H. Lieb and H. Tasaki, *Physical Review Letters* **59**, 799 (1987).
- I. Affleck, T. Kennedy, E. H. Lieb and H. Tasaki, *Communications Mathematical Physics* **115**, 477 (1988).
- M. den Nijs and K. Rommelse, *Physical Review* B **40**, 4709 (1989).
- T. Kennedy and H. Tasaki, *Physical Review* B **45**, 304 (1992).

The generalization to integer spins of arbitrary value is considered in:

- M. Oshikawa, *Journal Physics: Condensed Matter* **4**, 7469 (1992).
- F. Pollmann, E. Berg, A. M. Turner and M. Oshikawa, *Physical Review* B **85**, 075125 (2012).

About Berry phase using twisted boundary conditions and its application to spin systems:

- D. J. Thouless, M. Kohmoto, M. P. Nightingale and M. den Nijs, *Physical Review Letters* **49**, 405 (1982).
- T. Hirano, H. Katsura and Y. Hatsugai, *Physical Review* B **77**, 094431 (2008).

On the Kosterlitz-Thouless transition:

- J. M. Kosterlitz and D. J. Thouless. *Journal Physics* C **5**, L124 (1972).
- J. M. Kosterlitz and D. J. Thouless. *Journal Physics* C **6**, 1181 (1973).
- J. M. Kosterlitz, *Journal Physics* C *Solid State Phys.*, **7**(6), 1046, (1974).

On duality and topology:

- P. D. Sacramento and V. R. Vieira, *Annals of Physics* **391**, 216 (2018).

Chapter 7

Photonic systems with topological properties

7.1 Topological phases in photonic systems

The existence of topological phases in photonic systems is of great interest. The transmission of signals through waveguides is affected, in general, by significative problems due to collisions and energy loss. Having in mind the bulk-edge correspondence in a topological system, the existence of topological phases would lead to edge states topologically protected (for instance, edge states in a two-dimensional system where a wave can propagate), that would not be affected by collisions that reverse the propagation direction, such as, for example, changes in the electric or magnetic properties due to impurities.

Having as inspiration electronic topological states, it is reasonable to expect that a periodic structure that has a band structure similar to bands in a solid, with gaps that forbid propagation in certain frequency regimes (in which the photonic system would behave as a mirror to those frequencies), may have edge states localized in these gaps, that may be protected from impurities or imperfections.

Such idea was proposed by Haldane and Raghu considering a non-reciprocal system with one hexagonal lattice of dielectric tubes, with a Faraday effect resulting from an imaginary part of the dielectric permittivity tensor in the form $\epsilon_{xy} = -\epsilon_{yx} = i\epsilon_0\eta(\mathbf{r}, \omega)$, where $\eta(\mathbf{r}, \omega) = \eta_0(\omega) + \eta_1(\omega)V_G(\mathbf{r})$, with $\eta_0(\omega)$ and $\eta_1(\omega)$ real and odd functions of ω and V_G periodic with the periodicity of the hexagonal lattice. Edge states appear in the frequency spectrum of unidirectional waveguides, that allow the transmission of energy along a given direction, since the Faraday effect breaks time reversal symmetry.

7.2   Edge modes with time reversal symmetry breaking

In this section we discuss the propagation of electromagnetic waves in a waveguide, in the presence of a periodic potential obtained placing periodically ferrite tubes on the surface of the waveguide. First, we briefly review the propagation modes in a rectangular waveguide. Next, we consider some properties of the magnetic permeability of the ferrite tubes, organized in a periodic manner, and how the solutions of Maxwell equations are affected by the periodicity and how topological properties emerge.

7.2.1   *Waveguides*

Let us consider a finite box in the plane xy with sides of lengths a along x and b along y, and infinite length along z. The box is filled with a material with dielectric permittivity ϵ and magnetic permeability μ and its walls are metallic, made of a very good conductor. Consider a wave that is propagating along $-z$, as $e^{-ik_z z + i\omega t}$. We may obtain, from Maxwell's equations,

$$\frac{\partial E_z}{\partial y} + ik_z E_y = -i\omega\mu H_x\,,$$

$$-\frac{\partial E_z}{\partial x} - ik_z E_x = -i\omega\mu H_y\,,$$

$$\frac{\partial E_y}{\partial x} - \frac{\partial E_x}{\partial y} = -i\omega\mu H_z\,, \tag{7.1}$$

and

$$\frac{\partial H_z}{\partial y} + ik_z H_y = i\omega\epsilon E_x\,,$$

$$-\frac{\partial H_z}{\partial x} - ik_z H_x = i\omega\epsilon E_y\,,$$

$$\frac{\partial H_y}{\partial x} - \frac{\partial H_x}{\partial y} = i\omega\epsilon E_z\,. \tag{7.2}$$

These equations may be rewritten as

$$H_x = \frac{i}{k_c^2}\left(\omega\epsilon\frac{\partial E_z}{\partial y} - k_z\frac{\partial H_z}{\partial x}\right)\,,$$

$$H_y = -\frac{i}{k_c^2}\left(\omega\epsilon\frac{\partial E_z}{\partial x} + k_z\frac{\partial H_z}{\partial y}\right)\,,$$

$$E_x = -\frac{i}{k_c^2}\left(\omega\mu\frac{\partial H_z}{\partial y} + k_z\frac{\partial E_z}{\partial x}\right)\,,$$

$$E_y = \frac{i}{k_c^2}\left(\omega\mu\frac{\partial H_z}{\partial x} - k_z\frac{\partial E_z}{\partial y}\right)\,, \tag{7.3}$$

where $k_c^2 = k^2 - k_z^2$ e $k^2 = \omega^2 \mu \epsilon$. These equations show that if we determine E_z and H_z, then, the other components are exclusive functions of these components.

There are, now, different types of solutions. The transverse electric solution (TE) is such that $E_z = 0, H_z \neq 0$; another type of solution, transverse magnetic (TM), is such that $E_z \neq 0, H_z = 0$. There is another solution, transverse electric and magnetic (TEM), such that $E_z = 0, H_z = 0$, but it does not occur for a rectangular waveguide.

In the case of a TE mode we need to solve Helmholtz equation for H_z,

$$\nabla^2 H_z + k^2 H_z = 0 \,, \tag{7.4}$$

and obtain the other components knowing that

$$H_x = -i\frac{k_z}{k_c^2}\frac{\partial H_z}{\partial x} \,,$$

$$H_y = -i\frac{k_z}{k_c^2}\frac{\partial H_z}{\partial y} \,,$$

$$E_x = -i\frac{\mu\omega}{k_c^2}\frac{\partial H_z}{\partial y} \,,$$

$$E_y = i\frac{\mu\omega}{k_c^2}\frac{\partial H_z}{\partial x} \,. \tag{7.5}$$

In the case of TM modes ($H_z = 0, E_z \neq 0$) we solve first the Helmholtz equation for E_z,

$$\nabla^2 E_z + k^2 E_z = 0 \,, \tag{7.6}$$

and obtain the other components using

$$H_x = i\frac{\omega\epsilon}{k_c^2}\frac{\partial E_z}{\partial y} \,,$$

$$H_y = -i\frac{\omega\epsilon}{k_c^2}\frac{\partial E_z}{\partial x} \,,$$

$$E_x = -i\frac{k_z}{k_c^2}\frac{\partial E_z}{\partial x} \,,$$

$$E_y = -i\frac{k_z}{k_c^2}\frac{\partial E_z}{\partial y} \,. \tag{7.7}$$

In order to solve the Helmholtz equation we need to specify the boundary conditions. In general, the tangential components of the electric field are continuous at the interface. Since inside the conductor (considered as ideal) the electric field vanishes, the tangential component of the electric field in the waveguide also vanishes. Let us consider first the TM modes for which $E_z \neq 0$. The boundary conditions are written as

$$E_z(0, y, z) = E_z(a, y, z) = E_z(x, 0, z) = E_z(x, b, z) = 0. \qquad (7.8)$$

Note that Helmholtz equation reduces to a two-dimensional equation,

$$\left(\frac{\partial^2}{\partial x^2} + \frac{\partial^2}{\partial y^2} + k_c^2 \right) E_z = 0. \qquad (7.9)$$

The solutions are of the type

$$E_{zmn}(x, y, z) = e_{mn} \sin \left(\frac{m\pi x}{a} \right) \sin \left(\frac{n\pi y}{b} \right) e^{-ik_z z}, \qquad (7.10)$$

which implies that $k_c = \sqrt{\left(\frac{m\pi}{a} \right)^2 + \left(\frac{n\pi}{b} \right)^2}$. The numbers m and n are natural numbers. We may see immediately that

$$H_x, E_y \sim \sin \left(\frac{m\pi x}{a} \right) \cos \left(\frac{n\pi y}{b} \right) e^{-ik_z z},$$

$$H_y, E_x \sim \cos \left(\frac{m\pi x}{a} \right) \sin \left(\frac{n\pi y}{b} \right) e^{-ik_z z}.$$

The boundary condition for the tangential component of the electric field is not directly useful for the TE modes, since these involve the solution of Helmholtz equation for H_z. But we know that

$$E_x = -i \frac{\mu\omega}{k_c^2} \frac{\partial H_z}{\partial y}, \qquad E_y = i \frac{\mu\omega}{k_c^2} \frac{\partial H_z}{\partial x}. \qquad (7.11)$$

Let us look, then, for a general solution for H_z in the form

$$H_z(x, y) = [A \cos(k_x x) + B \sin(k_x x)] [C \cos(k_y y) + D \sin(k_y y)]. \qquad (7.12)$$

Then,

$$E_x = -i \frac{\mu\omega}{k_c^2} [A \cos(k_x x) + B \sin(k_x x)] [-C k_y \sin(k_y y) + D k_y \cos(k_y y)],$$

$$E_y = i \frac{\mu\omega}{k_c^2} [-A k_x \sin(k_x x) + B k_x \cos(k_x x)) (C \cos(k_y y) + D \sin(k_y y)].$$

$$(7.13)$$

Using the boundary conditions $E_x(x, 0) = E_x(x, b) = 0$ and $E_y(0, y) = E_y(a, y) = 0$, we obtain that $D = 0$ and $B = 0$. Therefore,

$$H_{zmn}(x, y, z) = h_{mn} \cos \left(\frac{m\pi x}{a} \right) \cos \left(\frac{n\pi y}{b} \right) e^{-ik_z z}. \qquad (7.14)$$

It may be shown that the same solution is obtained using directly boundary conditions for H_z in the form

$$\frac{\partial H_z}{\partial x}(0, y, z) = \frac{\partial H_z}{\partial x}(a, y, z) = \frac{\partial H_z}{\partial y}(x, 0, z) = \frac{\partial H_z}{\partial y}(x, b, z) = 0. \qquad (7.15)$$

Note that, in this case, $m = 0, 1, 2, \cdots$ and $n = 0, 1, 2, \cdots$, but we must exclude the point $(m, n) = (0, 0)$. For these modes we have that

$$E_x, H_y \sim \cos(m\pi x/a) \sin(n\pi y/b),$$
$$E_y, H_x \sim \sin(m\pi x/a) \cos(n\pi y/b).$$

Recall that

$$k_z = k_{zmn} = \sqrt{k^2 - k_{c,mn}^2} \tag{7.16}$$

and must be real for a propagating wave. Therefore, it must hold true that

$$\omega^2 \mu\epsilon > \left(\frac{m\pi}{a}\right)^2 + \left(\frac{n\pi}{b}\right)^2. \tag{7.17}$$

Therefore, there is a frequency threshold in order to have a given mode, both for TE and TM modes, given by

$$f_{c,mn} = \frac{1}{2\pi\sqrt{\mu\epsilon}}\sqrt{\left(\frac{m\pi}{a}\right)^2 + \left(\frac{n\pi}{b}\right)^2}, \tag{7.18}$$

therefore, there is a gap at low frequencies.

7.2.2 *Ferrite tubes*

Materials are called reciprocal if the response function between two space points is symmetric in the exchange of their locations. Otherwise, they are called non-reciprocal. A usual example of non-reciprocity is obtained applying a magnetic field. We will analyse its influence on the dielectric permittivity and on the magnetic permeability.

Let us begin by considering a simple model for a dielectric, assuming that it is composed by electrons bound harmonically and isotropically to a given center of force, and let us apply an electric field and a magnetic field, taking into account the presence of friction on the motion of the electrons. The equation of motion of one of these electrons of charge q and mass m is written, in a semi-classical approximation, as

$$m\frac{d^2\mathbf{r}}{dt^2} = q\mathbf{E} + q\frac{d\mathbf{r}}{dt} \times \mathbf{B} - m\nu_c\frac{d\mathbf{r}}{dt} - m\omega_0^2\mathbf{r}. \tag{7.19}$$

Looking for a solution of the type $\mathbf{r} \propto \mathbf{r}_0 e^{i\omega t}$, we obtain

$$-m\omega^2\mathbf{r} = q\mathbf{E} + i\omega q\mathbf{r} \times \mathbf{B} - m\nu_c i\omega\mathbf{r} - m\omega_0^2\mathbf{r}. \tag{7.20}$$

The relation between the coordinates and the electric field may be written in a matrix form,

$$A \begin{pmatrix} x \\ y \\ z \end{pmatrix} = q \begin{pmatrix} E_x \\ E_y \\ E_z \end{pmatrix}, \tag{7.21}$$

where the matrix A is given by

$$\begin{pmatrix} m(\omega_0^2 - \omega^2) + m\nu_c i\omega & -qi\omega B_z & qi\omega B_y \\ qi\omega B_z & m(\omega_0^2 - \omega^2) + m\nu_c i\omega & -qi\omega B_x \\ -qi\omega B_y & qi\omega B_x & m(\omega_0^2 - \omega^2) + m\nu_c i\omega \end{pmatrix} .$$

$$(7.22)$$

The dipolar moment is given by $\mathbf{p} = q\mathbf{r}$, the polarization $\mathbf{P} = n_q q\mathbf{r} = \epsilon_0 \chi_e \mathbf{E}$, and also $\epsilon = \epsilon_0 (1 + \chi_e)$. From here we obtain that the dielectric tensor is given by

$$\epsilon = \epsilon_0 \left(1 + \frac{n_q q^2}{\epsilon_0} A^{-1} \right) . \tag{7.23}$$

The inverse of the matrix A is given by

$$A^{-1} = \frac{1}{a(a^2 - \tilde{B}^2)} \begin{pmatrix} a^2 - \tilde{B}_x^2 & -\tilde{B}_x\tilde{B}_y + ia\tilde{B}_z & -\tilde{B}_x\tilde{B}_z - ia\tilde{B}_y \\ -\tilde{B}_x\tilde{B}_y - ia\tilde{B}_z & a^2 - \tilde{B}_y^2 & -\tilde{B}_y\tilde{B}_z + ia\tilde{B}_x \\ -\tilde{B}_x\tilde{B}_z + ia\tilde{B}_y & -\tilde{B}_y\tilde{B}_z - ia\tilde{B}_x & a^2 - \tilde{B}_z^2 \end{pmatrix} ,$$

$$(7.24)$$

where $a = A_{11} = A_{22} = A_{33}$ and $\tilde{B}_j = B_j q\omega$. We can see that the matrix A is antisymmetric due to the presence of a term that breaks time reversal invariance, the magnetic field. Considering for instance a magnetic field along the z axis, and a transverse wave ($E_z = 0$) propagating along this direction, the matrix structure of the dielectric tensor ensures that the index of refraction has two distinct eigenvalues, that lead to two different indices of refraction and the Faraday effect.

A similar result is obtained if we consider a ferrimagnetic material, considering now the magnetic permeability. Examples are iron oxides (YIG) and various systems such as aluminum, cobalt, manganese and nickel. In these systems, the magnetic anisotropy of the material is induced applying a magnetic field, as above.

Let us consider, now, a magnetic material where there are magnetic dipoles. The magnetic moment and the angular momentum are related by $\mathbf{m} = -\gamma\mathbf{s}$, where γ is the gyromagnetic ratio and $\mathbf{s}$ the spin angular momentum. Applying a magnetic field along z, for instance, a torque (force moment) arises on the magnetic moment that is given by

$$\boldsymbol{\tau} = \mathbf{m} \times \mathbf{B}_0 = \mu_0 \mathbf{m} \times \mathbf{H}_0 = -\mu_0 \gamma \mathbf{s} \times \mathbf{H}_0 . \tag{7.25}$$

Since the torque is equal to the time variation of the angular momentum, we obtain

$$\frac{d\mathbf{m}}{dt} = -\mu_0 \gamma \mathbf{m} \times \mathbf{H}_0 . \tag{7.26}$$

Therefore, the magnetization also obeys the same equation of motion,

$$\frac{d\mathbf{M}}{dt} = -\mu_0\gamma\mathbf{M}\times\mathbf{H}\,,\tag{7.27}$$

where $\mathbf{H}$ is the internal magnetic field. Let us consider, now, the effect of adding a small magnetic field, $\mathbf{h}$, that changes with time. The total field and total magnetization are , then, given by

$$\mathbf{H}_t = H_0\mathbf{e}_z + \mathbf{h}\,,$$
$$\mathbf{M}_t = M_s\mathbf{e}_z + \mathbf{M}\,,\tag{7.28}$$

where M_s is the saturation magnetization, due to the external field, and the additional magnetization is taken in the plane xy. The equations of motion for the magnetization components are given, in the limit of a small changing field, by

$$\frac{dM_x}{dt} = -\omega_0 M_y + \omega_m h_y\,,$$
$$\frac{dM_y}{dt} = \omega_0 M_x - \omega_m h_x\,,$$
$$\frac{dM_z}{dt} = 0\,,\tag{7.29}$$

where $\omega_0 = \mu_0\gamma H_0$ and $\omega_m = \mu_0\gamma M_s$. We obtain that the components satisfy

$$\frac{d^2 M_x}{dt^2} + \omega_0^2 M_x = \omega_m\frac{dh_y}{dt} + \omega_0\omega_m h_x\,,$$
$$\frac{d^2 M_y}{dt^2} + \omega_0^2 M_y = -\omega_m\frac{dh_x}{dt} + \omega_0\omega_m h_y\,.\tag{7.30}$$

If the field $\mathbf{h}$ has a time dependence of the form $e^{i\omega t}$, then we obtain

$$\left(\omega_0^2 - \omega^2\right)M_x = \omega_0\omega_m h_x + i\omega\omega_m h_y\,,$$
$$\left(\omega_0^2 - \omega^2\right)M_y = -i\omega\omega_m h_x + \omega_0\omega_m h_y\,.\tag{7.31}$$

These relations may be condensed in a matrix form,

$$M = \chi H = \begin{pmatrix} \chi_{xx} & \chi_{xy} & 0 \\ \chi_{yx} & \chi_{yy} & 0 \\ 0 & 0 & 0 \end{pmatrix} h\,,\tag{7.32}$$

where the matrix elements are given by

$$\chi_{xx} = \chi_{yy} = \frac{\omega_0\omega_m}{\omega_0^2 - \omega^2}\,,$$
$$\chi_{xy} = -\chi_{yx} = \frac{i\omega\omega_m}{\omega_0^2 - \omega^2}\,.\tag{7.33}$$

This implies that the magnetic permeability tensor $B = \mu_t h$ is given by

$$\mu_t = \begin{pmatrix} \mu & i\kappa & 0 \\ -i\kappa & \mu & 0 \\ 0 & 0 & \mu_0 \end{pmatrix} . \tag{7.34}$$

The elements of the permeability tensor are given by

$$\mu = \mu_0 \left(1 + \chi_{xx}\right) = \mu_0 \left(1 + \chi_{yy}\right) = \mu_0 \left(1 + \frac{\omega_0 \omega_m}{\omega_0^2 \omega^2}\right) ,$$

$$\kappa = -i\mu_0 \chi_{xy} = i\mu_0 \chi_{yx} = \mu_0 \frac{\omega \omega_m}{\omega_0^2 - \omega^2} . \tag{7.35}$$

A material that has a permeability of this form is called gyrotropic. The matrix is also antisymmetric due to the presence of the magnetic field. Changing the direction of the magnetic fields, the signal of μ remains invariant but κ changes sign. These results assume that there are no losses. One may include the effect of losses in the usual way taking $\omega_0 \to \omega_0 - i\alpha\omega$.

7.2.3 *Waves in a periodic system: photonic crystals*

The Maxwell equations in a material take the form

$$\nabla \cdot \mathbf{B} = 0 ,$$

$$\nabla \cdot \mathbf{D} = \rho ,$$

$$\nabla \times \mathbf{E} + \frac{\partial \mathbf{B}}{\partial t} = 0 ,$$

$$\nabla \times \mathbf{H} - \frac{\partial \mathbf{D}}{\partial t} = \mathbf{J} . \tag{7.36}$$

Let us consider a material where the relative dielectric permittivity and the relative magnetic permeability depend on the space location, in such a way that $\mathbf{D}(\mathbf{r}) = \epsilon_0 \epsilon(\mathbf{r}) \mathbf{E}(\mathbf{r})$ and $\mathbf{B}(\mathbf{r}) = \mu_0 \mu(\mathbf{r}) \mathbf{H}(\mathbf{r})$. We will study the propagation of modes of angular frequency ω, described by functions $\mathbf{E}(\mathbf{r}, t) = \mathbf{E}(\mathbf{r}) e^{i\omega t}$ and $\mathbf{H}(\mathbf{r}, t) = \mathbf{H}(\mathbf{r}) e^{i\omega t}$. We may consider that the material is transparent and for which $\epsilon(\mathbf{r}, \omega = 0)$ is real and positive. Usually $\mu \sim 1$ but we will also consider a more general case. Then, in the absence of free charges and free currents, Maxwell's equations take the form

$$\nabla \cdot [\mu_0 \mu(\mathbf{r}) \mathbf{H}] = 0 ,$$

$$\nabla \cdot [\epsilon_0 \epsilon(\mathbf{r}) \mathbf{E}] = 0 ,$$

$$\nabla \times \mathbf{E} + i\omega \mu_0 \mu(\mathbf{r}) \mathbf{H} = 0 ,$$

$$\nabla \times \mathbf{H} - i\omega \epsilon_0 \epsilon(\mathbf{r}) \mathbf{E} = 0 . \tag{7.37}$$

We may eliminate, for instance, the electric field to obtain the so-called master equation,

$$\nabla \times \left(\frac{1}{\epsilon(\mathbf{r})} \nabla \times \mathbf{H}(\mathbf{r}) \right) = \left(\frac{\omega}{c} \right)^2 \mu(\mathbf{r}) \mathbf{H}(\mathbf{r}), \tag{7.38}$$

where $c = 1/\sqrt{\epsilon_0 \mu_0}$. This equation may be solved using the condition that $\nabla \cdot [\mu(\mathbf{r})\mathbf{H}(\mathbf{r})] = 0$. If $\mathbf{H}(\mathbf{r}) = \mathbf{H}_0 e^{i\mathbf{k}\cdot\mathbf{r}}$ and $\mu(\mathbf{r}) \sim 1$ we obtain the usual transversality condition for the magnetic field $\mathbf{k} \cdot \mathbf{H}_0 = 0$. Having solved the equation for the magnetic field, we may next calculate the electric field using

$$\mathbf{E}(\mathbf{r}) = \frac{-i}{\omega \epsilon_0 \epsilon(\mathbf{r})} \nabla \times \mathbf{H}(\mathbf{r}). \tag{7.39}$$

The transversality condition for the electric displacement field, that takes the form $\nabla \cdot [\epsilon(\mathbf{r})\mathbf{E}(\mathbf{r})] = 0$, is guaranteed, since the divergence of the rotational vanishes.

We may also eliminate the magnetic field and obtain the master equation for the electric field

$$\nabla \times \left(\frac{1}{\mu(\mathbf{r})} \nabla \times \mathbf{E}(\mathbf{r}) \right) = \left(\frac{\omega}{c} \right)^2 \epsilon(\mathbf{r}) \mathbf{E}(\mathbf{r}). \tag{7.40}$$

The master equation for the magnetic field or for the electric field may be seen as a generalized eigenvalue equation

$$\hat{O}_H \mathbf{H}(\mathbf{r}) = \left(\frac{\omega}{c} \right)^2 \mu(\mathbf{r}) \mathbf{H}(\mathbf{r}), \tag{7.41}$$

or

$$\hat{O}_E \mathbf{E}(\mathbf{r}) = \left(\frac{\omega}{c} \right)^2 \epsilon(\mathbf{r}) \mathbf{E}(\mathbf{r}), \tag{7.42}$$

where

$$\hat{O}_H = \nabla \times \left(\frac{1}{\epsilon(\mathbf{r})} \nabla \times \right),$$

$$\hat{O}_E = \nabla \times \left(\frac{1}{\mu(\mathbf{r})} \nabla \times \right). \tag{7.43}$$

If $\mu(\mathbf{r}) \sim 1$ ($\epsilon(\mathbf{r}) \sim 1$) and, therefore, are constants, the operator $1/\mu(\mathbf{r})\hat{O}_H$ ($1/\epsilon(\mathbf{r})\hat{O}_E$) is hermitian and the eigenvalues are real.

Let us suppose now that the system has a periodic structure reflected in its various physical quantities, such as the permittivity and the permeability, that are consequently periodic. Then, Bloch's theorem is valid, like in electronic systems. Representing the basis vectors of the periodic structure (lattice) by $\mathbf{a}_1, \mathbf{a}_2, \mathbf{a}_3$, any physical quantity has a periodicity, under the action of translations by $\mathbf{R} = l\mathbf{a}_1 + m\mathbf{a}_2 + n\mathbf{a}_3$. For example, we have that $\epsilon(\mathbf{r}) = \epsilon(\mathbf{r}+\mathbf{R})$. Associated to the basis vectors, we may define a reciprocal lattice and a wave vector in the usual way as $\mathbf{k} = k_1\mathbf{b}_1 + k_2\mathbf{b}_2 + k_3\mathbf{b}_3$, where $\mathbf{b}_i$ are the basis vectors of the reciprocal lattice. The magnetic field satisfies the Bloch theorem, and can, therefore, be written as

$$\mathbf{H_k}(\mathbf{r}) = e^{i\mathbf{k}\cdot\mathbf{r}}\mathbf{u_k}(\mathbf{r})\,,\tag{7.44}$$

where $u_\mathbf{k}(\mathbf{r}) = u_\mathbf{k}(\mathbf{r}+\mathbf{R})$. Substituting this expression in the generalized eigenvalue problem, we obtain

$$\hat{O}_\mathbf{k}u_\mathbf{k}(\mathbf{r}) = \left(\frac{\omega}{c}\right)^2\mu(\mathbf{r})u_\mathbf{k}(\mathbf{r})\,,\tag{7.45}$$

where the operator

$$\hat{O}_\mathbf{k} = (i\mathbf{k}+\nabla)\times\frac{1}{\epsilon(\mathbf{r})}(i\mathbf{k}+\nabla)\times\tag{7.46}$$

and the transversality condition may be written as

$$(i\mathbf{k}+\nabla)\cdot[\mu(\mathbf{r})\mathbf{u}_k(\mathbf{r})] = 0\,.\tag{7.47}$$

The solution of this eigenvalue problem leads to a band structure for $\omega_n(\mathbf{k})$, in a way similar to the band structure in electronic systems. In general, a numerical solution is required, due to the complexity of the operator.

7.2.4   *TM modes in a periodic lattice*

We may solve the master equation for the TM modes considering

$$\nabla\times[\mu^{-1}(\mathbf{r})\nabla\times\mathbf{E}] = \omega^2\epsilon(\mathbf{r})\mathbf{E}\,,\tag{7.48}$$

(we are using units in which $c = 1$) where the tensor μ takes the form of the equation(7.34). Its inverse is

$$\mu^{-1} = \begin{pmatrix} \tilde{\mu}^{-1} & i\eta & 0 \\ -i\eta & \tilde{\mu}^{-1} & 0 \\ 0 & 0 & \mu_0^{-1} \end{pmatrix}\,,\tag{7.49}$$

with

$$\tilde{\mu}^{-1} = \frac{\mu}{\mu^2-\kappa^2}\,,$$

$$\eta = -\frac{\kappa}{\mu^2-\kappa^2}\,.\tag{7.50}$$

For the specific example of the YIG ferrite tubes in the presence of a magnetic induction $B = 1600$ Gauss and a frequency $f = 4.28$ GHz, one takes $\kappa = 12.4\mu_0$, $\mu = 14\mu_0$.

As mentioned above, the solution of the master equation may be seen as a generalized eigenvalue problem. Let us consider TM modes characterized by $E_x = E_y = 0$ (for $k_z = 0$). We obtain the equation for E_z,

$$\left[-\nabla^2 + (\nabla \ln \tilde{\mu} - i\tilde{\mu}\mathbf{e}_z \times \nabla \eta) \cdot \nabla - \tilde{\mu}\epsilon\omega^2\right] E_z = 0.$$ (7.51)

Defining $\psi = E_z/\sqrt{\tilde{\mu}}$, we obtain

$$\left(-|\nabla + i\tilde{A}(\mathbf{r})|^2 + \tilde{V}(r)\right)\psi = 0,$$ (7.52)

with a vector potential

$$\tilde{A} = \frac{\tilde{\mu}}{2}\mathbf{e}_z \times \nabla \eta,$$ (7.53)

and a scalar potential

$$\tilde{V} = \frac{1}{4}\left(|\nabla \ln \tilde{\mu}|^2 + |\tilde{\mu}\nabla \eta|^2\right) - \frac{1}{2}\nabla^2 \ln \tilde{\mu} - \tilde{\mu}\epsilon\omega^2.$$ (7.54)

By similarity with a Schrödinger equation, we may interpret that this equation describes zero energy modes of a particle in the presence of periodic vector and scalar potentials. The amplitude of the scalar potential depends on the value of ω. This equation may be solved numerically at the interface between a normal photonic crystal and a magneto-optical photonic crystal. At the interface edge states appear due to the bulk-edge correspondence, signalling the existence of non-trivial topology. Note that if $\kappa = 0$ and, as a consequence, $\eta = 0$, the vector potential $\tilde{A} = 0$, and there is no time reversal symmetry breaking.

7.2.5 Effective model for quadratic bands

Even though, in general, the problem has to be solved numerically, some approximations may be carried out that correctly describe the behavior of the bands near points where the low energy modes are located. These may be points where the spectrum is linear (Dirac points), as in the lattice considered by Haldane and Raghu, or near bands with quadratic dispersion relations, degenerate in a given point in momentum space, that arise in a square lattice with ferrite tubes. If the system is time reversal invariant, the bands touch. The degeneracy is raised if we break the invariance for time reversal or parity. In the relevant cases, the subspace around the degeneracy point involves only two bands.

An effective model may be introduced to describe the behavior of the two bands near the degeneracy point. The model depends on some parameters, that are adjusted using the numerical solution, guaranteeing that the description is sufficiently accurate at low energies. Breaking parity, states appear inside the gap that are not topologically protected and that may be removed by appropriate choices of the parameters. In the case the time reversal symmetry is broken, edge states appear that are robust, guaranteeing the topological nature, that may be confirmed by finite Chern numbers, as for electronic systems.

Let us consider a square lattice whose basis vectors are designated by $\mathbf{a}_1 = a(1,0)$, $\mathbf{a}_2 = a(0,1)$, where a is the lattice constant. In a square lattice of dielectric tubes there is a quadratic dispersion at the degeneracy point at the corner of the Brillouin zone, that we may call $\mathbf{k}_M$. We want now to consider several perturbations that break the square symmetry.

One way consists to consider the result of a shear force of the basis vectors in the form $\mathbf{a}_1 = a(\cos\theta, \sin\theta)$, $\mathbf{a}_2 = a(\sin\theta, \cos\theta)$. This perturbation breaks the symmetry under rotations C_4 and reflections around the axes x and y. Another possibility consists in performing distortions along the x or y axes that break the symmetry under the action of rotations or reflections around the lines $y = \pm x$. Including gyromagnetic effects, that break both parity and time reversal symmetry, a gap appears in this point. The degeneracy may be lifted introducing a term in the magnetic permeability of the form $\mu_{xy} = i\kappa$. In this case, if the system is finite, a chiral edge state appears.

Let us consider a generic case such that the Hamiltonian in the subspace of these two bands is written as

$$H = \lambda_0 \left(\sum_{i=1}^{3} \alpha_i \sigma_i + \beta \left(\kappa_x^2 - \kappa_y^2 \right) \sigma_1 + 2\kappa_x \kappa_y \sigma_3 + \gamma |\kappa|^2 \right), \quad (7.55)$$

where $\kappa = \mathbf{k} - \mathbf{k}_M$ is the shift of the degeneracy point in the momentum space, and σ_i are Pauli matrices. λ_0 is a multiplicative factor that determines the energy scale and β and γ fix the curvatures of the bands along different directions, α_1 the relative length of the different basis vectors, α_2 is proportional to the permeability component that breaks parity, and α_3 is proportional to the twist angle θ of the basis axes. We consider that the parameters responsible by the symmetry breaking are small (for instance, $\theta \ll 1$) being, therefore, of zeroth order in κ. This Hamiltonian may be written in matrix form,

$$\frac{H}{\lambda_0} = \begin{pmatrix} \alpha_3 + \gamma|\kappa|^2 + 2\kappa_x\kappa_y & \beta\left(\kappa_x^2 - \kappa_y^2\right) + \alpha_1 - i\alpha_2 \\ \beta\left(\kappa_x^2 - \kappa_y^2\right) + \alpha_1 + i\alpha_2 & -\alpha_3 + \gamma|\kappa|^2 - 2\kappa_x\kappa_y \end{pmatrix}. \quad (7.56)$$

Let us consider first $\alpha_1 = \alpha_3 = 0$. The eigenvalues of the effective Hamiltonian are given by

$$\frac{\lambda_\pm(\kappa)}{\lambda_0} = \gamma|\kappa|^2 \pm \sqrt{|\kappa|^4 + (\beta^2 - 1)(\kappa_x^2 - \kappa_y^2)^2 + \alpha_2^2}. \tag{7.57}$$

Choosing $\beta = 1$, the eigenvalues have the simplified form

$$\frac{\lambda_\pm(\kappa)}{\lambda_0} = \gamma|\kappa|^2 \pm \sqrt{|\kappa|^4 + \alpha_2^2}, \tag{7.58}$$

and if the imaginary non-diagonal part is absent, $\alpha_2 = 0$, we obtain

$$\frac{\lambda_\pm(\kappa)}{\lambda_0} = (\gamma \pm 1)|\kappa|^2. \tag{7.59}$$

These two eigenvalues are degenerate at $\kappa = 0$. A gap appears if $\alpha_2 \neq 0$.

If, now, $\alpha_3 > 0$, the points where the degeneracy occurs change, and are now:

$$\mathbf{k}_\pm = \pm\sqrt{\frac{\alpha_3}{2}} \; (1, -1) \; .$$

The same effect is found if $\alpha_1 \neq 0$, or a combination of the two. We may expand the moment around the new degeneracy points defining

$$q_1 = \frac{1}{2}(\kappa_x + \kappa_y) \; ,$$

$$q_2 = \frac{1}{2}(-\kappa_x + \kappa_y) \pm \sqrt{\frac{\alpha_3}{2}} \; . \tag{7.60}$$

Then, the Hamiltonian expanded to the first order in q_1 and q_2 takes the form

$$\frac{H_\pm(\mathbf{q})}{\lambda_0} \approx \gamma\left(\alpha_3 \mp \sqrt{8\alpha_3}q_2\right) \pm \sqrt{8\alpha_3}\left(\beta q_1\sigma_1 + q_2\sigma_3\right) + \alpha_2\sigma_2 \; . \tag{7.61}$$

When $\gamma = 0$ and $\beta = 1$, this expression reduces to a two-dimensional Dirac Hamiltonian, near each degeneracy point. Here α_2 takes the role of a mass term, opening a gap $2\lambda_0\alpha_2$. If $\gamma \neq 0$ the Dirac Hamiltonian gets distorted.

When $\alpha_2 \neq 0$, there is a gap and the effective bands have a finite Chern number. The lower bands have Chern numbers that are given by $C = \text{sgn}(\alpha_2)$ and for the upper bands, $C = -\text{sgn}(\alpha_2)$. Recalling the bulk-edge correspondence, this result leads to edge modes that are topologically protected, as expected.

The calculation of the Chern number may be carried out in the usual way. Focusing the attention on the lower band and considering, for simplicity, $\alpha_1 = \alpha_3 = 0$, we may write

$$\frac{H}{\lambda_0} = \begin{pmatrix} \gamma|\kappa|^2 + 2\kappa_x\kappa_y & \beta\left(\kappa_x^2 - \kappa_y^2\right) - i\alpha_2 \\ \beta\left(\kappa_x^2 - \kappa_y^2\right) + i\alpha_2 & \gamma|\kappa|^2 - 2\kappa_x\kappa_y \end{pmatrix} . \tag{7.62}$$

The term proportional to γ does not contribute to the Chern number because it is of the form of the identity operator. Therefore, we may ignore it. Considering $\beta = 1$ leads to

$$\frac{H}{\lambda_0} = \begin{pmatrix} 2\kappa_x\kappa_y & \left(\kappa_x^2 - \kappa_y^2\right) - i\alpha_2 \\ \left(\kappa_x^2 - \kappa_y^2\right) + i\alpha_2 & -2\kappa_x\kappa_y \end{pmatrix}. \tag{7.63}$$

Using polar coordinates we write that $\kappa_x = \kappa\cos(\phi), \kappa_y = \kappa\sin(\phi)$. Then,

$$\frac{H}{\lambda_0} = \begin{pmatrix} \kappa^2\sin(2\phi) & \kappa^2\cos(2\phi) - i\alpha_2 \\ \kappa^2\cos(2\phi) + i\alpha_2 & -\kappa^2\sin(2\phi) \end{pmatrix}. \tag{7.64}$$

The eigenstate of the lower band may be written as

$$|\psi^-(\kappa)\rangle = \frac{1}{\sqrt{2\left[\kappa^4 + \alpha_2^2 + \kappa^2\sqrt{\kappa^4 + \alpha_2^2}\sin(2\phi)\right]}} \begin{pmatrix} -\kappa^2\cos(2\phi) + i\alpha_2 \\ \kappa^2\sin(2\phi) + \sqrt{\kappa^4 + \alpha_2^2} \end{pmatrix} \tag{7.65}$$

and the corresponding Berry connection is given by

$$\mathbf{A}^-(\kappa) = i\langle\psi^-(\kappa)|\nabla_\kappa|\psi^-(\kappa)\rangle = -\frac{\alpha_2\kappa(\cos(2\phi)\mathbf{e}_\kappa - \sin(2\phi)\mathbf{e}_\phi)}{\kappa^4 + \alpha_2^2\sqrt{\kappa^4 + \alpha_2^2}\sin(2\phi)}. \tag{7.66}$$

The Chern number is given by

$$C^- = \frac{1}{2\pi}\oint d\kappa \cdot \mathbf{A}^-(\kappa)$$
$$= \frac{2\alpha_2}{\pi}\int_{-\pi/4}^{\pi/4} \frac{\kappa_0^2\sin(2\phi)d\phi}{\kappa_0^4 + \alpha_2^2 + \kappa_0^2\sqrt{\kappa_0^4 + \alpha_2^2}\sin(2\phi)}, \tag{7.67}$$

where κ_0 is the radius of the circumference that contains the degeneracy point. Making the substitution $\sin(2\phi) = \tanh(u)$ we obtain that

$$|C^-| = \text{sgn}(\alpha_2) - \frac{\alpha_2}{\sqrt{\kappa_0^4 + \alpha_2^2}} \to \text{sgn}(\alpha_2) \tag{7.68}$$

for $|\alpha_2| << \kappa_0^2$.

Separating the Dirac points due to $\alpha_3 \neq 0$ and/or due to $\alpha_1 \neq 0$, it can be shown that the contribution of each Dirac point to the Chern number is $\text{sgn}(\alpha_2)/2$.

7.2.6 *Experimental implementation*

The proposal by Haldane and Raghu was implemented experimentally, as illustrated in figure 7.1. In the experiment, a rectangular waveguide, narrow along a direction designated vertical direction, is filled by a square

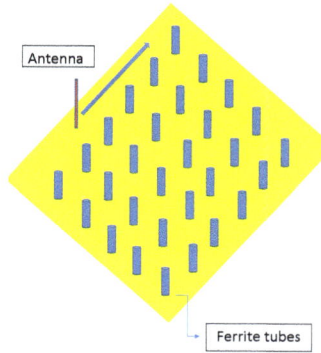

Fig. 7.1   Waveguide with unidirectional edge mode with topological protection in the presence of a lattice of ferrite tubes.

lattice of ferrite tubes, a gyromagnetic material. Applying a magnetic field along the vertical direction, a degeneracy between bands is lifted and, since there is time reversal symmetry breaking, gives origin to gaps that are filled, between the second and third bands, by a chiral edge mode, as result of the non-trivial Chern numbers of the two bands. Without the magnetic field, these bands with quadratic momentum dispersion relation are connected at a point where the degeneracy takes place, being composed by a pair of Dirac cones. The magnetic field gives origin to non-diagonal terms, antisymmetric and imaginary, in the magnetic permeability tensor. In the gap between the lower frequency band and the second band there is no mode, since the sum of the Chern numbers below a frequency in the gap vanishes. This sum equals 1 in the gap between the second and third bands and, therefore, the edge mode appears. An antenna is placed on the metallic border of the waveguide and a mode is detected along the same border, even if an imperfection, of arbitrary shape and length, is placed on the path of the propagating mode. In this experiment transverse magnetic (TM) modes are considered, while in the original proposal by Haldane and Raghu transverse electrical (TE) modes were considered. Numerical simulations of the collision process with the imperfection have been confirmed experimentally and show that the amplitudes of the transmitted wave and reflected wave differ considerably, by more than five orders of magnitude, and this asymmetry is independent of the size of the obstacle, placed on the path of the edge mode.

7.3 Systems with time reversal symmetry

Similarly to electronic systems, it is also possible to have topological modes in the case there is time reversal symmetry. This is discussed in this section.

7.3.1 *Scattering of a particle by a potential*

Let us consider a beam of free particles that impinges on a region where they interact with a target of finite extension. The beam goes through that region and emerges such that the particles may be considered again as free. Neglecting the interactions between the beam particles, we may study the scattering process considering one particle at a time.

Let us define, then, three spatial regions (see figure 7.2): a region I, for negative values of x and where we may take the potential as vanishing; a region III for positive values of x and where the particles are also free, and the interaction region, called region II, where the interaction with a given potential V takes place. We will consider a one-dimensional problem. We may write with generality the wave functions in the regions I and III, since the potential is zero. We consider, therefore, solutions as:

$$\psi_I = Ae^{ikx} + Be^{-ikx},$$
$$\psi_{III} = Ce^{ikx} + De^{-ikx}. \tag{7.69}$$

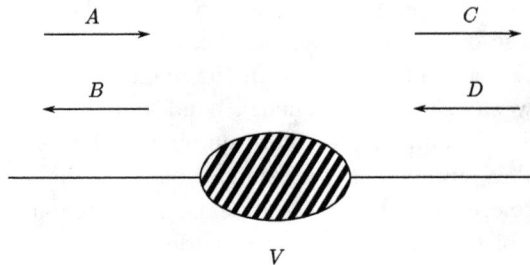

Fig. 7.2   Generic potential scattering experiment. Figure reproduced from the book *Mecânica Quântica*, Coleção Ensino da Ciência e da Tecnologia, Vol. 50, IST Press (2017).

Since the probability current is conserved, we have that

$$|A|^2 + |D|^2 = |C|^2 + |B|^2.$$

The waves with coefficients A and D are incoming waves, and the waves with coefficients C and B are outgoing waves. We define the S matrix that relates the outgoing states with the incoming states,

$$\begin{pmatrix} B \\ C \end{pmatrix} = S \begin{pmatrix} D \\ A \end{pmatrix}.$$

Taking the adjoint, we obtain that

$$\begin{pmatrix} B^* & C^* \end{pmatrix} = \begin{pmatrix} D^* & A^* \end{pmatrix} S^\dagger.$$

The probability current conservation can be expressed as

$$\begin{pmatrix} B^* & C^* \end{pmatrix} \begin{pmatrix} B \\ C \end{pmatrix} = \begin{pmatrix} D^* & A^* \end{pmatrix} \begin{pmatrix} D \\ A \end{pmatrix},$$

from which we obtain $S^\dagger S = 1$, therefore, the S matrix is unitary. The unitary property is consequence of the probability conservation.

Let us consider now Schrödinger's equation and its complex conjugate:

$$i\hbar \frac{\partial \psi}{\partial t} = -\frac{\hbar^2}{2m} \frac{\partial^2 \psi}{\partial x^2} + V(x)\psi(x, t),$$

$$-i\hbar \frac{\partial \psi^*}{\partial t} = -\frac{\hbar^2}{2m} \frac{\partial^2 \psi^*}{\partial x^2} + V(x)\psi^*(x, t),$$

$$(7.70)$$

in the case of a real potential. Under time reversal, $t \to -t$, we have that if $\psi(x, t)$ is a solution, then $\psi^*(x, -t)$ is also a solution, That is, if

$$\psi_I = A e^{ikx} + B e^{-ikx},$$

$$\psi_{III} = C e^{ikx} + D e^{-ikx} \qquad (7.71)$$

are solutions, then:

$$A^* e^{-ikx} + B^* e^{ikx},$$

$$C^* e^{-ikx} + D^* e^{ikx} \qquad (7.72)$$

are also solutions with the same energy. Under time reversal, the incoming and outgoing waves are interchanged. We may write the relation between the incoming and outgoing amplitudes in the time reversal state, $\psi^*(x, -t)$, using once more the S matrix,

$$\begin{pmatrix} D^* \\ A^* \end{pmatrix} = S \begin{pmatrix} B^* \\ C^* \end{pmatrix}.$$

Taking the complex conjugate of this matrix equation, we obtain

$$\begin{pmatrix} D \\ A \end{pmatrix} = S^* \begin{pmatrix} B \\ C \end{pmatrix} = S^* S \begin{pmatrix} D \\ A \end{pmatrix},$$

which implies that $S^* S = 1$, as a result of the time reversal. Using the unitarity of S, we conclude that $S^\dagger = S^*$ and, therefore, the S matrix is symmetric. Note that repeating the time reversal leads us back to the initial state. Therefore, calling $\mathcal{T}$ the time reversal operator, we have that $\mathcal{T}^2 = 1$. The same happens if we have bosonic systems, including the case of light propagation described by Maxwell's equations. Therefore, in the case of light, the scattering matrix is supposed to be symmetric, as well.

7.3.2   *S matrix for the scattering of electromagnetic waves*

Maxwell's equations (7.37) may be rewritten in a compact way as

$$\hat{N}\mathbf{f} = \omega\mathbf{g}, \tag{7.73}$$

with

$$\hat{N} = \begin{pmatrix} 0 & i\nabla \times 1_{3\times3} \\ -i\nabla \times 1_{3\times3} & 0 \end{pmatrix}, \tag{7.74}$$

and where

$$\mathbf{f} = \begin{pmatrix} \mathbf{E} \\ \mathbf{H} \end{pmatrix}, \qquad \mathbf{g} = \begin{pmatrix} \mathbf{D} \\ \mathbf{B} \end{pmatrix} \tag{7.75}$$

are the electric and magnetic fields, electric displacement field and magnetic induction and ω is the oscillation frequency. Also $\mathbf{g} = M\mathbf{f}$ with

$$M = \begin{pmatrix} \epsilon_0\epsilon & \frac{1}{c}\xi \\ \frac{1}{c}\zeta & \mu_0\mu \end{pmatrix}. \tag{7.76}$$

The adimensional tensors $\epsilon(\omega), \mu(\omega), \xi(\omega), \zeta(\omega)$ are the permittivity, permeability and relative magnetoelectric tensors, respectively. The time reversal operator, that we represent by $\mathcal{T}$, is defined by $\mathcal{T} = K\sigma_z$, where K is complex conjugation and

$$\sigma_z = \begin{pmatrix} 1_{3\times3} & 0 \\ 0 & -1_{3\times3} \end{pmatrix}$$

and acts on the fields such as $\mathbf{f} \to \mathcal{T}\mathbf{f}$, and $\mathbf{g} \to \mathcal{T}\mathbf{g}$. The terms that involve the electric field transform into their complex conjugates, while the magnetic part acquires in addition a minus sign. Therefore, the Poynting vector gets its direction altered. According to this definition, it is clear that $\mathcal{T}^2 = 1$. We saw in chapter 2 that for fermions with spin we get $\mathcal{T}^2 = -1$.

The sign of $\mathcal{T}^2$ has profound consequences on the symmetry of the S matrix. Let us consider a situation of scattering of electromagnetic waves from a configuration similar to that shown in figure 7.2. Let us generalize the scattering of N propagating modes, both on the left side and on the right side of the collision potential. Let us call the incoming modes from the left as $\mathbf{f}_l^+$, with $l = 1, \cdots, N$ and the incoming modes from the right as $\mathbf{f}_l^+$, with $l = N+1, \cdots, 2N$. Then, the incoming wave that propagates towards the potential region (or of the junction of two waveguides) may be written as

$$\mathbf{f}^+ = \sum_{l=1}^{N} a_l^+ \mathbf{f}_l^+, \tag{7.77}$$

on the left side, and

$$\mathbf{f}^+ = \sum_{l=N+1}^{2N} a_l^+ \mathbf{f}_l^+ \tag{7.78}$$

on the right side, where a_l^+ are the amplitudes of the incoming modes. Assuming that the wires that transport the waves are invariant under time reversal, the outgoing waves may be written as

$$\mathbf{f}^- = \sum_{l=1}^{N} a_l^- \mathcal{T} \mathbf{f}_l^+ \tag{7.79}$$

on the left side and

$$\mathbf{f}^- = \sum_{l=N+1}^{2N} a_l^- \mathcal{T} \mathbf{f}_l^+ \tag{7.80}$$

on the right side, since we must have $\mathbf{f}_l^- = \mathcal{T} \mathbf{f}_l^+$, given that the energy flux direction, determined by the direction of the Poynting vector, gets reversed.

As before, we may now define a S matrix, of dimension $2N$, that relates the amplitudes of the incoming and outgoing modes, $[a_l^-] = S[a_l^+]$. The column vector $[a]$ has as elements the components a_l of the modes. Acting with the time reversal operator we also obtain that $\mathcal{T} \mathbf{f}_l^- = \pm \mathbf{f}_l^+$ since $\mathcal{T}^2 = \pm 1$, depending if the system under consideration is photonic or electronic. The time reversal operator converts the incoming wave in outgoing waves, as we saw before. But for electronic systems there is an additional minus sign, since the operator is antiunitary. We obtain, for electronic systems,

$$[(a_l^+)^*] = -S[(a_l^-)^*]. \tag{7.81}$$

Therefore, for electronic system, we obtain

$$S^{-1} = -S^*,$$

in contrast with the result for photons where $S^{-1} = S^*$. Using the fact that the S matrix is unitary we obtain

$$S = \pm S^T \begin{cases} \text{photons}, \\ \text{electrons}. \end{cases} \tag{7.82}$$

For photons the matrix is symmetric, and for electrons the matrix is antisymmetric. Let us call the left side as 1 and the right side as 2. The S matrix may be written in block form

$$\begin{pmatrix} a_1^- \\ \cdots \\ a_N^- \\ a_{N+1}^- \\ \cdots \\ a_{2N}^- \end{pmatrix} = \begin{pmatrix} (S_{11})_{N \times N} & (S_{12})_{N \times N} \\ (S_{21})_{N \times N} & (S_{22})_{N \times N} \end{pmatrix} \begin{pmatrix} a_1^+ \\ \cdots \\ a_N^+ \\ a_{N+1}^+ \\ \cdots \\ a_{2N}^+ \end{pmatrix}. \tag{7.83}$$

Due to the antisymmetry of the matrix in the fermionic case, the blocks S_{11} and S_{22} are also antisymmetric. Therefore, $\det(S_{11}) = (-1)^N \det(S_{11})$, and the same for S_{22}, since they are matrices of dimension $N \times N$. As a consequence, if N is odd, the determinant vanishes. Since this is given by the product of the matrix eigenvalues, this implies that it is possible to choose an incoming wave such that there is no outgoing wave, at least for a given mode. Therefore, if there is energy conservation, there is a mode for which there is complete transmission (since there is no back scattering). For $N = 1$ the mode itself satisfies this condition. In this case, the blocks only have one element and, when they vanish, there is not the mode a^- corresponding to the mode a^+. We found in chapter 3 this absence of backscattering in electronic systems, that leads to the spin Hall effect, where the states are protected by the time reversal symmetry .

These results suggest that in a photonic system with time reversal symmetry there should not be protected states, since $\mathcal{T}^2 = 1$. However, it is possible to identify other symmetries, or so-called generalized time reversal operators, that satisfy the condition that their square equals -1. Such operator could, for example, be the product of parity, time reversal and a dual operator. The resulting scattering matrix would be antisymmetric[1].

Further reading:

The proposal of topology in photonic systems:

- F. D. M. Haldane and S. Raghu, *Physical Review Letters* **100**, 013904 (2008).
- S. Raghu and F. D. M. Haldane, *Physical Review* A **78**, 033834 (2008).

Experimental results on the existence of protected edge modes:

- Zheng Wang, Yidong Chong, J. D. Joannopoulos and Marin Soljacic, *Nature Letters* **461**, 772 (2009).
- Ling Lu, John D. Joannopoulos and Marin Soljacic, *Nature Photonics* **8**, 821 (2014).

---

[1] *Physical Review* B **95**, 035153 (2017).

Generic discussion on aspects of electromagnetic theory and its applications on photonic crystals:

- J. D. Jackson, *Classical Electrodynamics*, Wiley.
- J. D. Joannopoulos, S. G. Johnson, J. N. Winn and R. D. Meade, *Photonic crystals: Molding the flow of light*, Princeton University Press (2008).
- D. M. Pozar, *Microwave Engineering*, Wiley (2012).

On specific calculations of topological properties and topological invariants, including systems with no time reversal symmetry breaking:

- Y. D. Chong, Xiao-Gang Wen and Marin Soljacic, *Physical Review* B **77**, 235125 (2008).
- Mário G. Silveirinha, *Physical Review* B **95**, 035153 (2017).

Chapter 8

Quantum information and topological systems

Various quantum information methods have been used to signal phase transitions in alternative to the more conventional methods in condensed matter, such as correlation functions. These techniques have been used to detect both finite temperature transitions, and quantum phase transitions at zero temperature. A natural extension is their use in topological systems, with particular emphasis on transitions between different topological regimes. In topological systems, the Landau description of a phase transition, in terms of a local order parameter, is not valid in general and quantities associated with degrees of freedom of long range are needed to describe the different topological phases. In addition, the non-existence of an usual order parameter raises the question of which entities may be used to replace the concept of an order parameter. As mentioned before, topological invariants or the existence of edge states are usually chosen. An alternative is provided by entanglement measures designed to identify or detect global correlations. Recently, it was also proposed that entanglement measures may be used to construct entities, interpreted as order parameters, that may distinguish various topological phases, or, at least, signal the transitions between them.

8.1 Entanglement

Consider a simple system of two spin-1/2 operators. The four possible states that result from the coupling between the spin angular momenta

189

may be separated in a triplet state of total spin 1,

$$| \uparrow_1 \rangle | \uparrow_2 \rangle \quad (m = 1),$$
$$| \downarrow_1 \rangle | \downarrow_2 \rangle \quad (m = -1),$$
$$\frac{1}{\sqrt{2}} (| \uparrow_1 \rangle | \downarrow_2 \rangle + | \downarrow_1 \rangle | \uparrow_2 \rangle) \quad (m = 0), \tag{8.1}$$

and in a singlet state of total spin 0 and $m = 0$,

$$\frac{1}{\sqrt{2}} (| \uparrow_1 \rangle | \downarrow_2 \rangle - | \downarrow_1 \rangle | \uparrow_2 \rangle). \tag{8.2}$$

A state of more than one spin is called *separable* if it can be written as the tensor product of pure states of its parts. Otherwise it is called *entangled*. The states with $m = \pm 1$ of the triplet states may be seen as separable, while the states with $m = 0$, both from the triplet or the singlet, may be seen as entangled.

We may consider linear combinations of these states selecting four entangled states, as

$$| \psi^\pm \rangle = \frac{1}{\sqrt{2}} (| \downarrow_1 \rangle | \uparrow_2 \rangle \pm | \uparrow_1 \rangle | \downarrow_2 \rangle),$$
$$| \phi^\pm \rangle = \frac{1}{\sqrt{2}} (| \downarrow_1 \rangle | \downarrow_2 \rangle \pm | \uparrow_1 \rangle | \uparrow_2 \rangle). \tag{8.3}$$

In this basis, all states, called Bell states, are entangled. Note that the states $| \psi^\pm \rangle$ mix states with different total spin values.

Entangled states are a general characteristic of quantum states of a many-body system. If we have n separate systems, the Hilbert space is the tensor product and the total state may be written as a superposition

$$| \psi \rangle = \sum_{i_1, \cdots, i_n} c_{i_1, \cdots, i_n} | i_1 \rangle \otimes | i_2 \rangle \otimes \cdots \otimes | i_n \rangle, \tag{8.4}$$

where i_j index the states of each part of the system. The state $| \psi \rangle$ can not be written, in general, as a product of states of the individual subsystems: $| \psi \rangle \neq | \psi_1 \rangle \otimes | \psi_2 \rangle \otimes \cdots \otimes | \psi_n \rangle$.

The problem simplifies if there are only two subsystems. A state of the total system may be written as above

$$| \psi \rangle = \sum_{m,n} A_{m,n} | \psi_m^1 \rangle | \psi_n^2 \rangle, \tag{8.5}$$

where $| \psi_m^1 \rangle$ e $| \psi_n^2 \rangle$ are orthonormal basis functions in the two Hilbert spaces. The matrix A may be written as a product, UDV, where U is a unitary matrix, D is rectangular diagonal and the lines of V are orthonormal. This is the so-called singular value decomposition. Forming new bases combining the states m using U and the states n using V, we obtain the *Schmidt decomposition*,

$$|\psi\rangle = \sum_n \lambda_n |\Phi_n^1\rangle |\Phi_n^2\rangle, \qquad (8.6)$$

where λ_n are the elements of the diagonal matrix D. The state $|\psi\rangle$ is normalized if the sum of the squares of the absolute values of λ_n equals unity. Writing the state in this form, we see that the state is, in general, entangled. Unless all λ_n vanish except one, the state can not be written as a separable state, and, therefore, is entangled. The distribution of the values of λ_n quantifies somehow the entanglement. Clearly, the state is maximally entangled if all the λ_n are equal. A convenient way to quantify the entanglement is to introduce an entropy similar to that of Shannon, designated in this context as von Neumann entropy.

8.1.1 *von Neumann entropy*

Let us consider a system at zero temperature, whose density matrix describes the groundstate. Being this a pure state, the density matrix is simply

$$\rho = |\Phi\rangle\langle\Phi|, \qquad (8.7)$$

where $|\Phi\rangle$ denotes the groundstate. Let us consider now, that the system is divided in two parts, A and B, and let us define

$$\rho_A = \text{Tr}_B \rho. \qquad (8.8)$$

This quantity is the reduced density matrix of subsystem A, since the states associated with the degrees of freedom of the rest of the system, that is of B, have been integrated over. However, the information on the correlations intrinsic to B, and the correlations between the two subsystems remains implicit. We may, now, define the von Neumann entropy as the information contained in the part A of the system. In the usual way,

$$S_A = -\text{Tr}\rho_A \ln \rho_A. \qquad (8.9)$$

This quantity is a possible measure of the entanglement between the subsystem A and the rest of the system.

Let us then consider the Schmidt decomposition of the state of the system. The reduced density matrices may be easily obtained:

$$\rho = |\psi\rangle\langle\psi| = \sum_{m,n} \lambda_n \lambda_m^* |\Phi_n^A\rangle |\Phi_n^B\rangle \langle\Phi_m^B| \langle\Phi_m^A|. \qquad (8.10)$$

The reduced density matrix of subsystem A is given by

$$\rho_A = \mathrm{Tr}_B \rho$$
$$= \sum_{n,m} \lambda_n \lambda_m^* \mathrm{Tr}_B \left(|\Phi_n^A\rangle |\Phi_n^B\rangle \langle \Phi_m^B| \langle \Phi_m^A| \right)$$
$$= \sum_{n,m} \lambda_n \lambda_m^* |\Phi_n^A\rangle \mathrm{Tr}_B \left(|\Phi_n^B\rangle \langle \Phi_m^B| \right) \langle \Phi_m^A|$$
$$\Leftrightarrow \rho_A = \sum_n |\lambda_n|^2 |\Phi_n^A\rangle \langle \Phi_n^A|, \tag{8.11}$$

and in a similar way for ρ_B,

$$\rho_B = \sum_n |\lambda_n|^2 |\Phi_n^B\rangle \langle \Phi_n^B|, \tag{8.12}$$

showing that the two reduced density matrices have the same spectrum, designated *entanglement spectrum*, given by the squares of the absolute values of the coefficients in the Schmidt decomposition. Then, the von Neumann entropy is simply the Shannon entropy of these eigenvalues $S = -\sum_n |\lambda_n|^2 \log |\lambda_n|^2$. The entropy vanishes if and only if one of the eigenvalues is 1 and is maximal if they are all equal.

Let us consider a finite size system, and, as subsystems, two system regions. The subsystems may have different sizes. The simplest case corresponds to choose A as a single site and B the remaining $N-1$ sites. In a system with translational invariance, the entropy S_A is the same for each site, but in a non-homogeneous system the entropy is site dependent. It is also interesting to consider that the subsystem A is composed by two sites (chosen arbitrarily in the lattice) and B are the remaining $N-2$ sites.

8.1.2 *Relation with correlation functions*

In many-particle electronic systems, a description in terms of wave functions is very complex and is more convenient to use second quantization to carry out calculations on the system properties. The matrix density is, however, naturally defined in terms of the states of the system and, therefore, in terms of the wave functions. Alternatively, the elements of the density matrix may be expressed in terms of correlation functions of the system. For example, in the case of the entanglement of one site, one may use the basis $|0\rangle, |\uparrow, \downarrow\rangle, |\uparrow\rangle, |\downarrow\rangle$, that refers to the possible four states: unoccupied, doubly occupied, single occupancy with spin $\uparrow$, and single occupancy with spin $\downarrow$, respectively. The first two basis states are associated with the charge part and the remaining two with the spin part.

The expression for the reduced density matrix may be obtained writing the single site density matrix as

$$\rho = \sum_{n,m} |n\rangle \rho_{nm} \langle m|, \qquad (8.13)$$

where $|n\rangle, |m\rangle$ are the four states described above. Consider, now, an operator O defined in terms of electronic creation and destruction operators, and let us calculate

$$\mathrm{Tr}\{O\rho\} = \sum_{n,m} \langle n|O|m\rangle \rho_{m,n}. \qquad (8.14)$$

To determine the matrix element ρ_{nm} of the density matrix is enough to look, by inspection, the operator O such that only the matrix element $\langle n|O|m\rangle$ is non-vanishing. This will let us know which correlation function corresponds to which density matrix element. Alternatively, we may express the operator $|m\rangle\langle n|$ in terms of the creation and annihilation operators.

The density matrix is then, given by

$$\rho_A = \begin{pmatrix} \langle (1-n_\uparrow)(1-n_\downarrow) \rangle & \langle c_\uparrow^\dagger c_\downarrow^\dagger \rangle & 0 & 0 \\ \langle c_\downarrow c_\uparrow \rangle & \langle n_\uparrow n_\downarrow \rangle & 0 & 0 \\ 0 & 0 & \langle n_\uparrow (1-n_\downarrow) \rangle & \langle c_\downarrow^\dagger c_\uparrow \rangle \\ 0 & 0 & \langle c_\uparrow^\dagger c_\downarrow \rangle & \langle (1-n_\uparrow)n_\downarrow \rangle \end{pmatrix}. \qquad (8.15)$$

These matrix elements are defined in the subspace of the one-site states. However, since we have integrated over the remaining sites, we may now replace the matrix elements by the matrix elements over the full system (direct product with the states of the remaining $N-1$ sites).

The charge and spin parts decouple. The spin part couples the two spin orientations and the charge part couples the empty and doubly occupied sites. The matrix diagonal terms describe the number of empty sites, sites with doubly occupancy, the number of $\uparrow$ spins and the number of $\downarrow$ spins, respectively. This matrix is easily diagonalizable and the von Neumann entropy is obtained in a simple way. The sum of the diagonal terms equals unity due to normalization. In the case of a system with no interactions or quadratic, the correlation functions are solved easily, using a representation of the electronic operators in terms of the spectrum of the quasi-particle operators.

The case of the entanglement of two sites is more complex. We may introduce two sites, i and j, and use the same state basis for each site. The basis states are given by the direct product of these two basis states and the reduced density matrix has dimension $(4^2) \times (4^2) = 16 \times 16$. It is convenient to organize this basis in even-even and odd-odd states, in the number of electrons on the sites i and j, respectively. These two sets of states decouple and the problem is reduced to the diagonalization of two 8×8 matrices. The calculation of the various correlation functions in these matrices is now complex. Note that some correlation functions involve double occupancy on the two sites, therefore, products of eight fermionic operators. In general, the density matrix for N sites has dimension $(4^N) \times (4^N)$. However, if the problem is quadratic the calculation simplifies and it is possible to reduce the problem to the diagonalization of a matrix of the order of $(gN) \times (gN)$, where g is the number of degrees of freedom on each lattice site. Defining a correlation matrix in a subspace S of lattice sites, χ_S, with matrix elements given by

$$\chi_{i,j} = \langle c_i c_j^\dagger \rangle, \tag{8.16}$$

the von Neumann entropy of the reduced density matrix, ρ_S, associated to this subsystem S of lattice points, may be obtained as

$$S(\rho_S) = -\frac{1}{2}\mathrm{Tr}\left((1 - \chi_S)\ln(1 - \chi_S) + \chi_s \ln \chi_S\right). \tag{8.17}$$

8.1.3   *Impurity in a conventional superconductor*

As example of the utility of the von Neumann entropy to detect a quantum phase transition, let us consider the transition induced by the variation of the coupling between a magnetic impurity and the spin density of electrons in a conventional superconductor with s-wave symmetry, described by the Hamiltonian of equation (4.188). This transition was discussed in detail in section 4.5. Even though this transition is not topological, similar treatments will be considered in this chapter to detect topological transitions.

   In order to clarify the meaning of the von Neumann entropy, we may compare the entanglement entropy of one site in three cases, namely, a system of free electrons in a tight-binding model with no superconductivity or magnetic impurity, a superconductor with no impurity and a superconductor with magnetic impurity. The entropy is basically uniform in the free case and in the case of the pure superconductor. In the free case the entropy has higher value, as expected, due to the ordered nature of the superconductor. Introducing the impurity the entropy changes in its vicinity but it does not change the value in the bulk, far from the impurity.

   In figure 8.1 we show the single site entropy, for different values of the chemical potential, in a system of dimensions 15×15. The entropy is calculated using the BCS theory to obtain the elements of the reduced density

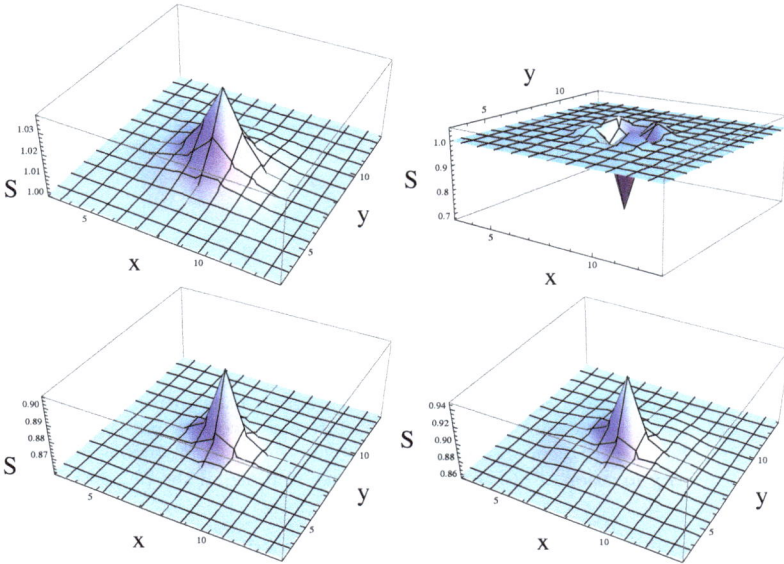

Fig. 8.1   von Neumann entropy of one site $S_A(1)$ for $J = 1, 2$ and different chemical potentials $\epsilon_F = -1, -2$. The size of the system is 15×15. In the top row $\epsilon_F = -1, J = 1$ (left column) and $\epsilon_F = -1, J = 2$ (right column) and in the bottom row $\epsilon_F = -2, J = 1$ (left column) and $\epsilon_F = -2, J = 2$ (right column).

matrix. The various correlation functions are obtained solving the BdG equations. If the system is large enough, the states induced by the impurity are localized and, therefore, are not affected by the presence of the border. Also, since in this problem the transition is first order, there are no long-range correlations near the transition, whose presence would imply considering larger systems, to reduce finite size effects. In the case of chemical potential or Fermi energy $\epsilon_F = -1$, the quantum phase transition is signalled by the change of the entropy near the impurity location. Before the quantum phase transition occurs there is a maximum and, after the transition occurs, the entropy develops a local minimum. Before the transition is reached, the impurity induces a local increase of the entropy, since it acts against the superconducting order: $\Delta S_{imp} > 0$. At the transition point an electron is captured by the impurity, which gives rise to a local decrease of the entropy: $\Delta S_{imp} < 0$. In the case $\epsilon_F = -2$, a value of $J = 2$ is not sufficiently large to cross the phase transition.

In figure 8.2 we show various contributions to the single-site entropy, signalling the location of the phase transition. The largest change of the

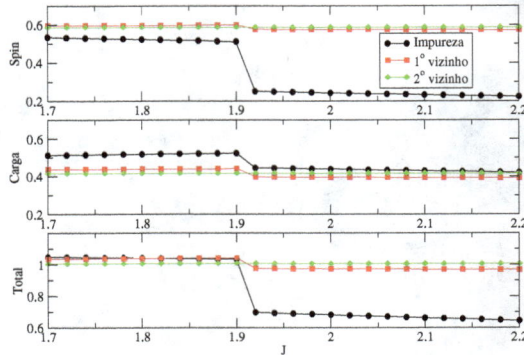

Fig. 8.2　Charge and spin contributions, and their sum, to the entropy of one site, as a function of the coupling value. The black and white curve corresponds to the impurity location and the other two curves correspond to two points distant by one and two sites from the impurity location. Note the discontinuities that signal the phase transition.

entropy is related with the spins degrees of freedom, as expected, due to the interaction between the impurity spin and the electrons spin. However, since the trace of the reduced density matrix is normalized, the charge part also presents a discontinuity at the phase transition. It is also interesting to note that, as expected, at the impurity location, and for values of J larger than the critical value, the spin contribution for the entropy is smaller than the charge contribution.

8.2　Entanglement spectrum

8.2.1　*Entanglement in real space*

In addition to the von Neumann entropy, the eigenvalues of the reduced density matrix of a bipartite system, with a real space partition, also provide valuable information on the system. If the two partitions are decoupled and the global state is a state that is the product of the parts, only one eigenvalue is finite, due to the normalization. Writing the reduced density matrix in the form

$$\rho = e^{-H},$$

(as a Boltzmann factor with effective temperature $T = 1$), we may expect that, if the entanglement is weak, one of the eigenvalues is larger than the remaining. Defining eigenvalues analogous to "energy", in the form $e^{-\xi_i}$, the set of ξ_i is called entanglement spectrum of the reduced density matrix (or logarithmic spectrum). If the entanglement is weak, we expect a state of low "energy" (or a set of low "energy" states) separated by a gap of high value from the remaining eigenvalues. Therefore, the entanglement spectrum separates in two sectors, one of small eigenvalues separated from the rest by a large gap. This gap goes to infinite if only one eigenvalue has a significant value. We expect, therefore, that the appearance of a gap in the entanglement spectrum provides a characterization of the degree of entanglement, in a more detailed way and with more information than the von Neumann entropy, given by the sum over all eigenvalues, $S = \sum_i \xi_i e^{-\xi_i}$.

In the case of the fractional Hall effect, it has been shown that the entanglement spectrum separates in two sets of eigenvalues. The group of states with the lower values of the logarithmic spectrum is separated by a gap from the higher eigenvalues. It was found that there is an agreement between the state sequence with those of a conformal theory of the low energy states, associated to the edge modes, of topological origin. As a consequence, it was proposed that the entanglement spectrum may be a signature of topology.

It may be expected that the states that characterize the topological properties are associated to the gapless edge states, at the border of the system. A group of eigenvalues of the entanglement spectrum may be associated with the states in the bulk of the system and the other group associated with the edges states of low energy. We expect that the partition of the system leads to a entanglement spectrum that may provide information on states along the partition, and with a structure similar to the edge states of a finite system.

In the context of spin systems, an example is a spin ladder, selecting a partition such that block A is one of the chains and block B is the other chain. Each chain is the border of the other one. Calculating the entanglement spectrum it was found that the spectrum has a set of states that is equivalent to the states of one chain.

Extending these ideas to other systems, it may be shown that the method gives us a way to identify and understand edge states, that for a topological system have a topological nature. For example, in the case of the Sato and Fujimoto model, equation (4.183), it is possible to identify contributions to the entropy that are the result of the edge states, that have a fractional entropy characteristic of Majorana modes, as expected in that topological model.

8.2.2 *Momentum space entanglement*

Describing an Hamiltonian in momentum space, we may consider a partition in this space. The corresponding entanglement spectrum and eigenvectors of the reduced density matrix reveal interesting information on the degree of entanglement.

In a multiband system at low temperature, the density matrix spectrum in $\mathbf{k}$ space has, in the lowest positive energy band, an eigenvalue near unity and all others near zero (except in degenerate points when the eigenvalue is the inverse of the degeneracy). A simple example is the Sato and Fujimoto model, equation (4.183), for a superconductor of spinfull electrons, where pairings with symmetries s and p are considered. Since the Hamiltonian is separable in momentum space, the density matrix operator for a given momentum $\mathbf{k}$ may be defined in the usual way

$$\hat{\rho}_\mathbf{k} = \frac{e^{-\beta \hat{H}_\mathbf{k}}}{Z_\mathbf{k}}. \tag{8.18}$$

This expression applies at finite temperatures. Later we will take the limit of very small temperatures $T \to 0$. In the diagonal basis it is written as

$$\rho_\mathbf{k} = \langle n | \hat{\rho}_\mathbf{k} | n \rangle = \frac{e^{-\beta \langle n | \hat{H}_\mathbf{k} | n \rangle}}{Z_\mathbf{k}}. \tag{8.19}$$

We may use the basis of occupation numbers for a given momentum and its symmetrical value, and the two spin components. We consider the representation

$$\widetilde{H}_\mathbf{k} = \langle n_{\mathbf{k}\uparrow} n_{-\mathbf{k}\uparrow} n_{\mathbf{k}\downarrow} n_{-\mathbf{k}\downarrow} | \, \hat{H}_\mathbf{k} \, | n_{\mathbf{k}\uparrow} n_{-\mathbf{k}\uparrow} n_{\mathbf{k}\downarrow} n_{-\mathbf{k}\downarrow} \rangle. \tag{8.20}$$

The eigenvalue problem of this Hamiltonian can be written as

$$\widetilde{H}_\mathbf{k} \boldsymbol{Q}_{\mathbf{k},n} = \lambda_{\mathbf{k},n} \boldsymbol{Q}_{\mathbf{k},n} \quad ; \quad n = 1, \ldots, 16. \tag{8.21}$$

In the same basis the density matrix may be written as

$$\rho_\mathbf{k} = \frac{e^{-\beta \widetilde{H}_\mathbf{k}}}{Z_\mathbf{k}} \tag{8.22}$$

and the eigenvalue problem of the density matrix is written as $\rho_\mathbf{k} \boldsymbol{Q}_{\mathbf{k},n} = \Lambda_{\mathbf{k},n} \boldsymbol{Q}_{\mathbf{k},n}$, where

$$\Lambda_{\mathbf{k},n} = \frac{e^{-\beta \lambda_{\mathbf{k},n}}}{\sum_{n'} e^{-\beta \lambda_{\mathbf{k},n'}}}. \tag{8.23}$$

The entropy for each momentum may be obtained using

$$S_\mathbf{k} = - \sum_{n=1}^{16} \Lambda_{\mathbf{k},n} \ln(\Lambda_{\mathbf{k},n}) \tag{8.24}$$

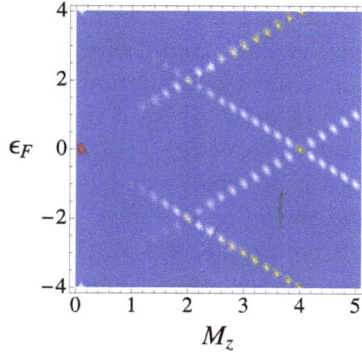

Fig. 8.3   Total entropy as a function of the chemical potential and magnetization for $t = 1$, $\alpha = 0.6$, $d = 0.6$, $\Delta_s = 0.1$ and temperature $T = 10^{-3}$. The transitions occur in three momenta sets: $\boldsymbol{k} = (0, 0)$ ($\epsilon_F = -4$), $\boldsymbol{k} = (\pi, 0)$ ($\epsilon_F = 0$) and $\boldsymbol{k} = (\pi, \pi)$ ($\epsilon_F = 4$).

and the entropy may be obtained summing over the Brillouin zone

$$S = \sum_{\mathbf{k}} S_{\mathbf{k}} \,. \qquad (8.25)$$

The results for the entropy as a function of the chemical potential and magnetization (Zeeman term) are presented in figure 8.3, for a temperature $T = 10^{-3}$, where $\beta = 1/T$ ($k_B = 1$). At very low temperatures, the entropy is $S = \ln \Omega$, where Ω is the groundstate degeneracy. In general, the groundstate is non-degenerate and the entropy vanishes. At the transition points the spectrum gap vanishes, the groundstate becomes degenerate and the entropy is finite. Integrating over the momenta, we sum all the points where there is no gap. As we can see from the figure 8.3, the entropy detects very well the topological transitions. On the other hand, except at the transitions the entropy does not present singular features and does not distinguish the various topological phases.

A detailed analysis of the entropy as a function of momentum, $S_{\mathbf{k}}$, may be carried out in a way to determine the origin of the peaks in figure 8.3. For $M_z \to 0$ three peaks arise associated with different momentum values. These are special points, degenerate at zero temperature, with entropy values $\ln(4)$, $2\ln(4)$ and $\ln(4)$, respectively. These peaks are located at the momenta $\boldsymbol{k} = (0, 0)$, $\boldsymbol{k} = (\pi, 0)$, $\boldsymbol{k} = (\pi, \pi)$, respectively, and at their equivalent points in the Brillouin zone. At very low temperatures the peaks become very narrow. The relevance of the zero energy points extends beyond these particular values of the chemical potential.

The eigenvectors provide additional information on the phase of the system, and in particular on the phase transitions. In general, each eigenvector expressed in this basis is a linear combination of the 16 states of the used basis $|n_{\mathbf{k}_\uparrow} n_{-\mathbf{k}_\uparrow} n_{\mathbf{k}_\downarrow} n_{-\mathbf{k}_\downarrow}\rangle$. In most cases, of the 16 coefficients, only a few are non-vanishing for each eigenvector. Their physical interpretation is, in several cases, clear, and it is possible to understand the physical content of the dominant states, and the origin of a given transition.

The dominant states are: *i)* the states $|1100\rangle$ and $|0011\rangle$ corresponding to the triplet pairing; *ii)* the state $|0000\rangle$ corresponding to an empty state in momentum and spin state; *iii)* the state $|1111\rangle$ corresponding to a basis state fully occupied with four electrons (for a given momentum and its symmetric and both spin components); *iv)* the states $|1001\rangle, |0110\rangle$ corresponding to a singlet pairing. Other states are also present, but their coefficients are typically very small. The detailed study of the various components may be found in the literature.

8.3   Fidelity

8.3.1   *Pure states*

Consider a system given by an Hamiltonian, $\hat{H}(q)$, that depends on a set of parameters, q. At zero temperature, the system is in the pure ground-state $|\Phi(q)\rangle$. The fidelity between two eigenstates for two values of the parameters q is defined as

$$F(q, \tilde{q}) = |\langle \Phi(q) | \Phi(\tilde{q}) \rangle| \qquad (8.26)$$

and is the absolute value of the overlap between two groundstates $|\Phi(q)\rangle$ and $|\Phi(\tilde{q})\rangle$. The fidelity quantifies the distinguishability between two states. It is frequently useful to determine the overlap between two states characterized by two parameters close to each other q and $\tilde{q} = q + \delta q$, where δq is small. This overlap may, for example, indicate points where a phase transition occurs. The idea is simple: two quantum states defining different phases should be easily distinguishable, in comparison to two states in the same phase. In the case of pure systems, the density matrix for $T = 0$ is simply the projector to that state $\rho = |\Phi(q)\rangle\langle\Phi(q)|$. If two states are identical, the fidelity is simply the state normalization, taken as being 1.

Since the fidelity is a measure of the distinguishability between two states, it may be considered a generalized order parameter, encompassing, in some way, all the order parameters that one may define. Therefore, regions of non-analyticity in parameter space may signal the critical behavior of a system, with no need to consider specific order parameters and/or its associated correlated functions.

8.3.2 Fidelity between partial states

Two mixed quantum states, a and b, may be defined by their density matrices (or density operators) $\hat{\rho}_a$ e $\hat{\rho}_b$. Various distinguishability measures between the states have been introduced in the literature. The fidelity is one of the most used, being defined by the expression

$$F(\hat{\rho}_a, \hat{\rho}_b) = \text{Tr}\sqrt{\sqrt{\hat{\rho}_a}\hat{\rho}_b\sqrt{\hat{\rho}_a}}\,. \tag{8.27}$$

The numerical value of the fidelity varies between $F = 0$, in the case of two states that are fully distinguishable, to $F = 1$, when the two states are indistinguishable (that is, they are identical). The introduction of this expression for the quantum fidelity (where we compare quantum states) is motivated by an analogous definition in classical systems.

The classical fidelity F_c, a measure of the distinguishability between two probability distributions $\{p_a(i)\}$ and $\{p_b(i)\}$ is defined as $F_c(p_a, p_b) = \sum_i \sqrt{p_a(i)p_b(i)}$. Measuring one observable $\hat{A}$ in states $\hat{\rho}_a$ and $\hat{\rho}_b$, we obtain probability distributions $\{p_a^{\hat{A}}(i)\}$ and $\{p_b^{\hat{A}}(i)\}$), respectively, whose mutual distinguishability is given by the classical fidelity $F_c(p_a^{\hat{A}}, p_b^{\hat{A}})$. It may be shown that for any observable $\hat{A}$, the inequality $F(\hat{\rho}_a, \hat{\rho}_b) \le F_c(p_a^{\hat{A}}, p_b^{\hat{A}})$ is valid and there is always an optimal observable, $\hat{A}_{op}$, for which the equality holds. In other words, the quantum fidelity, $F(\hat{\rho}_a, \hat{\rho}_b)$, gives the optimal value for the distinguishability between two quantum states when compared through probability distributions $\{p_a^{\hat{A}}(i)\}$) and $\{p_b^{\hat{A}}(i)\}$. Defining the modulus of an operator $\hat{R}$ as $|\hat{R}| = (\hat{R}\hat{R}^\dagger)^{1/2}$, we obtain an alternative expression for the fidelity:

$$F(\hat{\rho}_a, \hat{\rho}_b) = \text{Tr}\left|\sqrt{\hat{\rho}_a}\sqrt{\hat{\rho}_b}\right| = \text{Tr}\sqrt{\sqrt{\hat{\rho}_a}\sqrt{\hat{\rho}_b}(\sqrt{\hat{\rho}_a}\sqrt{\hat{\rho}_b})^\dagger} = \text{Tr}\sqrt{\sqrt{\hat{\rho}_a}\hat{\rho}_b\sqrt{\hat{\rho}_a}}\,. \tag{8.28}$$

Instead of considering global states that describe a system, we consider partial quantum states of a subsystem. Therefore, these are the states associated with the reduced density matrix defined for this subsystem. If the global system $\mathcal{S}$ is divided in two parts A and B, we may consider the fidelity between the mixed states $F(\hat{\rho}_A(q), \hat{\rho}_A(\tilde{q}))$, where $\hat{\rho}_A(q) = \text{Tr}_B|\Phi(q)\rangle\langle\Phi(q)|$, and analogously for $\hat{\rho}_A(\tilde{q})$.

8.3.3   *Two-level system*

Let us consider a two-level system. The Hamiltonian may be written in terms of Pauli matrices as

$$H = \boldsymbol{h} \cdot \boldsymbol{\sigma} \,. \tag{8.29}$$

A general state may be written as a two-component spinor

$$\chi = \begin{pmatrix} c_1 \\ c_2 \end{pmatrix} \,. \tag{8.30}$$

We may define a density matrix associated with this state as

$$\rho = \chi\chi^{\dagger} = \begin{pmatrix} c_1 \\ c_2 \end{pmatrix} \begin{pmatrix} c_1^* & c_2^* \end{pmatrix} = \begin{pmatrix} |c_1|^2 & c_1 c_2^* \\ c_2 c_1^* & |c_2|^2 \end{pmatrix} \,. \tag{8.31}$$

This definition satisfies the usual condition for the average value of an operator A:

$$\langle \chi | A | \chi \rangle = \mathrm{Tr}\,(A\rho) = \mathrm{Tr}\,(A\chi\chi^{\dagger}) \tag{8.32}$$

An alternative way to write the density matrix (8.31) is

$$\rho = \frac{1}{2}\,(I + \boldsymbol{P} \cdot \boldsymbol{\sigma}) \,, \tag{8.33}$$

where I is the identity matrix of dimension 2. We may now determine the vector $\boldsymbol{P}$. The relation (8.33) satisfies the normalization condition, $\mathrm{Tr}\rho = 1$, since the trace of any Pauli matrix vanishes. The vector $\boldsymbol{P}$ may be determined calculating the average of any Pauli matrix:

$$\mathrm{Tr}\,(\rho\boldsymbol{\sigma}) = \frac{1}{2}\mathrm{Tr}\,(I\boldsymbol{\sigma} + \boldsymbol{P} \cdot \boldsymbol{\sigma}\boldsymbol{\sigma}) \,. \tag{8.34}$$

We then obtain,

$$\mathrm{Tr}\,(\rho\sigma_\alpha) = \frac{1}{2}\sum_{\beta} P_\beta \mathrm{Tr}\,(\sigma_\alpha\sigma_\beta) = P_\alpha \,, \tag{8.35}$$

therefore,

$$\rho = \frac{1}{2}\,(I + \mathrm{Tr}\,(\rho\boldsymbol{\sigma}) \cdot \boldsymbol{\sigma}) \,. \tag{8.36}$$

We may also see that

$$
\begin{aligned}
P_x &= 2Re\,(c_1^* c_2) \,, \\
P_y &= 2Im\,(c_1^* c_2) \,, \\
P_z &= (|c_1|^2 - |c_2|^2) \,.
\end{aligned}
\tag{8.37}
$$

When diagonalized, the density matrix may be written as

$$\rho = \rho_+|\psi_+\rangle\langle\psi_+| + \rho_-|\psi_-\rangle\langle\psi_-|, \tag{8.38}$$

where

$$\begin{aligned} H|\psi_+\rangle &= |\boldsymbol{h}||\psi_+\rangle, \\ H|\psi_-\rangle &= -|\boldsymbol{h}||\psi_-\rangle, \end{aligned} \tag{8.39}$$

therefore, the density matrix may also be written as

$$\rho = \frac{1}{2}\left(I - \frac{H}{|\boldsymbol{h}|}\right). \tag{8.40}$$

since it acts as a projector onto the lowest energy state. Note that the equality $(\boldsymbol{\sigma}\cdot\boldsymbol{n})^2 = 1$, for a unit vector $\boldsymbol{n}$, leads to $\rho^2 = \rho = \sqrt{\rho}$.

We may, now, calculate the fidelity between two states described by two density matrices $\rho(H_1)$ and $\rho(H_2)$,

$$\begin{aligned} F &= \mathrm{Tr}\sqrt{\sqrt{\rho(H_1)}\rho(H_2)\sqrt{\rho(H_1)}} \\ &= \mathrm{Tr}\sqrt{\frac{1}{2}\left(I - \frac{H_1}{|\boldsymbol{h}_1|}\right)\frac{1}{2}\left(I - \frac{H_2}{|\boldsymbol{h}_2|}\right)\frac{1}{2}\left(I - \frac{H_1}{|\boldsymbol{h}_1|}\right)}, \end{aligned} \tag{8.41}$$

which is also valid when the density matrix represents a pure state for which $\mathrm{Tr}(\rho^2) = \mathrm{Tr}(\rho)$. Using $H_1 H_2 + H_2 H_1 = 2\boldsymbol{h}_1\cdot\boldsymbol{h}_2 I$, and recalling the definition of the versor $\hat{\boldsymbol{h}} = \boldsymbol{h}/|\boldsymbol{h}|$, we may obtain

$$\begin{aligned} F &= \mathrm{Tr}\sqrt{\frac{1}{2}\left(1 + \hat{\boldsymbol{h}}_1\cdot\hat{\boldsymbol{h}}_2\right)\frac{1}{2}\left(I - \frac{H_1}{|\boldsymbol{h}_1|}\right)} \\ &= \sqrt{\frac{1}{2}\left(1 + \hat{\boldsymbol{h}}_1\cdot\hat{\boldsymbol{h}}_2\right)}\,\mathrm{Tr}\rho(H_1) \\ \Leftrightarrow F &= \sqrt{\frac{1}{2}\left(1 + \hat{\boldsymbol{h}}_1\cdot\hat{\boldsymbol{h}}_2\right)}. \end{aligned} \tag{8.42}$$

8.3.4 *States of a superconductor with magnetic impurities*

At $T = 0$, the system is in a pure state $|\Phi\rangle$ and has a density operator $\hat{\rho} = |\Phi\rangle\langle\Phi|$. Dividing the system in two subsystems, A and B, the mixed partial state is given by the reduced density operator, $\hat{\rho}_A$, for the subsystem A. Let us calculate the matrix elements of mixed states of one lattice site i for different values of the coupling between the spins, J. In order to infer the point of the quantum phase transition, let us calculate the fidelity between states of one site corresponding to two close values J and $J + \delta J$.

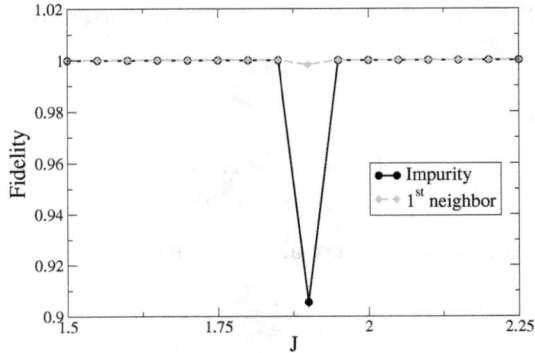

Fig. 8.4   Fidelity for states of one lattice site between two states with J and $J + \delta J$ ($\delta J = 0.05$), for the impurity site (black line) and a nearest neighbor (green line). Note the fast decrease of the fidelity for the impurity close to the point where the phase transition takes place.

To understand better the influence of the transition on the states structure, let us also examine the fidelity between two states, with the same coupling constant, J, but at different lattice locations.

Let us consider the fidelity between states of the reduced density matrix of one site in a two-dimensional lattice, of size 15×15. First, we calculate the fidelity (8.27) between two states, one respecting to site i and the other to site j, and two close values of the coupling parameter J and $J + \delta J$, with $\delta J = 0.05$, such that $\hat{\rho}_a = \hat{\rho}(J; i)$ and $\hat{\rho}_b = \hat{\rho}(J + \delta J; j)$.

For $i = j = l_c$, when we consider the reduced density matrix respecting to the impurity site, the results for the fidelity of one site $F_1(\hat{\rho}_a, \hat{\rho}_b)$, as a function of J, are shown in figure 8.4. We can see a clear decrease of the common value $F_1(\hat{\rho}_a, \hat{\rho}_b) \simeq 1$ at the critical value $J_0 \simeq 1.9$ of the parameter J. When we consider one of the states away from the impurity site, the decrease of the fidelity F_1 around the quantum transition point, becomes smaller; for the first neighbors of the impurity the decrease of the fidelity is still visible, but for second neighbors and further neighbors it is negligible. We see, then, that even for a very small subsystem, the fidelity is changed significantly in relation to the value in which both states refer to two sites far from the impurity, as long as we are in the vicinity of the impurity.

A more detailed study of the fidelity between the states of one site is presented in figure 8.5, where the charge and spin contributions to the change of the partial states are presented. The charge and spin contributions differ

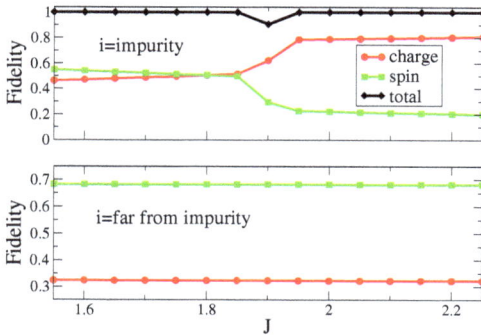

Fig. 8.5   Charge and spin contributions to the fidelity of one site as a function of the parameter J, with $\delta J = 0.05$, for the site at the impurity (top figure) and a site located in the bulk of the system (bottom figure). Note the discontinuity for both contributions around the phase transition, for the site located on the impurity.

from each other, and we may see that both display a discontinuity in the vicinity of the phase transition point.

8.3.5   *Fidelity spectrum and phase transitions in quantum systems*

As we know, the fidelity between two quantum states characterized by two density matrices, ρ_1 and ρ_2, is defined as the trace of an operator,

$$F(\rho_1, \rho_2) = \mathrm{Tr}\sqrt{\sqrt{\rho_1}\rho_2\sqrt{\rho_1}} \equiv \mathrm{Tr}\mathcal{F}\,. \qquad (8.43)$$

Let us now study the spectrum of the fidelity operator, $\mathcal{F}(\rho_1, \rho_2)$. The set of its eigenvalues λ_i is called spectrum of the fidelity operator; the set of logarithms $-\ln \lambda_i$ is called fidelity spectrum. The spectrum may provide additional information in comparison to the fidelity (given by its trace) in a way similar to the information obtained from the entanglement spectrum in comparison to the von Neumann entropy. In the case of two equal mixed states the fidelity is simply

$$F(\rho, \rho) = \mathrm{Tr}\rho = 1\,, \qquad (8.44)$$

and the operator $\mathcal{F}$ in this case has a set of eigenvalues, $\lambda_i = r_i$, such that $-\ln r_i$ is the entanglement spectrum, as seen previously.

While the entanglement spectrum has some relation with the energy spectrum of the edge states, or even states inside the system, the fidelity spectrum contains information over which eigenvalues of the fidelity operator have a larger contribution for the distinguishability between quantum states, that is, the elements of the fidelity spectrum that take smaller values as we approach a transition.

We will next analyse the fidelity spectrum in topological superconductors, with particular emphasis on the vicinity of phase transitions, as well as on the properties characterizing their phases.

8.3.6 *Fidelity spectrum of a topological superconductor*

The fidelity operator, $\mathcal{F}$, may be studied using different basis associated with different representations, such as position, momentum, energy or charge and spin. Rewriting the fidelity operator in these representations allows us to look more directly on the specific relevant modes that participate more actively in the critical phenomena that are involved in the phase transition.

Be $\rho_{1\mathbf{k}}$ and $\rho_{2\mathbf{k}}$ two density matrices that result from a partition in momentum space, for two points in the parameter space, defined as in equation (8.18). We may write

$$\rho_{1\mathbf{k}}\boldsymbol{Q}_{1\mathbf{k}} = \boldsymbol{Q}_{1\mathbf{k}}\boldsymbol{\Lambda}_{1\mathbf{k}} \quad ; \quad \rho_{2\mathbf{k}}\boldsymbol{Q}_{2\mathbf{k}} = \boldsymbol{Q}_{2\mathbf{k}}\boldsymbol{\Lambda}_{2\mathbf{k}} \,. \tag{8.45}$$

As mentioned earlier, the fidelity between two states, characterized by $\rho_{1\mathbf{k}}$ and $\rho_{2\mathbf{k}}$, is given by the trace of the fidelity operator $\mathcal{F}_{\mathbf{k}}$,

$$F_{\mathbf{k}}(\rho_{1\mathbf{k}}, \rho_{2\mathbf{k}}) = \mathrm{Tr}\mathcal{F}_{\mathbf{k}} = \mathrm{Tr}\sqrt{\sqrt{\rho_{1\mathbf{k}}}\rho_{2\mathbf{k}}\sqrt{\rho_{1\mathbf{k}}}} \,. \tag{8.46}$$

The square root of the density matrix may be written as

$$\sqrt{\rho_{1\mathbf{k}}} = \boldsymbol{Q}_{1\mathbf{k}}\sqrt{\boldsymbol{\Lambda}_{1\mathbf{k}}}\boldsymbol{Q}_{1\mathbf{k}}^{-1} \,, \tag{8.47}$$

and

$$\mathcal{F}_{\mathbf{k}}^{\,2} = \sqrt{\rho_{1\mathbf{k}}}\rho_{2\mathbf{k}}\sqrt{\rho_{2\mathbf{k}}} \,. \tag{8.48}$$

Diagonalizing,

$$\mathcal{F}_{\mathbf{k}} = \boldsymbol{U}_{\mathbf{k}}\boldsymbol{f}_{\mathbf{k}}\boldsymbol{U}_{\mathbf{k}}^{-1} \,, \tag{8.49}$$

we write

$$\mathcal{F}_{\mathbf{k},n} = \boldsymbol{U}_{\mathbf{k},n}\boldsymbol{f}_{\mathbf{k},n}\boldsymbol{U}_{\mathbf{k},n}^{-1} \quad ; \quad n = 1,\dots,16 \,. \tag{8.50}$$

Therefore, the k-fidelity, $F_{\mathbf{k}}$, is obtained as

$$F_{\mathbf{k}} = \mathrm{Tr}(\mathcal{F}_{\mathbf{k}}) = \sum_n f_{\mathbf{k},n} \tag{8.51}$$

and the total fidelity

$$F = \prod_{\mathbf{k}} F_{\mathbf{k}} \,. \tag{8.52}$$

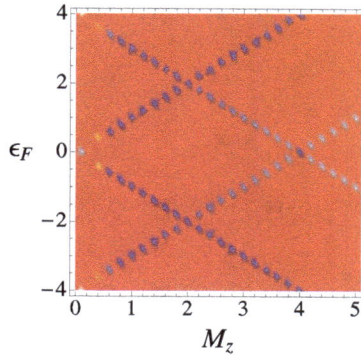

Fig. 8.6   Total fidelity as a function of the chemical potential and magnetization of the Sato and Fujimoto model, equation (4.183). The fidelity is calculated using two points M_z and $M_z + \delta M_z$, with $\delta M_z = 0.01$. Almost everywhere the fidelity is close to unity. Along the transition lines the fidelity has minima.

Let us consider that $\rho_{1\mathbf{k}}, \rho_{2\mathbf{k}}$, are defined for two points that are close in parameter space. Since the fidelity measures the distinguishability, if one of the points is on one site of the transition and the other on the other side, the fidelity has a minimum.

In figure 8.6 we present results for the total fidelity as a function of the chemical potential and magnetization in the Sato and Fujimoto model. Far from the transition points, the fidelity approaches unity, since the two states are very similar. The minima in the fidelity identify correctly the transitions considered and discussed before. In the same way as the von Neumann entropy, taking a magnetization value such that $M_z \to 0$, the k-fidelity identifies the momenta values that are singular.

The transitions may also be signalled by the decrease of the highest eigenvalue of the fidelity spectrum (k-fidelity), that is, the spectrum of the operator $\mathcal{F}_{\mathbf{k}}$, at the transition point, when the two density matrices are calculated in points of the phase diagram around the transition.

Complementar and interesting information may be obtained considering two points that are far from each other in the phase diagram and using the respective density matrices. This allows to compare different phases, and not specifically detect the location of the transition or transitions in the phase diagram, obtaining, for instance, interesting information on the momenta that are responsible for the transitions and also to describe the distinguishability between two phases in momentum space.

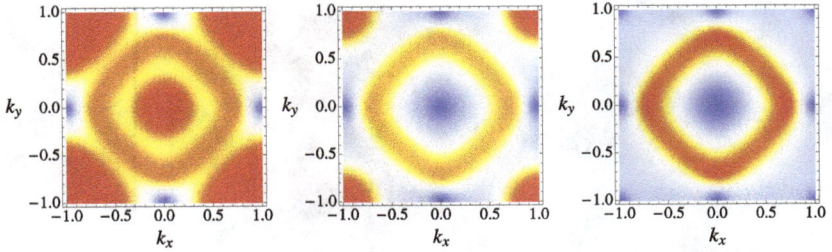

Fig. 8.7   Spectrum of the operator k-fidelity, $\mathcal{F}_\mathbf{k}$, for ρ_A corresponding to $\varepsilon_F = -1, M_z = 0.5$ where the Chern number $C = 0$, and ρ_B corresponding to $\varepsilon_F = -1$ and $M_z = 2, C = -2$, $M_z = 4, C = -1$ and $M_z = 5.5, C = 0$, respectively from left to right. In the first figure, the fidelity approaches unity for small momentum values and near the limits of the Brillouin zone; in the middle figure, the fidelity maximum near the limits of the Brillouin zone is maintained, while in the right figure the fidelity tends to zero in this region.

In figure 8.7 we present results for the spectrum of the operator k-fidelity, where the two density matrices define two points in parameter space that are localized inside different topological phases. In figure 8.7 we consider a density matrix in a state specified by a point in a trivial phase with $C = 0$ (specifically $\mu = -1, M_z = 0.5$) and the other density matrix corresponds to states with different magnetization, along the vertical line of figure 8.3, for which the Chern numbers take the values $C = -2, C = -1, C = 0$ and the magnetization takes the values $M_z = 2, M_z = 4, M_z = 5.5$, respectively. As we move along this line in parameter space, we cross, sequentially, transition lines characterized by momenta $\mathbf{k} = (\pi, 0), \mathbf{k} = (0, 0), \mathbf{k} = (\pi, \pi)$, respectively. The k-fidelity highlights the differences between the various phases. The deviations from unity along the Brillouin zone are significant, even when the phases have the same Chern number, as a consequence of the different electronic occupation numbers (different band-fillings). The points where the transitions occur give rise to zeros in the fidelity spectrum. As shown on the left panel, the fidelity spectrum in momentum space vanishes at the momenta $\mathbf{k} = (\pi, 0)$ and equivalent points that differ by multiples of $(\pm 2\pi, \pm 2\pi)$ (two of them are independent). The middle panel also shows a zero at the center of the Brillouin zone $\mathbf{k} = (0, 0)$. The last panel shows the existence of zeros in all the transition points. The fidelity operator in momentum space signals, therefore, the minimal number of transitions that may occur when one goes from the trivial phase to another, also trivial phase.

8.3.7 Quantum phase transition in Kitaev model

The momentum space fidelity may also be used to detect transition lines in the one-dimensional Kitaev model, equation (4.59). Let us consider, for instance, the line in parameter space $\Delta = t = 1$ as a function of the chemical potential μ. Let us focus, for instance, our attention on the transition that takes place at $\mu = 2t = 2$. We may calculate the fidelity associated with a given momentum value, k. The total fidelity is the product of the fidelities for each momentum k, as $F = \prod_k F_k$. For a given momentum value, we obtain

$$\boldsymbol{h} = (0, 2\sin k, -2\cos k - \mu) \ . \tag{8.53}$$

Defining $\tilde{\mu} = \mu/2$, we may obtain the fidelity, in momentum space, between two states with different chemical potential values, using the equation 8.42 for a given momentum value,

$$F_k = \sqrt{\frac{1}{2}\left(1 + \frac{1 + \cos k(\tilde{\mu}_1 + \tilde{\mu}_2) + \tilde{\mu}_1\tilde{\mu}_2}{\sqrt{1 + 2\tilde{\mu}_1\cos k + (\tilde{\mu}_1)^2}\sqrt{1 + 2\tilde{\mu}_2\cos k + (\tilde{\mu}_2)^2}}\right)} \ . \tag{8.54}$$

The energy eigenvalues are given by $\pm|\boldsymbol{h}|$. Therefore, the gap vanishes for the momentum $k = \pi$. The fidelity for this momentum value is given by

$$F_{k=\pi} = \sqrt{\frac{1}{2}\left(1 + \frac{1 - (\tilde{\mu}_1 + \tilde{\mu}_2) + \tilde{\mu}_1\tilde{\mu}_2}{|1 - \tilde{\mu}_1||1 - \tilde{\mu}_2|}\right)} \ . \tag{8.55}$$

It is now easy to see that, if we select two values of the chemical potential on the same side of the transition, then $F_{k=\pi} = 1$. But if we select now two values of the chemical potential, one on each side of the transition, $F_{k=\pi} = 0$. This result only occurs for this momentum value, but since the total fidelity is the product over all momenta, the fidelity vanishes. Therefore, a state in a topological region is orthogonal to a state in a trivial region, in agreement with the fact that the fidelity measures the distinguishability between states. Clearly this result occurs in general for quantum transitions, even if there are no topological properties associated. In general, if gapless points occur we may expect a similar result. But since in a topological transition a gap should vanish, the result should be generic for topological transitions.

8.3.8 Fidelity susceptibility

Let us return to the case where two close points in parameter space are considered. Let us recall the expression for the fidelity in the form

$$F = \frac{1}{\sqrt{2}}\sqrt{1 + \boldsymbol{n}_1 \cdot \boldsymbol{n}_2} \ , \tag{8.56}$$

where the vectors are unitary. If we consider two points very close to each other, then the unit vectors are also very close to each other and we may write $n_2 = n_1 + \delta n$. Therefore, we obtain

$$F = \frac{1}{\sqrt{2}} \sqrt{2 + n \cdot \delta n} \,, \tag{8.57}$$

where we omit the vector index. The unit vector depends, in general, on the various parameters in the Hamiltonian. Supposing that the transition is obtained varying a single parameter, h, we write $n = n(h)$. Then, the fidelity is given by

$$F = \frac{1}{\sqrt{2}} \sqrt{1 + n(h) \cdot n(h + \delta h)} \,. \tag{8.58}$$

The Taylor expansion,

$$n(h + \delta h) = n(h) + \delta h \frac{\partial n}{\partial h} + \frac{1}{2} (\delta h)^2 \frac{\partial^2 n}{\partial h^2} + \cdots \tag{8.59}$$

allows to se that the fidelity may be approximated by

$$F = 1 - \frac{(\delta h)^2}{2} \chi_F \,, \tag{8.60}$$

where

$$\chi_F = -\frac{1}{4} n(h) \cdot \frac{\partial^2 n}{\partial h^2} \,. \tag{8.61}$$

The linear term vanishes because the normalization of n implies that $n \cdot (\partial n)/(\partial h) = 0$. The quantity χ_F is called fidelity susceptibility.

The fidelity susceptibility has a singularity on the transition line and may be used efficiently to detect quantum phase transitions.

To illustrate, let us consider the simple case of a zero temperature transition involving only pure states, instead of the cases described by a density matrix. The fidelity may be written as

$$F = |\langle \psi_0(h) | \psi_0(h + \delta h) \rangle| \,, \tag{8.62}$$

that is, the absolute value of the overlap between two groundstates, for the two parameters h and $h + \delta h$. We may use perturbation theory to write the wave function $\psi_0(h + \delta h)$ from $\psi_0(h)$. Assuming that the variation of h can be seen as a perturbation, H_I, then

$$|\psi_0(h + \delta h)\rangle = |\psi_0(h)\rangle + \delta h \sum_{n \neq 0} \frac{\langle \psi_n(h) | H_I | \psi_0(h) \rangle}{E_0(h) - E_n(h)} |\psi_n(h)\rangle \,. \tag{8.63}$$

The state (8.63) is not yet normalized. Indeed,

$$\langle \psi_0(h + \delta h) | \psi_0(h + \delta h) \rangle = 1 + (\delta h)^2 \sum_{n \neq 0} \frac{|\langle \psi_n(h) | H_I | \psi_0(h) \rangle|^2}{(E_n(h) - E_0(h))^2}. \qquad (8.64)$$

Normalizing the wave function and inserting in the fidelity, we obtain

$$F^2 = 1 - (\delta h)^2 \sum_{n \neq 0} \frac{|\langle \psi_n(h) | H_I | \psi_0(h) \rangle|^2}{(E_n(h) - E_0(h))^2}. \qquad (8.65)$$

The fidelity susceptibility

$$\chi_F(h) = -\frac{\partial^2 F}{\partial (\delta h)^2}$$

$$= \sum_{n \neq 0} \frac{|\langle \psi_n(h) | H_I | \psi_0(h) \rangle|^2}{(E_n(h) - E_0(h))^2}. \qquad (8.66)$$

This expression clearly shows that a singularity arises in $\chi_F(h)$ if the gap closes, i.e., when some eigenenergy coincides with the energy of the ground-state.

The calculation of the fidelity susceptibility for Kitaev model, when $\mu \approx 2t$ and keeping $\Delta = t$, leads to the result

$$\chi_F(k = \pi) \sim \frac{1}{(\frac{\mu}{2} - 1)^2}, \qquad (8.67)$$

displaying a divergence at the critical point.

8.4   Non-abelian permutation of Majorana fermions

We saw in chapter 4 that the Majorana fermions have interesting properties, such as being their own antiparticle, have zero energy and enjoy topological protection. The coupling of two Majorana fermions leads to a normal fermionic mode intrinsically non-local. This characteristic implies that a local transformation should not affect the non-local nature of the fermionic mode and implies, as a consequence, their robustness to perturbations that do not break the symmetry, or symmetries, involved in their topological nature. We will see next that the permutation of two Majorana fermions has a non-abelian nature. This is the result of the degeneracy of the normal fermion constructed from the two Majoranas, since the occupation states of the fermion are degenerate. The robustness of the Majorana fermions, due to the topological protection, allows to consider the possibility of non-trivial operations on these entities, that may be considered as gates on a set of operations associated with quantum computation. The robustness avoids the problem of errors and lack of coherence usually associated with operations on quantum systems. The Majorana states allow, as a consequence, a potentially interesting way for quantum computation. We will introduce here some basic concepts, since the details are beyond the scope of this introductory text.

8.4.1 *Products of Majorana fermions*

Considering systems with more than one Majorana fermion, one is naturally lead to products of different Majoranas. We saw previously that it is possible to represent one spin-1/2 operator by a bilinear expression of Majorana fermions. At the very least, one needs three Majorana fermions, although it is also possible to enlarge the operator space and consider four Majorana fermions. On the other hand, a normal fermion may be represented by two Majorana fermions, therefore four Majorana fermions may be interpreted as two normal fermions.

It is convenient to define operators $\mathscr{S}_i = \frac{-i}{2}\epsilon_{ijk}\gamma_j\gamma_k$, that are related with the usual spin operators, $\mathbf{S} = \frac{1}{2}\mathscr{S}$. With three Majorana fermions it is possible to define the operator $I_3 = -i\gamma_1\gamma_2\gamma_3$, that is also of Majorana type, since is an odd product of Majoranas.

We may, then, construct rules for products of Majorana fermions,

$$\gamma_i\gamma_j = \delta_{ij}I + i\epsilon_{ijk}\mathscr{S}_k \,, \tag{8.68}$$

where I is the identity operator, obtaining that $\gamma^2 = 3I$. We may write the multiplications of Majorana operators by spin operators,

$$\gamma_i\mathscr{S}_j = \delta_{ij}I_3 + i\epsilon_{ijk}\gamma_k \,, \tag{8.69}$$

$$\mathscr{S}_i\gamma_j = \delta_{ij}I_3 + i\epsilon_{ijk}\gamma_k \,, \tag{8.70}$$

and the product of spin operators,

$$\mathscr{S}_i\mathscr{S}_j = \delta_{ij}I + i\epsilon_{ijk}\mathscr{S}_k \,, \tag{8.71}$$

and also $I_3^2 = I$, $I_3\gamma_i = \gamma_iI_3 = \mathscr{S}_i$, $I_3\mathscr{S}_i = \mathscr{S}_iI_3 = \gamma_i$, $\mathscr{S}^2 = 3I$ and $\gamma \cdot \mathscr{S} = \mathscr{S} \cdot \gamma = 3I_3$.

We may now consider operations that involve these operators. In particular, unitary operations of various types. Two interesting examples are:

(a) $U = \frac{1}{\sqrt{2}}(1 + i\mathbf{n} \cdot \mathscr{S})$, with $|\mathbf{n}| = 1$. The action of this operator may be seen in the following way:

$$U\mathscr{S}U^\dagger = \mathbf{n}(\mathbf{n} \cdot \mathscr{S}) + \mathbf{n} \times \mathscr{S} \,,$$

$$U\gamma U^\dagger = \mathbf{n}(\mathbf{n} \cdot \gamma) + \mathbf{n} \times \gamma \,. \tag{8.72}$$

This is a canonical transformation, since it does not change the nature of the operators.

(b) $U = \frac{1}{\sqrt{2}}\left(1 + i\mathbf{n}\cdot\boldsymbol{\gamma}\right)$, with $|\mathbf{n}| = 1$. The action of this operator leads to

$$U\boldsymbol{\gamma}U^\dagger = \mathbf{n}\left(\mathbf{n}\cdot\boldsymbol{\gamma}\right) + \mathbf{n}\times\mathscr{S}\,,$$
$$U\mathscr{S}U^\dagger = \mathbf{n}\left(\mathbf{n}\cdot\mathscr{S}\right) + \mathbf{n}\times\boldsymbol{\gamma}\,, \qquad (8.73)$$

and since it mixes the nature of the operators, is a non-canonical transformation.

The unitary transformations may also be written in the usual way associated with transformations in $SU(2)$:

$$e^{i\frac{\theta}{2}\mathbf{n}\cdot\boldsymbol{\gamma}} = \cos\frac{\theta}{2}I + i\sin\frac{\theta}{2}\mathbf{n}\cdot\boldsymbol{\gamma}\,, \qquad (8.74)$$

$$e^{i\frac{\theta}{2}\mathbf{n}\cdot\mathscr{S}} = \cos\frac{\theta}{2}I + i\sin\frac{\theta}{2}\mathbf{n}\cdot\mathscr{S}\,, \qquad (8.75)$$

with $\mathbf{n}^2 = 1$.

8.4.2 *Flux quantization and Majoranas permutations*

The non-abelian statistics of the Majorana fermions was shown for the example of a bidimensional superconductor with triplet pairing, where vortices may arise that have an excitation spectrum with a Majorana fermion. Unlike usual fermions, that follow anticommutation relations, the Majorana fermions have a different statistics and are *anyons*. This term was introduced in the context of the fractional quantum Hall effect, where the excitations obey a statistics that is neither fermionic nor bosonic.

Let us consider a vortex in a two-dimensional superconductor and let us recall that the magnetic flux associated with the vortex is quantized. The superconducting state may be understood as a set of pairs of electrons (Cooper pairs). In a way similar to the case of a boson condensate, these may be represented by a macroscopic wave function $\psi(\mathbf{r})$, that has the meaning of the probability amplitude of finding a pair of electrons centered in the position $\mathbf{r}$, and may also be taken as the superconducting order parameter $\Delta(\mathbf{r})$. The concentration of the pairs is, then, $n_p(\mathbf{r}) = \psi^*(\mathbf{r})\psi(\mathbf{r})$. We may represent the wave function as

$$\psi(\mathbf{r}) = \sqrt{n_p}\,e^{i\theta(\mathbf{r})} \qquad (8.76)$$

and approximate the concentration of the pairs by a constant. In a magnetic field we know that the velocity of one particle is given by the generalization of the equation (6.36),

$$\mathbf{v} = \frac{1}{m}\left(\mathbf{p} - q\mathbf{A}\right) = \frac{1}{m}\left(\frac{\hbar}{i}\nabla - q\mathbf{A}\right), \tag{8.77}$$

where q is the charge of the Cooper pair, m its mass and $\mathbf{A}$ the vector potential. This result implies that

$$\psi^{*}\mathbf{v}\psi = \frac{n_{p}}{m}\left(\hbar\nabla\theta(\mathbf{r}) - q\mathbf{A}\right) \tag{8.78}$$

and the current is

$$\mathbf{J} = q\psi^{*}\mathbf{v}\psi = \frac{n_{p}q}{m}\left(\hbar\nabla\theta - q\mathbf{A}\right). \tag{8.79}$$

Consider, now, that the magnetic field penetrates the superconductor in a given spatial point. Inside the superconductor, sufficiently far from this region, we have that $\mathbf{B} = 0, \mathbf{J} = 0$, due to the Meissner effect. Since the current vanishes, we obtain that

$$\hbar\nabla\theta = q\mathbf{A}. \tag{8.80}$$

Therefore, choosing a contour C in the bulk and around the line of the magnetic field, we obtain

$$\int_{C}\nabla\theta\cdot d\mathbf{l} = \theta_{2} - \theta_{1} = \frac{q}{\hbar}\int_{C}\mathbf{A}\cdot d\mathbf{l} = \frac{q}{\hbar}\Phi, \tag{8.81}$$

where Φ is the magnetic flux through the contour C. The difference $\theta_{2} - \theta_{1}$ is a multiple of 2π, therefore, the flux is quantized as

$$\Phi = \frac{h}{q}n = \Phi_{0}n, \tag{8.82}$$

where

$$\Phi_{0} = \frac{h}{q} \tag{8.83}$$

denotes the flux quantum and n is an integer. Recall that the particles are, actually, pairs, with mass $m = 2m_{e}$ and charge $q = 2e$. This result leads us to conclude that the flux through the magnetic field lines that penetrate the superconductor of type-*II* is quantized.

In the case of the superconducting pairing studied in chapter 4 the wave function is given by [see equation (4.27)]

$$\psi(\mathbf{r}) = e^{i\varphi}\left[d_{x}\left(|\uparrow\uparrow\rangle + |\downarrow\downarrow\rangle\right) + id_{y}\left(|\uparrow\uparrow\rangle - |\downarrow\downarrow\rangle\right) + d_{z}\left(|\uparrow\downarrow\rangle + |\downarrow\uparrow\rangle\right)\right]$$
$$\left(k_{x} + ik_{y}\right), \tag{8.84}$$

where the vector $\mathbf{d}$ characterizes the triplet pairing. In a vortex, the pairing order parameter has a spatial dependence: the phase of the condensate, φ, as the vector **d** changes in real space, i.e., are functions of **r**. Obviously, the order parameter, $\psi(\mathbf{r})$, has to resume its value when we carry out one turn around the center of the vortex.

Note that $\psi(\mathbf{r})$ remains invariant under the transformation $(\varphi, \mathbf{d}) \rightarrow (\varphi + \pi, -\mathbf{d})$. What characterizes a half *quantum* flux vortex is that the phase φ varies by π and the vector **d** rotates to $-\mathbf{d}$ in a turn around the vortex. This property can not be eliminated by a continuous transformation (homotopy) of the superconducting order parameter.

A way to satisfy the described situation corresponds to the choice $d_z = 0$ and $d_x + i d_y = d e^{i\varphi(\mathbf{r})}$. Let us denote by r the distance to the vortex center. At the center, the order parameter should vanish, therefore, $d = d(r)$ and $d(0) = 0$. Then, we see in the expression (8.84) that the two spin directions acquire different phases, and only the $\uparrow$ spin component has a dependence on φ. Specifically,

$$\psi(\mathbf{r}) = d(r) \left(e^{2i\varphi(\mathbf{r})} | \uparrow\uparrow\rangle + | \downarrow\downarrow\rangle \right) (k_x + i k_y) . \tag{8.85}$$

The expression (8.85) shows that a vortex exists associated with the spin $\uparrow$ component while for the other component the vortex does not exist. For the $\uparrow$ spin component low energy excitations are found, located at the vortex center. These low energy modes include a zero energy mode, of Majorana. We may ignore the spectrum associated to the $\downarrow$ spin component, since it has associated energies that are, in general, finite. The problem is therefore, reduced to a problem of effectively spinless fermions, described by the Hamiltonian

$$\hat{H} = \int d\mathbf{r} \left[\hat{\psi}^\dagger(\mathbf{r}) \left(-\frac{\hbar^2 \nabla^2}{2m} - \mu \right) \hat{\psi}(\mathbf{r}) + \frac{\Delta e^{-i\theta}}{2} \hat{\psi}(\mathbf{r}) \left(-\partial_x + i\partial_y \right) \hat{\psi}(\mathbf{r}) \right.$$

$$\left. + \frac{\Delta e^{i\theta}}{2} \hat{\psi}^\dagger(\mathbf{r}) \left(\partial_x + i\partial_y \right) \hat{\psi}^\dagger(\mathbf{r}) \right], \tag{8.86}$$

where θ is the polar angular coordinate. This Hamiltonian was studied in chapter 4 (but where the phase of the condensate was spatially constant). Recall that in chapter 4 it was shown that there was a Majorana mode at the edge with a normal fluid region. In the case of a vortex, this region is the center of the vortex, where $\Delta(r) \rightarrow 0$, which suggests the existence of a Majorana mode spatially localized. Indeed, solving the problem for the vortex, modes with energies $\epsilon_n = n\omega_0$ arise (with $\omega_0 \sim \Delta^2/\mu$), where n is an integer, being the zero energy mode a Majorana fermion at the center of the vortex. The solutions are obtained performing a usual transformation to new operators $\hat{\gamma}^\dagger = u\hat{\psi}^\dagger + v\hat{\psi}$. A change of phase by a value ϕ implies that $(u, v) \rightarrow (ue^{i\phi/2}, ve^{-i\phi/2})$, as we can see from equation (4.22). If $\phi = 2\pi$ we obtain that $\gamma \rightarrow -\gamma$. If we have two vortices, each one having an associated Majorana mode, and one of them is transported around the other, the rotation exchanges the signal of the fermion associated with the vortex that was transported (this "feels" the phase θ of the condensate of the other), leaving invariant the Majorana fermion that was at rest. If we denote the Majorana fermion that was moved by γ_1 and the fixed one by γ_2, we obtain the transformation

$$\gamma_1 \rightarrow -\gamma_2 \,,$$
$$\gamma_2 \rightarrow \gamma_1 \,. \tag{8.87}$$

The non-abelian statistics leads to the following: in a system with various vortices the groundstate is degenerate and the permutation of two vortices leads to a state that is the linear combination of the degenerate states, instead of a simple multiplication of the wave function by a factor.

A unitary transformation that describes this process may be constructed considering the second transformation of equation (8.72) or the second transformation of equation (8.75), taking $\mathbf{n} = \mathbf{e}_z$ and $\theta = \pi/2$. That is, choosing the transformation

$$U_{12} = e^{\frac{\pi}{4}\gamma_1\gamma_2} = \frac{1}{\sqrt{2}} (1 + \gamma_1\gamma_2) \,, \tag{8.88}$$

it is easy to see that, considering two Majorana fermions 1 and 2,

$$U_{12}\gamma_1 U_{12}^\dagger = -\gamma_2$$
$$U_{12}\gamma_2 U_{12}^\dagger = \gamma_1 \,. \tag{8.89}$$

The two Majorana fermions give origin to a usual fermion, $f = 1/2(\gamma_1 + i\gamma_2)$, and as we saw previously, $\gamma_1\gamma_2 = i(1 - 2f^\dagger f)$. This operator has two eigenstates, with different fermionic parities, $|n_f = 0, 1\rangle$, where n_f is the occupation number of the operator f. The action of the unitary operator on these states is given by

$$U_{12}|n_f\rangle = e^{i\frac{\pi}{4}(1-2n_f)}|n\rangle \tag{8.90}$$

and, therefore, is simply a phase factor and the parity is conserved in the process.

Consider, now, two pairs of Majorana fermions to which we may associate two usual fermionic operators, $f_1 = 1/2(\gamma_1 + i\gamma_2)$ and $f_3 = 1/2(\gamma_3 + i\gamma_4)$, to which we associate eigenstates $|n_1\rangle$ and $|n_3\rangle$. The action of the unitary operators U_{12} and U_{34} is given by

$$U_{12}|n_1 n_3\rangle = e^{i\frac{\pi}{4}(1-2n_1)}|n_1 n_3\rangle \,,$$
$$U_{34}|n_1 n_3\rangle = e^{i\frac{\pi}{4}(1-2n_3)}|n_1 n_3\rangle \,. \tag{8.91}$$

On the other hand, the permutation of Majoranas of different pairs takes the form

$$U_{23}|n_1 n_3\rangle = \frac{1}{\sqrt{2}} \left(|n_1 n_3\rangle + i|1 - n_1, 1 - n_3\rangle \right), \qquad (8.92)$$

therefore, mixes different states, even though it preserves the fermionic parity. We see that, if we consider other Majorana pairs, the permutation operation is non-trivial and non-commutative. We may check, for instance, that

$$[U_{12}, U_{23}] = \gamma_1 \gamma_3 . \qquad (8.93)$$

The commutator is non-vanishing and the order of the fermionic permutation operations affects the final result. Recalling the form of the permutation operator, this may be interpreted as a rotation by an angle $\pi/2$ in a space of spin-1/2, therefore is not a sufficiently general transformation. Specifically, note that

$$U_{12} = e^{i\frac{\pi}{4}\mathscr{S}_z},$$
$$U_{23} = e^{i\frac{\pi}{4}\mathscr{S}_x}, \qquad (8.94)$$

(the unitary operator U_{34} is of the same form as U_{12}) and, therefore, corresponds to the algebra of one qubit (a spin-1/2 operator). Even though promising, the permutation does not allow directly universal computation. This may be obtained adding other types of gates (unitary operations); the property of topological protection may, however, be lost.

The permutations between the Majoranas may be obtained in different ways, typically displacing the Majoranas physically, such as spatially moving their physical supports (as the one-dimensional systems that support the edge states), or using the motion of domain walls.

Further reading:

A generic discussion of entanglement is presented in:

- L. Amico, R. Fazio, A. Osterloh and V. Vedral, *Reviews Modern Physics* **80**, 517 (2008).

An application to a phase transition in a superconductor with an impurity is presented in:

- P. D. Sacramento, P. Nogueira, V. R. Vieira and V. K. Dugaev, *Physical Review* B **76**, 184517 (2007).

On the entanglement spectrum:

- H. Li and F. D. M. Haldane, *Physical Review Letters* **101**, 010504 (2008).
- N. Regnault, B. A. Bernevig and F. D. M. Haldane, *Physical Review Letters* **103**, 016801 (2009).
- D. Poilblanc, *Physical Review Letters* **105**, 077202 (2010).
- T. P. Oliveira and P. D. Sacramento, *Physical Review* B **89**, 094512 (2014).
- T. P. Oliveira, P. Ribeiro and P. D. Sacramento, *Journal of Physics: Condensed Matter* **26**, 425702 (2014).

On the notion of fidelity and its use in many particle systems:

- P. Zanardi and N. Paunković, *Physical Review* E **74**, 031123 (2006).
- Damian F. Abasto, Alioscia Hamma and Paolo Zanardi, *Physical Review* A **78**, 010301R (2008).
- A. Hamma, W. Zhang, S. Haas and D. A. Lidar, *Physical Review* B **77**, 155111 (2008).
- Shi-Jian Gu, *International Journal of Modern Physics* B **24**, 4371 (2010).

The notion of partial state fidelity was introduced in:

- N. Paunković, P. D. Sacramento, P. Nogueira, V. R. Vieira and V. K. Dugaev, *Physical Review* A **77**, 052302 (2008).
- H. Q. Zhou, *arXiv*:0704.2945.

On the fidelity spectrum:

- P. D. Sacramento, N. Paunković and V. R. Vieira, *Physical Review* A **84**, 062318 (2011).
- P. D. Sacramento, B. Mera and N. Paunković, *Annals of Physics* **401**, 40 (2019).

On the entanglement of Majorana fermions;

- D. A. Ivanov, *Physical Review Letters* **86**, 268 (2001).
- R. Aguado, *Rivista Nuovo Cimento* **40**, 523 (2017).

On topological quantum computation:

- C. Nayak, S. H. Simon, A. Stern, M. Freedman and S. Das Sarma, *Reviews Modern Physics* **80**, 1083 (2008).
- S. Das Sarma, M. Freedman and C. Nayak, *npj Quantum Information* **1**, 15001 (2015).
- V. T. Lahtinen and J. K. Pachos, *SciPost Physics* **3**, 021 (2017).

On the transport and manipulation of Majorana fermions:

- J. Alicea, Y. Oreg, G. Refael, F. von Oppen and M. P. A. Fisher, *Nature Physics* **7**, 412 (2011).
- Ching-Kai Chiu, M. M. Vazifeh and M. Franz, *Europhysics Letters* **110**, 10001 (2015).
- J. Li, T. Neupert, B. A. Bernevig and A. Yazdani, *Nature Communications* **7**, 10395 (2016).

Chapter 9

Out of equilibrium topological systems

9.1 Sudden quantum transformations

9.1.1 *Survival probability and Loschmidt echo*

Let us consider a system in a given initial state, $|\psi_m(\xi)\rangle$, where ξ represents a set of parameters upon which the Hamiltonian depends. As we change the parameters, the system may cross various phases that, in the case of a topological system, may have different topological properties. Let us suppose that at time $t = 0$ the parameters are abruptly changed to the set ξ'. After this sudden change, the system evolves in time under the action of a different Hamiltonian, $H(\xi')$. Since the Hamiltonian is changed, the initial states, calculated at $t = 0^-$, are no longer eigenstates and get mixed. The time evolution is dictated by

$$|\psi_m(\xi, t)\rangle = e^{-iH(\xi')t}|\psi_m(\xi)\rangle . \tag{9.1}$$

Representing the eigenstates of the Hamiltonian $H(\xi')$ by $|\psi_n(\xi')\rangle$ and the eigenvalues by $E_n(\xi')$, we may write the time evolved state as

$$|\psi_m(\xi, t)\rangle = \sum_n e^{-iE_n(\xi')t}|\psi_n(\xi')\rangle\langle\psi_n(\xi')|\psi_m(\xi)\rangle . \tag{9.2}$$

Additionally, we may calculate the superposition of this time evolved state with the initial state,

$$A_m(t) = \langle\psi_m(\xi)|\psi_m(\xi, t)\rangle . \tag{9.3}$$

This result may be expressed as

$$A_m(t) = \sum_n |\langle\psi_m(\xi)|\psi_n(\xi')\rangle|^2 e^{-iE_n(\xi')t} \tag{9.4}$$

and involves the superposition between this initial state and all the eigenstates of the final Hamiltonian. One can define the survival probability of the initial state as

$$P_m(t) = |A_m(t)|^2 . \tag{9.5}$$

The states considered may be either many-body states or single-particle states.

In the case of a many-particle system, it is usual to define an equivalent quantity, called *Loschmidt echo*. This quantity compares the time evolution of a same state $|\psi_0\rangle$ under the action of two different Hamiltonians, $H(\xi_0)$ and $H(\xi_1)$. The Loschmidt echo is defined by the square of the fidelity between the two evolved states, that is,

$$L(t) = |\langle \psi_0 | e^{iH(\xi_1)t} e^{-iH(\xi_0)t} | \psi_0 \rangle|^2 . \tag{9.6}$$

This expression simplifies if the groundstate is an eigenstate of one of the Hamiltonians. For example, if it is an eigenstate of the Hamiltonian $H(\xi_1)$ then (9.6) reduces to

$$L(t) = \langle \psi_0 | e^{-iH(\xi_0)t} | \psi_0 \rangle|^2 , \tag{9.7}$$

that may be calculated exactly for bilinear Hamiltonians. In the case of a system of many particles it has to be calculated numerically, in general.

Exemplifying with a bilinear Hamiltonian, the us consider the one-dimensional XX Heisenberg model, with coupling constant J, and a transverse field that includes an uniform component, h, and a component that alternates between sites, h_{st}:

$$H = J \sum_j \frac{1}{2} \left(S_j^+ S_{j+1}^- + S_{j+1}^+ S_j^- \right) + \sum_j \left(h + (-1)^j h_{st} \right) S_j^z . \tag{9.8}$$

Applying a Jordan-Wigner transformation, we arrive at a spinless fermion problem,

$$H = -\frac{J}{2} \left(\sum_{j=1}^{N-1} \left(c_j^\dagger c_{j+1} + c_{j+1}^\dagger c_j \right) - e^{i\pi M} \left(c_N^\dagger c_1 + c_1^\dagger c_N \right) \right)$$
$$+ \sum_{j=1}^N \left(h + (-1)^j h_{st} \right) \left(n_j - \frac{1}{2} \right) , \tag{9.9}$$

where M is the number of fermions and $n_j = c_j^\dagger c_j$. We may define a Fourier transform as

$$c_j = \frac{1}{\sqrt{N}} \sum_{n=1}^N c_{k_n} e^{-ij(k_n+a)} , \tag{9.10}$$

where $k_n = 2\pi n/N$, $a = \pi/N$ for even M (antiperiodic boundary conditions for the fermions) and $a = 0$ for odd M (periodic boundary conditions for the fermions). We are, then, lead to the Hamiltonian

$$H = \sum_{n=1}^N (\epsilon_{k_n} - h) c_{k_n}^\dagger c_{k_n} - h_{st} \sum_{n=1}^N c_{k_n}^\dagger c_{k_n+N/2} . \tag{9.11}$$

Using a reduced Brillouin zone and introducing new operators, $d_{k_n} = c_{k_n - \pi}$, we obtain

$$H = \sum_{n=1}^{N/2} \left((\epsilon_{k_n} - h) c_{k_n}^\dagger c_{k_n} - (\epsilon_{k_n} + h) d_{k_n}^\dagger d_{k_n} \right)$$

$$- h_{st} \sum_{n=1}^{N/2} \left(c_{k_n}^\dagger d_{k_n} + d_{k_n}^\dagger c_{k_n} \right) , \tag{9.12}$$

where $\epsilon_{k_n} = -J \cos(k_n + a)$. This Hamiltonian may be diagonalized by a canonical transformation to new operators, α_{k_n} e β_{k_n}, as

$$H = \sum_{n=1}^{N/2} \left(\left(\sqrt{\epsilon_{k_n}^2 + h_{st}^2} - h \right) \alpha_{k_n}^\dagger \alpha_{k_n} - \left(\sqrt{\epsilon_{k_n}^2 + h_{st}^2} + h \right) \beta_{k_n}^\dagger \beta_{k_n} \right) . \tag{9.13}$$

As example, let us consider a sudden transformation of an Hamiltonian with parameters $J = h = 0$ and alternating (staggered) field $h_{st} \neq 0$ to a final Hamiltonian with $h = h_{st} = 0$ and $J \neq 0$. The initial Hamiltonian is, then,

$$H_i = -h_{st} \sum_n \left(c_{k_n}^\dagger d_{k_n} + d_{k_n}^\dagger c_{k_n} \right)$$

$$= \sum_n \left(|h_{st}| \alpha_{k_n}^\dagger \alpha_{k_n} - |h_{st}| \beta_{k_n}^\dagger \beta_{k_n} \right) . \tag{9.14}$$

The groundstate of the Hamiltonian (9.14) has Néel order and is of the form

$$|\psi_0\rangle = \prod_n \beta_{k_n}^\dagger |0\rangle , \tag{9.15}$$

where $|0\rangle$ represents the vacuum. The final Hamiltonian has the form

$$H_f = \sum_n \epsilon_{k_n} \left(c_{k_n}^\dagger c_{k_n} - d_{k_n}^\dagger d_{k_n} \right) , \tag{9.16}$$

already in the diagonal form, since it is written in terms of the occupation number operators. Therefore, we may rewrite it as

$$H_f = \sum_n \left(|\epsilon_{k_n}| \alpha_{k_n}^\dagger \alpha_{k_n} - |\epsilon_{k_n}| \beta_{k_n}^\dagger \beta_{k_n} \right) . \tag{9.17}$$

In this case, it is simple to time evolve the initial state, $|\psi_0\rangle$. Introducing the variable $z = it$, we write the evolution in the form

$$e^{-zH_f} |\psi_0\rangle = e^{-z \sum_n \epsilon_{k_n} \left(c_{k_n}^\dagger c_{k_n} - d_{k_n}^\dagger d_{k_n} \right)} \prod_m \beta_{k_m}^\dagger |0\rangle . \tag{9.18}$$

The relation between the various operators is given by

$$c_{k_n} = \frac{1}{\sqrt{2}} \left(\alpha_{k_n} + \beta_{k_n} \right) ,$$

$$d_{k_n} = \frac{1}{\sqrt{2}} \left(\alpha_{k_n} - \beta_{k_n} \right) , \tag{9.19}$$

that leads to

$$e^{-zH_f} |\psi_0\rangle = e^{-z \sum_n \epsilon_{k_n} \left(c_{k_n}^\dagger c_{k_n} - d_{k_n}^\dagger d_{k_n} \right)} \prod_m \frac{1}{\sqrt{2}} \left(c_{k_m}^\dagger - d_{k_m}^\dagger \right) |0\rangle$$

$$= \prod_n \left(\frac{1}{\sqrt{2}} \left(e^{-z\epsilon_{k_n}} c_{k_n}^\dagger - e^{z\epsilon_{k_n}} d_{k_n}^\dagger \right) \right) |0\rangle . \tag{9.20}$$

Then, the Loschmidt amplitude is

$$\langle \psi_0 | e^{-zH} | \psi_0 \rangle = \langle 0 \prod_m \beta_{k_m} \prod_n \left(\frac{1}{\sqrt{2}} \left(e^{-z\epsilon_{k_n}} c_{k_n}^\dagger - e^{z\epsilon_{k_n}} d_{k_n}^\dagger \right) \right) |0\rangle$$

$$= \langle 0 | \prod_n \frac{1}{\sqrt{2}} \left(c_{k_m} - d_{k_m} \right)$$

$$\prod_n \left(\frac{1}{\sqrt{2}} \left(e^{-z\epsilon_{k_n}} c_{k_n}^\dagger - e^{z\epsilon_{k_n}} d_{k_n}^\dagger \right) \right) |0\rangle$$

$$= \prod_n \cosh \left(z\epsilon_{k_n} \right) = \prod_n \cos \left(t\epsilon_{k_n} \right) . \tag{9.21}$$

We see that at the times $t = \pi(2m + 1)/(2\epsilon_{k_n})$, with integer m, the state that results from the time evolution becomes orthogonal to the initial state.

9.1.2 *Energy non-conservation*

When we interact suddenly, slowly or in a time periodic way with a system, we exchange energy with the system. In the case of a periodic pertur-bation, energy is being transfered to the system, therefore, its energy is not-conserved in general. We expect, therefore, that the system will evolve to a regime characteristic of high temperature. This will tend to infinite for sufficiently long times, if the system is not in contact with a reservoir at a given temperature, that is, if the system is isolated (besides the energy ex-change that results from the periodic perturbation). This asymptotic limit may be avoided if the system is interactive, or if it is disordered. Then, it may reach a different regime, in which the states are localized. Another relevant case is an integrable system, since there is an infinite set of con-servation laws that lead to a final state that is qualitatively different from an infinite temperature regime.

Let us begin by considering a sudden transformation. A periodic perturbation will be considered later in this chapter. Let

$$H(\lambda) = H_0 + \lambda V \qquad (9.22)$$

the Hamiltonian, where λ is a parameter externally controlled and that we change abruptly at $t = 0$. For $t < 0$ the parameter takes the value $\lambda = \lambda_i$ and we assume that the system is at zero temperature, in its groundstate, that is assumed as non-degenerate, to simplify. We write

$$H(\lambda_i)|\psi_0\rangle = E_0(\lambda_i)|\psi_0\rangle. \qquad (9.23)$$

Let us, now, change abruptly $\lambda_i \to \lambda_f$. The new Hamiltonian has a different spectrum and may be written in terms of the projectors to its eigenstates, $|\Phi_m\rangle$, as

$$H(\lambda_f) = \sum_m E_m(\lambda_f)|\Phi_m\rangle\langle\Phi_m|, \qquad (9.24)$$

where $E_m(\lambda_f)$ are its eigenvalues.

For a generic transformation, either sudden or slow, we define the work performed on the system, when it is found in a final states m, as

$$W_m = E_m(\lambda_f) - E_0(\lambda_i). \qquad (9.25)$$

The final state is reached with a certain probability given by $p_m = |\langle\psi_0|\Phi_m\rangle|^2$.

We define the quantum work distribution as

$$P(W) = \sum_m p_m \delta(W - W_m). \qquad (9.26)$$

The first moment of this distribution is given by

$$\begin{aligned}
\langle W \rangle &= \int dW \, P(W) W \\
&= \langle\psi_0|H(\lambda_f)|\psi_0\rangle - \langle\psi_0|H(\lambda_i)|\psi_0\rangle \\
&= (\lambda_f - \lambda_i)\langle\psi_0|V|\psi_0\rangle \\
&= \delta\lambda \frac{\partial E_0}{\partial \lambda}\bigg|_{\lambda_i},
\end{aligned} \qquad (9.27)$$

where we used the Hellman-Feynman theorem, $V = \partial H/\partial\lambda$ and $\delta\lambda = \lambda_f - \lambda_i$. Therefore, the average work is related with the first derivative of the energy of the groundstate, in order to the parameter.

Thermodynamics demands that $\langle W \rangle \geq \Delta U$, where $\Delta U = E_0(\lambda_f) - E_0(\lambda_i)$ is the variation of the internal energy that results from the transformation. The equality is obtained in the case of a very slow and adiabatic transformation. But for a sudden transformation we expect that there is irreversibility and the work is defined by

$$\langle W_{irr} \rangle = \langle W \rangle - \Delta U \,. \tag{9.28}$$

If $\delta\lambda \ll 1$, the irreversible work may be expanded in $\delta\lambda$ as

$$\langle W_{irr} \rangle = \delta\lambda \frac{\partial E_0}{\partial\lambda}\bigg|_{\lambda_i} - E_0(\lambda_i + \delta\lambda) + E_0(\lambda_i)$$

$$\sim -\frac{(\delta\lambda)^2}{2} \frac{\partial^2 E_0}{\partial\lambda^2}\bigg|_{\lambda_i}, \tag{9.29}$$

and, therefore, is related with the second derivative of the groundstate energy.

This type of behavior depends on the type of quantum phase transition that we consider. If the transition is first order, there is a level crossing in the lowest energy states. If the transition is second order, the crossing is avoided. In the first case, the first energy derivative has a discontinuity at the transition point, where the second derivative has a Dirac delta function and a constant behavior away from the transition point. In the second case,

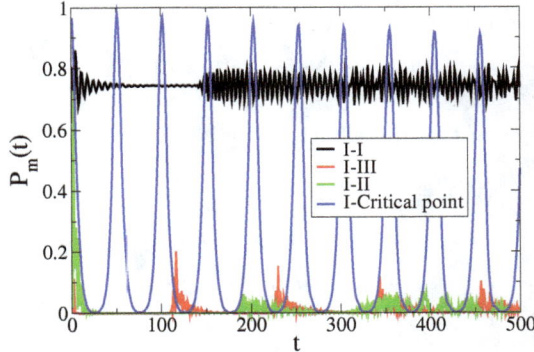

Fig. 9.1   Survival probability of a Majorana state of Kitaev model for different transitions across the phase diagram: i) transition between two points in phase I ($\mu = 0.5, \Delta = 0.6$) $\rightarrow$ ($\mu = 1.0, \Delta = 0.6$), ii) transition from the topological phase I to the trivial phase III ($\mu = 0.5, \Delta = 0.6$) $\rightarrow$ ($\mu = 2.2, \Delta = 0.6$), iii) transition from the topological phase I with positive Δ to the topological phase II with negative Δ ($\mu = 0.5, \Delta = 0.6$) $\rightarrow$ ($\mu = 0.5, \Delta = -0.6$), iv) transition between a point in phase I, to the quantum critical point ($\mu = 0, \Delta = 0.1$) $\rightarrow$ ($\mu = 0, \Delta = 0$), where the system has a gapless spectrum. The size of the system is 100 sites.

of a continuous transition, we expect a singularity at the transition point and a region of critical behavior.

9.1.3 *Kitaev model: stability of edge states*

The Hamiltonian of Kitaev model in equation (4.56) is quadratic, and therefore effectively non-interacting, and, therefore, a many-particle description is not needed. Note that in that equation the parameter t stands for the hopping amplitude and not time. Let us then take as initial state $|\psi_m(\xi)\rangle$ a state of lowest energy (eigenstate of the single-particle Hamiltonian for the set of parameters ξ), that may belong to a set of degenerate states. Remaining in the one excitation subspace after the transformation, the Hamiltonian associated with the unitary time evolution is the single particle Hamiltonian, for the set of parameters ξ'.

In figure 9.1 we show results for the survival probability of a Majorana mode for different sudden transformations of the Hamiltonian parameters of the Kitaev model. Recall that there are two topological phases, I and II and two trivial phases III, represented in figure 4.1. In the case of a transformation within the same phase, in this case phase I, the survival probability is finite. Since the parameters have changed, this is not unity and there is a decrease as a function of time that is due to the superposition with all the eigenstates of the chain with the new set of parameters. But it stabilizes on a given finite value after some oscillations. For larger times, oscillations appear again centered on that finite value. Therefore, the Majorana mode is robust to this transformation.

In the case of a transformation from the topological phase I to the trivial phase III, the behavior is quite different. The probability decays fast to values close to zero. After a certain time, it increases rapidly and repeats the decay. This behavior (renewal) is repeated for larger times. Similar results are found for a transformation between the two topological phases I and II. The renewal time scales with the system size. When the time is of the order of the renewal time, the wave function has a maximum around the center point of the chain and is the result of a mode that propagates along the chain with a given velocity and, therefore, scales with the size of the chain. In the limit of an infinite system, the renewal time diverges and the Majorana mode decays and is destroyed.

A qualitatively different case occurs when we study the transformation from the topological phase I to a critical point. At this point, the gap vanishes. In this case, the survival probability oscillates, the Majorana mode is periodically renewed, even in the limit of an infinite system.

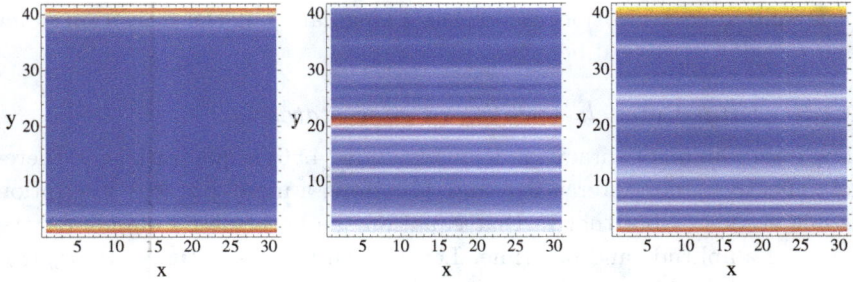

Fig. 9.2   Time evolution of $|u_\uparrow|^2$ as a result of the parameter change ($M_z = 2, \mu = -5$) $\rightarrow$ ($M_z = 0, \mu = -5$), to which corresponds $C = 1 \rightarrow C = 0$ (trivial) at the times $t = 0, t = 50, t = 62$, shown in (a), (b), (c), respectively, for the two-dimensional Sato and Fujimoto model. Note that, at these phase transitions, there are no edge modes in the final states. The horizontal axis (y) corresponds to the transverse direction and the vertical axis (x) to the longitudinal direction. The system size is 31 $\times$ 41.

9.1.4   *Sato and Fujimoto model: stability of edge modes*

In a two-dimensional topological system, of dimensions N_x by N_y, one finds edge states when we take open boundary conditions in a transversal dimension, y, and periodic boundary conditions in a longitudinal direction, x. The diagonalization of the Sato and Fujimoto model, given in equation (4.183), when expressed in real space involves the solution of an eigenvalue problem of dimension $(4N_x N_y) \times (4N_x N_y)$. Note, again, that in that equation the parameter t is the hopping amplitude between nearest neighbors and not time. The energy eigenstates include states in the bulk and states along the border of the system, and may be written in the form of a four component spinor,

$$\begin{pmatrix} u_n \\ v_n \end{pmatrix} = \begin{pmatrix} u_n(j_x, j_y, \uparrow) \\ u_n(j_x, j_y, \downarrow) \\ v_n(j_x, j_y, \uparrow) \\ v_n(j_x, j_y, \downarrow) \end{pmatrix}. \tag{9.30}$$

Here, j_x and j_y are the spatial coordinates along the directions x and y, respectively. Focusing our attention on a Majorana mode (that has zero energy), we present in figure 9.2 the time evolution of the absolute value of the component of the spinor $u_n(j_x, j_y, \uparrow)$ as a result of a sudden change of the parameters ($M_z = 2, \mu = -5$) to ($M_z = 0, \mu = -5$). To these parameters correspond Chern numbers $C = 1$ and $C = 0$ (trivial case), respectively. The evolution is shown for the characteristic instants $t = 0, t = 50, t = 62$, in panels (a), (b), (c), respectively. Time is measured in units of the inverse of the hopping amplitude between neighbors. The initial state has a significative maximum at the border of the system and decays fast inside the superconductor along the transverse direction. As time increases, the peaks move towards the center until they overlap at a later time, dependent on the transversal length of the system (as for the one-dimensional Kitaev model). Later, the peaks move away from the center, the wave functions become more extended, as evidence of the mixture with other eigenstates. For even longer times, the wave function eventually recovers a shape that is similar to the initial state. There is a partial renewal of the original state, therefore, an increase of the survival probability. The process repeats but the degree of coherence is lost for growing times. The other components of the spinor have similar behaviors.

9.1.5 *Evolution of the Chern numbers*

Consider, again, the Sato and Fujimoto model, equation (4.183), defined in momentum space. In equilibrium, the topology of each phase may be characterized by the Chern number. When the system evolves in time the wave functions change. Determining the evolution of the single-particle wave functions, in momentum space, we may calculate the Chern number (defined in terms of these functions that result from the time evolution) as a function of time and determine how the topology changes. Due to the fluctuations that result from the time evolution of the superpositions of one given state with the remaining states, the wave functions along the Brillouin zone oscillate considerably as time progresses.

In figure 9.3 we present the evolution of the Chern number of the Sato and Fujimoto model as a sudden transformation of the Hamiltonian parameters is carried out $(M_z = 2, \mu = -3) \rightarrow (M_z = 0, \mu = -3)$. Even though the edge modes evolve in time, as does the survival probability, the Chern number remains fixed in the initial value, until the Majorana mode reaches the center point of the system and, therefore, the topology is maintained. Beyond this point, the Chern number starts to fluctuate, which indicates that the gap opens and closes due to the time evolution. In the thermodynamic limit, the renewal times tend to infinite and the Chern number does not change, even though the edge modes decay. However, the Chern number may change due to the finite size of the system. The values of the Chern number at a given time may be quite high.

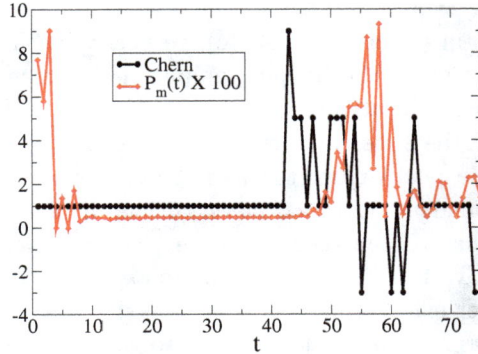

Fig. 9.3   Comparison of the time evolution of the survival probability and the Chern number for the case of a strong spin-orbit coupling for a sudden transition of parameters $(M_z = 2, \mu = -3) \rightarrow (M_z = 0, \mu = -3)$.

The fact that the Chern number remains invariant in the thermodynamic limit, if the evolution is unitary, may be understood, in a general way, as follows.

The Chern number may be obtained from

$$C = \frac{1}{2\pi} \sum_{n=1}^{N_F} \int d\mathbf{k} \, \Omega_n(\mathbf{k}), \qquad (9.31)$$

where the Berry curvature of band n may be written as

$$\Omega_n(\mathbf{k}) = -i\partial_{k_x} \langle u_n(\mathbf{k})|\partial_{k_y}|u_n(\mathbf{k})\rangle + i\partial_{k_y} \langle u_n(\mathbf{k})|\partial_{k_x}|u_n(\mathbf{k})\rangle, \qquad (9.32)$$

and the sum is taken over the occupied bands at zero temperature, until the Fermi level. If the system is infinite and has translational invariance, we may consider unitary transformations indexed by the momentum vector, that we write as $U(\mathbf{k})$. Under the action of this operator the states are changed as

$$|u_n(\mathbf{k})\rangle \rightarrow |u_n'(\mathbf{k})\rangle = U(\mathbf{k})|u_n(\mathbf{k})\rangle. \qquad (9.33)$$

Particularizing to the time evolution, the unitary operator $U(\mathbf{k})$ translates the time evolution of the states and, therefore, depends on time. The Berry curvature after the transformation is given by

$$\begin{aligned} \Omega_n'(\mathbf{k}) &= -i\partial_{k_x} \langle U(\mathbf{k})u_n(\mathbf{k})|\partial_{k_y}|U(\mathbf{k})u_n(\mathbf{k})\rangle \\ &+ i\partial_{k_y} \langle U(\mathbf{k})u_n(\mathbf{k})|\partial_{k_x}|U(\mathbf{k})u_n(\mathbf{k})\rangle \\ &= \Omega_n(\mathbf{k}) - i\partial_{k_x} \langle u_n(\mathbf{k})|U^\dagger \left(\partial_{k_y} U(\mathbf{k})\right)|u_n(\mathbf{k})\rangle \\ &+ i\partial_{k_y} \langle u_n(\mathbf{k})|U^\dagger(\mathbf{k}) \left(\partial_{k_x} U(\mathbf{k})\right)|u_n(\mathbf{k})\rangle. \end{aligned} \qquad (9.34)$$

The Chern number may, therefore, be written as

$$C' = C + \frac{1}{2\pi} \sum_{n=1}^{N_F} \int \int dk_x dk_y \ (-i)\partial_{k_x} \langle u_n(\mathbf{k})|U^\dagger \left(\partial_{k_y} U(\mathbf{k})\right) |u_n(\mathbf{k})\rangle$$

$$+ \frac{1}{2\pi} \sum_{n=1}^{N_F} \int \int dk_x dk_y \ i\partial_{k_y} \langle u_n(\mathbf{k})|U^\dagger(\mathbf{k}) \left(\partial_{k_x} U(\mathbf{k})\right) |u_n(\mathbf{k})\rangle . \quad (9.35)$$

If the states and the unitary transformation depend continuously on the momentum, with no sudden changes, the two integrals vanish due to the periodicity of the Brillouin zone. Therefore, the Chern number does not change and is conserved under the action of a unitary transformation.

Let us consider explicitly the derivative of the Chern number in order to time. This is given by

$$\partial_t C' = \frac{1}{2\pi} \sum_{n=1}^{N_F} \int d\mathbf{k} \left(\partial_{k_y} \langle u'_n(\mathbf{k})| \left(\partial_{k_x} H(\mathbf{k})\right) |u'_n(\mathbf{k})\rangle \right.$$

$$\left. - \partial_{k_x} \langle u'_n(\mathbf{k})| \left(\partial_{k_y} H(\mathbf{k})\right) |u'_n(\mathbf{k})\rangle \right) , \quad (9.36)$$

where we have used the relations

$$i\partial_t U = HU$$
$$\partial_t \left(U^\dagger \left(\partial_{k_i} U\right)\right) = -iU^\dagger \left(\partial_{k_i} H\right) U , \quad (9.37)$$

where H is the Hamiltonian and U the time evolution operator.

Consider, now, the example of a two band Hamiltonian, such as a spin-1/2 in a magnetic field $H(\mathbf{k}) = -1/2\mathbf{B}(\mathbf{k}) \cdot \boldsymbol{\sigma}$, where σ_i are the Pauli matrices. A state may be represented by the density matrix

$$\rho(\mathbf{k}) = 1/2 \left(\sigma_0 + \mathbf{S}(\mathbf{k}) \cdot \boldsymbol{\sigma}\right), \quad (9.38)$$

where $\mathbf{S}$ is the polarization. Recall that $\langle u_n|O|u_n\rangle = Tr\left(\rho_n O\right)$, where O is a given operator. Since the trace of a Pauli matrix vanishes and $\text{Tr}[\sigma_i \sigma_j] = 2\delta_{ij}$, we obtain

$$\partial_t C' = \frac{1}{4\pi} \int d\mathbf{k} \left(\partial_{k_x} \left(\mathbf{S}(\mathbf{k}) \cdot \partial_{k_y} \mathbf{B}(\mathbf{k})\right) - \partial_{k_y} \left(\mathbf{S}(\mathbf{k}) \cdot \partial_{k_x} \mathbf{B}(\mathbf{k})\right)\right) . \quad (9.39)$$

If the functions vary slowly with momentum, the integration over the Brillouin zone vanishes applying Stokes theorem, due to the periodic conditions and absence of a limiting surface, and the Chern number is invariant, as seen above.

9.2   Periodic perturbations: Floquet systems

A periodic perturbation may originate new topological states. It may also lead to a more generalized form of the correspondence principle between the bulk and boundary properties of a system. This correspondence exhibits a richer structure than that for equilibrium systems. These results can be shown to be valid for both topological and superconducting insulators.

The Schrödinger equation for a time dependent Hamiltonian,

$$i\hbar\partial_t\psi(t) = \hat{H}(t)\psi(t)\,, \tag{9.40}$$

admits the general solution

$$\psi(t) = \mathcal{T}e^{-\frac{i}{\hbar}\int_0^t \hat{H}(t')dt'}\,\psi(0), \tag{9.41}$$

where

$$
\mathcal{T}e^{-\frac{i}{\hbar}\int_0^t \hat{H}(t')dt'} \equiv 1 - \frac{i}{\hbar}\int_0^t dt_1 \hat{H}(t_1)
$$
$$
+ \left(-\frac{i}{\hbar}\right)^2 \int_0^t dt_1 \int_0^{t_1} dt_2 \hat{H}(t_1)\hat{H}(t_2) + \dots \tag{9.42}
$$

Let $\hat{H}(t) = \hat{H}(t+T)$ where the period $T = 2\pi/\omega$. Then:

$$i\hbar\partial_t\psi(t) = \hat{H}(t)\psi(t) \implies i\hbar\partial_t\psi(t+T) = \hat{H}(t)\psi(t+T)\,. \tag{9.43}$$

Any periodic function $\phi(t) = \phi(t+T)$ obviously satisfies Schrödinger equation. But, more generally, a phase can be included as $\psi(t) = e^{-i\epsilon t/\hbar}\phi(t)$, and

$$
i\hbar\partial_t\psi(t) = \hat{H}(t)\psi(t) \implies i\hbar\partial_t\phi(t) = \left(\hat{H}(t) - \epsilon\right)\phi(t)
$$
$$
\Rightarrow \left[\hat{H}(t) - i\hbar\partial_t\right]\phi(t) = \epsilon\phi(t)\,, \tag{9.44}
$$

where the *quasi-energy* ϵ may vary continuously. Equation (9.44) is known as *Floquet equation*. From this equation a set of periodic solutions, $\phi(t)$, is obtained.

Equation (9.44) is invariant under the gauge transformation:

$$\epsilon \to \epsilon + \hbar\omega\,,$$
$$\phi(t) \to e^{i\omega t}\phi(t)\,. \tag{9.45}$$

Functions $\phi(t)$ corresponding to different values of ϵ are orthogonal in space-time: the operator $\mathcal{H} \equiv \hat{H}(t) - i\hbar\partial_t$ acts in space-time domain and is time periodic. One then must define a dot product that includes a time integration over one period:

$$\langle\langle\phi_1|\phi_2\rangle\rangle \equiv \frac{1}{T}\int_0^T dt \int d\mathbf{r}\phi_1^*\phi_2\,. \tag{9.46}$$

The Floquet states, $\phi(t)$, form an orthonormal basis.

Considering now the evolution operator, the equality $U(T)\psi(t) = \psi(t+T)$ implies that

$$U(T)\phi(t) = \exp\left[-i\frac{\epsilon T}{\hbar}\right]\phi(t)\,, \tag{9.47}$$

and this leads us to define the effective Hamiltonian, H_{eff}, as $U(T) \equiv e^{-\frac{i}{\hbar}H_{eff}T}$. The spectrum of H_{eff} are the energies ϵ. Note the ambiguity in the definition of H_{eff}: the latter is defined up to a multiple of $\hbar\omega$. The evolution operator, $U(T)$, is commonly referred to as *Floquet operator*.

One may define an operator, $P(t)$, that "corrects" the evolution obtained from H_{eff} as follows:

$$U(t) \equiv P(t)e^{-\frac{i}{\hbar}H_{eff}t} \;\Rightarrow\; P(t) \equiv U(t)e^{\frac{i}{\hbar}H_{eff}t} \quad \text{is unitary.} \tag{9.48}$$

Independently of the definition of H_{eff}, one may apply $i\hbar\partial_t$ on both sides of the equation:

$$i\hbar\partial_t U(t) = i\hbar\left(\partial_t P(t)\right)e^{-\frac{i}{\hbar}H_{eff}t} + P(t)H_{eff}e^{-\frac{i}{\hbar}H_{eff}t} \tag{9.49}$$

and on the left-hand side, $i\hbar\partial_t U(t) = H(t)U(t) = H(t)P(t)e^{-\frac{i}{\hbar}H_{eff}t}$. Isolating H_{eff}, we obtain

$$H_{eff} = P^{-1}(t)H(t)P(t) - iP^{-1}(t)\partial_t P(t)\,. \tag{9.50}$$

This result has been obtained in general form, whatever the choice for H_{eff}. We now note that $P(t)$ in (9.50) is time periodic. To prove this,

$$\psi(t) = U(t)\psi_0$$
$$\Leftrightarrow\; e^{-\frac{i\epsilon}{\hbar}t}\phi(t) = P(t)e^{-\frac{i}{\hbar}H_{eff}t}\phi_0 \quad \text{because } \psi_0 = \phi_0$$
$$\Leftrightarrow\; \phi(t) = P(t)\phi_0 \quad \text{which is periodic.} \tag{9.51}$$

Equation (9.50) shows that one can map the Hamiltonian $H(t)$ onto another constant Hamiltonian, H_{eff}, by using a time-periodic unitary operator, $P(t)$. It is often said that a change to a rotating frame has been performed.

Equivalently, one can use the Fourier series,

$$\phi(t) = \sum_n e^{-in\omega t}\phi_n\,, \qquad \hat{H}(t) = \sum_n \hat{h}_n e^{in\omega t}\,, \tag{9.52}$$

where ϕ_n and $\hat{h}_n$ do not depend on time. Inserting in Schrödinger equation (9.44), we obtain

$$(\epsilon + n\hbar\omega)\,\phi_n = \sum_{m \in \mathbb{Z}} \hat{h}_{m-n}\phi_m \,. \tag{9.53}$$

The gauge invariance (9.45) here takes the form:

$$\epsilon \to \epsilon + \hbar\omega \,,$$
$$\phi_n \to \phi_{n+1} \,\forall_n \,. \tag{9.54}$$

Another invariance property of (9.53) is that under an arbitrary time translation, $t \to t + \theta$, we have:

$$\phi_n \to \phi_n e^{-in\omega\theta} \,,$$
$$\hat{h}_n \to \hat{h}_n e^{in\omega\theta} \,\forall_n \,,$$
$$\epsilon \to \epsilon \,. \tag{9.55}$$

It often happens that the Fourier components, $\hat{h}_1 = \hat{h}_{-1}^\dagger \neq 0$, so,

$$\hat{H}(t) = \hat{h}_0 + \hat{h}_1 e^{i\omega t} + \hat{h}_1^\dagger e^{-i\omega t} \tag{9.56}$$
$$\implies (\epsilon + n\hbar\omega)\,\phi_n = \hat{h}_0\phi_n + \hat{h}_1\phi_{n+1} + \hat{h}_1^\dagger\phi_{n-1} \,. \tag{9.57}$$

The effective Hamiltonian takes the form of a tridiagonal block matrix. A wave function localized on a certain block h_0 (from $n = 0$) describes the same quantum state as another wave function localized at a block $n \neq 0$. They only differ through a gauge transformation.

When dealing with translationally invariant systems, we work with the Hamiltonian in momentum space, $H(\mathbf{k}, t)$, and Floquet states, $\Phi(\mathbf{k}, t)$.

Because the functions are periodic, they can be expanded as

$$\Phi(\mathbf{k}, t) = \sum_n \phi_n(\mathbf{k}) e^{in\omega t} \,. \tag{9.58}$$

Equation (9.53) takes the form

$$\sum_{n'} H_{nn'}(\mathbf{k})\phi_{n'}(\mathbf{k}) = \epsilon(\mathbf{k})\phi_n(\mathbf{k}) \,, \tag{9.59}$$

where the Hamiltonian matrix is given by

$$H_{nn'}(\mathbf{k}) = \delta_{nn'}n\omega + \frac{1}{T}\int_0^T dt\, e^{-in\omega t} H(\mathbf{k}, t) e^{in'\omega t} \,. \tag{9.60}$$

If the Hamiltonian has two terms, one of them time independent, $H(\mathbf{k})$, and the other describing a periodic perturbation with frequency ω,

$$H(\mathbf{k}, t) = H(\mathbf{k}) + f(\omega t)H_d(\mathbf{k}) \,, \tag{9.61}$$

with $f(\omega t) = \cos(\omega t)$, the contribution of H_d to the second term of the Hamiltonian matrix (9.60) reduces to $1/2\,(\delta_{n'+1,n} + \delta_{n'-1,n})$. This is equivalent to writing $\hat{h}_1 = \frac{1}{2}H_d(\mathbf{k}) = \hat{h}_1^\dagger$ in equation (9.56). The system (9.57) for the quasi-energies, $\epsilon(\mathbf{k})$, and functions $\phi_n(\mathbf{k})$ may be written in the form of a tridiagonal matrix,

$$
\begin{pmatrix}
\cdots & \cdots & & \cdots & & \cdots & & \cdots \\
\cdots & (n-1)\hbar\omega + H(\mathbf{k}) & \frac{1}{2}H_d(\mathbf{k}) & & 0 & & \cdots \\
\cdots & \frac{1}{2}H_d(\mathbf{k}) & n\hbar\omega + H(\mathbf{k}) & & \frac{1}{2}H_d(\mathbf{k}) & & \cdots \\
\cdots & 0 & & \frac{1}{2}H_d(\mathbf{k}) & (n+1)\hbar\omega + H(\mathbf{k}) & \cdots \\
\cdots & \cdots & & \cdots & & \cdots & & \cdots
\end{pmatrix} . \tag{9.62}
$$

This infinite matrix may be truncated for sufficiently high frequency. Only a few blocks close to $n = 0$ are necessary to obtain an approximate solution.

Because the quasi-energies are defined up to a multiple of $\omega = 2\pi/T$, one may restrict them to the interval $-\hbar\omega/2 \leq \epsilon \leq \hbar\omega/2$. The latter is called the *first Floquet zone*. States with quasi-energies $\epsilon = \hbar\omega/2$ and $\epsilon = -\hbar\omega/2$ are physically the same and, as such, are related through a gauge transformation.

9.2.1 *Dirac cone under circularly polarized radiation*

The vector potential for circularly polarized radiation may be written as

$$\mathbf{A} = A\left(\cos\left(\omega t\right), \sin\left(\omega t\right)\right).$$

In the absence of radiation field, the Hamiltonian is $H_0 = \hbar v\,(k_x\tau_1 + k_y\tau_2)$. Minimal coupling leads us to the time-periodic Hamiltonian:

$$
\hat{H}(t) = \hbar v\left[\left(k_x - \frac{e}{\hbar}A_x\right)\tau_1 + \left(k_y - \frac{e}{\hbar}A_y\right)\tau_2\right], \tag{9.63}
$$

and to Floquet equation:

$$
\left[\hbar v\begin{pmatrix} 0 & k_- - \frac{e}{\hbar}Ae^{-i\omega t} \\ k_+ - \frac{e}{\hbar}Ae^{i\omega t} & 0 \end{pmatrix} - i\hbar\partial_t\right]\phi = \epsilon\phi, \tag{9.64}
$$

where $k_\pm = k_x \pm ik_y$. At the point $\mathbf{k} = 0$, it admits a solution of the form:

$$
\phi = \begin{pmatrix} \alpha \\ \beta e^{i\omega t} \end{pmatrix}, \qquad \epsilon = \frac{\hbar\omega}{2} \pm \sqrt{\left(\frac{\hbar\omega}{2}\right)^2 + (veA)^2},
$$

$$
\alpha^2 = 1 - \beta^2 = \frac{(veA)^2}{\epsilon^2 + (veA)^2}. \tag{9.65}
$$

This means that the radiation opened up a gap in the Dirac cone spectrum (in the Floquet spectrum).

For finite **k** we may use the decomposition (9.52)-(9.57):

$$\hat{h}_0 + n\hbar\omega = \hbar \begin{pmatrix} n\Omega & vk_- \\ vk_+ & n\Omega \end{pmatrix}, \quad \hat{h}_1 = \begin{pmatrix} 0 & 0 \\ -veA & 0 \end{pmatrix}, \quad \hat{h}_{-1} = \hat{h}_1^\dagger, \qquad (9.66)$$

$$\left(\hat{h}_0 + n\hbar\Omega\right)\phi_n + \hat{h}_1\phi_{n-1} + \hat{h}_1^\dagger\phi_{n+1} = \epsilon\phi_n, \forall n \in \mathbb{Z}, \qquad (9.67)$$

which can be numerically solved after a suitable truncation for sufficiently large $|n|$ has been made.

9.2.2 *Magnus expansion*

We now solve equation (9.57) perturbatively. If the frequency is high, one may consider $\hat{h}_1/(\hbar\omega)$ a small perturbation and find the solution as a series expansion in $1/(\hbar\omega)$. For each **k** in the Brillouin zone, the matrices $\hat{h}_n$ are fixed. They have dimension 2, at least (like the Pauli matrices). We shall consider $\hbar\omega \to \infty$. The matrices $\hat{h}_n$ are of order unity.

We first write the system of equations that follow from (9.57) explicitly:

$$\epsilon\phi_0 = \hat{h}_0\phi_0 + \hat{h}_1\phi_1 + \hat{h}_1^\dagger\phi_{-1}, \qquad (9.68)$$

$$(\epsilon + \hbar\omega)\phi_1 = \hat{h}_0\phi_1 + \hat{h}_1\phi_2 + \hat{h}_1^\dagger\phi_0, \qquad (9.69)$$

$$(\epsilon - \hbar\omega)\phi_{-1} = \hat{h}_0\phi_{-1} + \hat{h}_1\phi_0 + \hat{h}_1^\dagger\phi_{-2}. \qquad (9.70)$$

We consider ϵ and ϕ_0 to be of order unity, and

$$\phi_{\pm 1} \propto \frac{1}{\hbar\omega}, \qquad \phi_{\pm 2} \propto \frac{1}{(\hbar\omega)^2}.$$

Considering the terms $\mathcal{O}(1)$ in the system of equations above, we obtain:

$$\epsilon_0\phi_0 = \hat{h}_0\phi_0 \qquad \hbar\omega\phi_1 = \hat{h}_1^\dagger\phi_0, \qquad -\hbar\omega\phi_{-1} = \hat{h}_1\phi_0. \qquad (9.71)$$

Hence we have

$$\phi_1 = \frac{1}{\hbar\omega}\hat{h}_1^\dagger\phi_0, \qquad \phi_{-1} = -\frac{1}{\hbar\omega}\hat{h}_1\phi_0. \qquad (9.72)$$

Inserting the expressions for $\phi_{\pm 1}$ in (9.68) we obtain, up to order $\frac{1}{\hbar\omega}$,

$$\epsilon\phi_0 = \left\{\hat{h}_0 + \frac{1}{\hbar\omega}\left[\hat{h}_1, \hat{h}_1^\dagger\right]\right\}\phi_0. \qquad (9.73)$$

The periodic Floquet function, $\phi(t)$, can be obtained from expression (9.52). The solution to Schrödinger equation (9.40) is $\psi(t) = e^{-i\epsilon t/\hbar}\phi(t)$ which, up to order $1/\omega$, can be written as:

$$\phi(t) = \left[1 + \frac{1}{\hbar\omega}\left(e^{-i\omega t}\hat{h}_1^\dagger - e^{i\omega t}\hat{h}_1\right)\right]\phi_0, \qquad (9.74)$$

$$\psi(t) = e^{-i\epsilon t/\hbar}\left[1 + \frac{1}{\hbar\omega}\left(e^{-i\omega t}\hat{h}_1^\dagger - e^{i\omega t}\hat{h}_1\right)\right]\phi_0. \qquad (9.75)$$

One can rewrite (9.73) using $\phi(0) = \psi(0)$ instead of ϕ_0. Indeed, we see from (9.74) that

$$\phi(0) = \left[1 + \frac{1}{\hbar\omega}\left(\hat{h}_1^\dagger - \hat{h}_1\right)\right]\phi_0\,, \tag{9.76}$$

Inverting (9.76) to order $1/\omega \to 0$, we get

$$\phi_0 = \left[1 - \frac{1}{\hbar\omega}\left(\hat{h}_1^\dagger - \hat{h}_1\right)\right]\phi(0)\,, \tag{9.77}$$

and we can rewrite (9.73) as

$$\epsilon\phi(0) = \left\{\hat{h}_0 + \frac{1}{\hbar\omega}\left[\hat{h}_1^\dagger - \hat{h}_1, \hat{h}_0\right] + \frac{1}{\hbar\omega}\left[\hat{h}_1, \hat{h}_1^\dagger\right]\right\}\phi(0)\,, \tag{9.78}$$

where we must bear in mind that ϵ and $\phi(0)$ are correct only to order $1/\omega \to 0$. The Hamiltonian in equation (9.78) differs from the Hamiltonian in equation (9.73) only by a rotation that changes from the basis ϕ_0 to $\phi(0)$.

The expansion of the effective Hamiltonian as a power series in $1/\omega \to 0$ is called *Magnus expansion*. The terms to first order can be seen in equation (9.78).

There is a systematic way to obtain the Magnus expansion to arbitrary order. One may expand the series (9.42) for the Floquet evolution operator:

$$U(T) = 1 - \frac{i}{\hbar}\int_0^T dt_1 \hat{H}(t_1) + \left(-\frac{i}{\hbar}\right)^2 \int_0^T dt_1 \int_0^{t_1} dt_2 \hat{H}(t_1)\hat{H}(t_2) + \dots$$
$$= e^{-\frac{i}{\hbar}\hat{H}_{eff}T}\,, \tag{9.79}$$

where the period $T = 2\pi/\omega \to 0$ and where $\hat{H}_{eff}$ is also to be expanded as a power series in $1/\omega$. Using (9.56) for $\hat{H}(t)$ and comparing equal powers of $1/\omega$, we recover the result (9.78):

$$\hat{H}_{eff} = \hat{h}_0 + \frac{1}{\hbar\omega}\left[\hat{h}_1^\dagger - \hat{h}_1, \hat{h}_0\right] + \frac{1}{\hbar\omega}\left[\hat{h}_1, \hat{h}_1^\dagger\right] + \mathcal{O}\left(\frac{1}{(\hbar\omega)^2}\right)\,. \tag{9.80}$$

9.2.3 Invariants: frequency space formulation

It is possible to construct different types of topological invariants for Floquet systems. One possibility is to consider a formulation in frequency space. Another involves a description in terms of the time evolution. We consider first the description in terms of the frequencies (or quasi-energies).

A given eigenstate may be written as

$$|\psi(\mathbf{k}, t)\rangle = \sum_\alpha \psi_{n\alpha}(\mathbf{k}, t)c_{\mathbf{k}\alpha}^\dagger|0\rangle\,, \tag{9.81}$$

where

$$\psi_{n\alpha}(\mathbf{k}, t) = e^{-i\epsilon_n(\mathbf{k})t} \sum_{m=-\infty}^{\infty} \phi_{n\alpha}^{(m)}(\mathbf{k})e^{im\omega t} . \tag{9.82}$$

The choice of a time dependent Hamiltonian, of the form $H(t) = H_0 + \Delta e^{i\omega t} + \Delta^\dagger e^{-i\omega t}$, leads to a tridiagonal matrix that may be solved truncating its infinite dimension, as discussed above. This process leads to a set of bands of the time independent Hamiltonian, H_0, that are mixed due to the time dependent perturbation. States of the original Hamiltonian are mixed if their energy differences coincide with the perturbation frequency, ω. In general, the effect of the perturbation decays when m increases. Therefore, truncating the matrix is, in general, possible, giving results that converge rapidly as one increases the matrix dimension. To illustrate the result of this procedure, we may consider only the values $m = 0, -1, 1$ and a non-perturbed Hamiltonian with two bands. If the perturbation is absent ($\Delta = 0$), the procedure leads to two extra copies of the original Hamiltonian and a total of six bands, displaced up and down by the frequency ω. Assuming that this is larger than the bandwidth, Λ, these are not superposed and we obtain two gaps. Turning on the coupling between the bands, $\Delta \neq 0$, the lower and upper bands are not very affected by the perturbation (at least if this is weak). This is the result that the amplitude $\phi_{n\alpha}^{(m)}(\mathbf{k})$ decays fast for $|m| >> \Lambda/\omega$, justifying the approximation of only considering small values of $|m|$. The Magnus expansion in section 9.2.2 illustrates this point. Considering that the original bands are characterized by a given Chern number, C_0, these upper and lower bands retain their original Chern numbers. On the other hand, for small quasi-energies, in the first Floquet zones around the zero quasi-energy, the Chern numbers may change to different values, C_F. The Chern number of the Floquet bands may be calculated in the same way as for time independent bands, calculating the Berry curvature of the Floquet eigenstates and integrating over the Brillouin zone, if the system is invariant to spatial translations.

In the case of time independent systems, the topological invariant is defined by the sum of the Chern numbers of the occupied bands (if they are not degenerate). In the time dependent problem, one assumes that there are gaps in the spectrum and we may, then, calculate the topological invariant associated with a given quasi-energy localized in a spectrum gap. That is, knowing the Chern numbers of the various bands, one can sum their values for energies below the chemical potential, localized in a gap.

This is not always clear, since if the perturbation frequency, ω, is small or the coupling is strong, the original bands get mixed and, in general, there are no clear gaps in the quasi-energy spectrum. However, let us suppose they are present.

In the time independent problem, we usually assume a finite number of bands and, as a consequence, a finite number of bands below the Fermi level. In the Floquet band structure there are, in general, infinite bands, both above and below the vanishing quasi-energy (since $-\infty < m < \infty$). As we truncate the matrix (9.57) or (9.62), we obtain a finite number of bands. Note, however, that there is a periodicity in frequency space. This means that, even though the matrix truncation gives us a finite number of bands, the band of smaller energy may be linked to the higher energy band if they touch the limits of the Floquet zone. Even if they do not touch, and considering an open system along one or more spatial directions, there may be edge states that connect the lower and upper bands due to the periodicity in quasi-energy space. This point raises some interesting questions on the correspondence between the bulk and the edge states.

Consider, for now, a system with periodic boundary conditions and no edge states. We may, then, define the Chern number of a given quasi-energy in a way similar to the static case: this is given by the sum of the Chern numbers of all bands with smaller quasi-energy. Usually we only define the Chern number of quasi-energies located in the first Floquet zone.

As seen before, the Chern number of a band is equal to the difference between the number of chiral modes that leave the upper part of the band, and the number of those that enter in the lower part. In the case of a static system, the spectrum is limited and there are no chiral modes that extend beyond the top of the upper band or below the lower band. Therefore, in a two band static system, the Chern number defines in a unique way the net number of chiral modes that cross the gap between the two bands. The sum of the Chern numbers of all bands vanishes.

In a Floquet system the periodicity in quasi-energy space allows the existence of edge states, even if the bands do not have finite Chern numbers. Considering the bands in the first Floquet zone, it is possible that edge states exist that enter through the upper band by smaller quasi-energy values, and others that leave from the upper part of the band, if these modes re-enter the first Floquet zone from the lower band. It is possible a situation in which all Floquet bands have zero Chern number and there is a chiral mode that is present in all gaps.

9.2.4 *Quasi-energy bands and creation of π modes*

We may apply a periodic perturbation on a topological insulator shining it with light. This leads to a perturbation in the way $\Delta e^{i\omega t} + \Delta^* e^{-i\omega t}$. The total perturbed Hamiltonian is, then, the non-perturbed Hamiltonian, H_0, plus the perturbation.

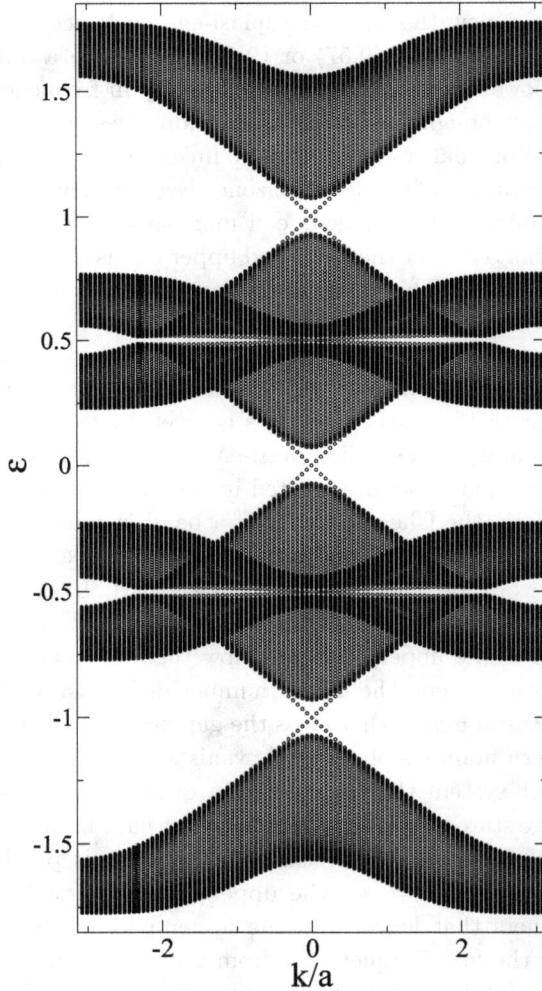

Fig. 9.4   Quasi-energy spectrum, in units of the frequency ω, of a topological insulator under the action of a periodic perturbation.

As first example, consider the Hamiltonian of a two-dimensional system in the form

$$H_0 = \mathbf{d}(\mathbf{k}) \cdot \boldsymbol{\sigma} , \qquad (9.83)$$

where $\boldsymbol{\sigma}$ are Pauli matrices. Taking

$$d_x(\mathbf{k}) = a \sin k_x ,$$
$$d_y(\mathbf{k}) = a \sin k_y ,$$
$$d_z(\mathbf{k}) = (\mu - J) - 2b(2 - \cos k_x - \cos k_y) + J \cos k_x \cos k_y , \quad (9.84)$$

the Hamiltonian has topological properties with zero energy modes (if we use open boundary conditions). The bands have Chern numbers $C_0 = \pm 1$.

Let us now add the perturbation and, to simplify, we may truncate the infinite series of equation (9.62) to the first three bands. If we choose the parameters $\mu = 1, J = b = 1.5, a = 4, \Delta = 1$ and the perturbation frequency $\omega = \Delta/0.07$, the perturbed system, beyond the zero energy modes, has now quasi-energies with additional states at finite energy $\pm\omega/2$. This result may be seen in figure 9.4. With the perturbation bands appear inside the Floquet zone that have vanishing Chern numbers, $C_F = 0$. The upper and lower bands are insensitive to the perturbation and retain their non-vanishing Chern numbers $C_0 = \pm 1$. Therefore, even though the Floquet bands have vanishing Chern numbers, edge states appear that connect the bands. If now we admit that the matrix (9.62) is infinite, the Floquet spectrum is limited to the interval $|\varepsilon| \leq \omega/2$ in figure 9.4 and is basically the same as the one represented in the figure, because truncating to three bands is already a good approximation in that interval.

In the case of a topological superconductor, the time dependent perturbation, $H_d(\mathbf{k})$, may be chosen turning a given term of the static Hamiltonian into a time dependent one, for example, the hopping amplitude, the chemical potential, the spin-orbit coupling, the pairing or the magnetization. The first four terms preserve the time reversal invariance (excluding non-unitary pairings or of type $p + ip$), while the magnetization breaks this invariance (if initially the magnetization is vanishing). Consider, as examples, a perturbation in the chemical potential or in the magnetization, since are easily achieved experimentally, applying a potential or a magnetic field, respectively. In this last case, it may be shown that, even though there are very low energy modes, these are not necessarily Majorana fermions, since the eigenvalues of the Floquet operator (i.e., the time evolution operator after a period in equation (9.47)) are not strictly ± 1, and, therefore, the quasi-energy is not vanishing. In the case the perturbation is associated to the chemical potential, zero energy modes are found, that are Majorana fermions. Due to the particle-hole symmetry of a superconductor, $\gamma_{-\epsilon} = \gamma_\epsilon^\dagger$ and to the equivalence between the energies $\epsilon = -\omega/2, \omega/2$, one expects a new type of Majorana modes (of finite quasi-energy) in addition to the zero energy usual modes. Since for $\epsilon = \omega/2$ we have $\epsilon T = \pi$, these modes of finite energy are, sometimes, called π modes. These finite energy modes occur in any Floquet system, including the topological insulators considered before, but in general are not Majorana modes but usual fermionic modes.

9.2.5 *Invariants: time formulation*

We may define a topological invariant in time space that has the additional advantage to predict the number of edge states in a two-dimensional system perturbed periodically in time.

We may write the spectral decomposition of the evolution operator,

$$U(\mathbf{k}, t) = \sum_{n=1}^{N} P_n(\mathbf{k}, t) e^{-i\Phi_n(\mathbf{k}, t)}, \qquad (9.85)$$

where n labels the eigenvalue, the eigenvalues are $e^{-i\Phi_n(\mathbf{k},t)}$, $P_n(\mathbf{k}, t) = |\chi_n\rangle\langle\chi_n|$ are the projectors on the eigenstates. The bands of phases $\Phi_n(\mathbf{k}, t)$ satisfy

$$\Phi_n(\mathbf{k}, T) = \epsilon_n(\mathbf{k})T, \qquad (9.86)$$

where ϵ_n are the quasi-energies associated with the eigenvalues of the Floquet operator. Also

$$\sum_n \epsilon_n = \frac{1}{T} \int_0^T dt \, \mathrm{Tr}(H). \qquad (9.87)$$

This result may be obtained using

$$\sum_n \epsilon_n(\mathbf{k}) = \frac{1}{T} \sum_n \Phi_n(\mathbf{k}, T)$$

$$= \frac{1}{T} \int_0^T dt \mathrm{Tr}\left(U^\dagger(\mathbf{k}, t) i \frac{\partial}{\partial t} U(\mathbf{k}, t) \right), \qquad (9.88)$$

and $i\partial U(\mathbf{k}, t)/\partial t = H(\mathbf{k}, t)U(\mathbf{k}, t)$. The equality between the first and the second line may be obtained in the following way. Using equation (9.85), and further that P_n satisfies $P_n^2 = P_n$, and $\sum_n P_n = 1$, we obtain

$$\frac{1}{T} \int_0^T dt \mathrm{Tr}\left(U^\dagger(\mathbf{k}, t) i \frac{\partial}{\partial t} U(\mathbf{k}, t) \right) = \frac{1}{T} \int_0^T dt \sum_n \mathrm{Tr}\left(P_n \frac{\partial \Phi_n}{\partial t} \right)$$

$$= \frac{1}{T} \sum_n \Phi_n(T), \qquad (9.89)$$

making the choice $\Phi_n(t = 0) = 0$.

Consider, now, a trivial Floquet operator such that $U(\mathbf{k}, T) = 1$ for all momenta. In a two-dimensional system $\mathbf{k} = (k_x, k_y)$. The three coordinates k_x, k_y, t are all defined on a circle: the momentum variables are defined in the first Brillouin zone with periodic boundary conditions and the time variable, due to the periodicity, is also defined on a circle of length T. Therefore, we have an application $S^1 \times S^1 \times S^1 \rightarrow U(N)$. As is known from topology, there is a topological invariant associated with this application, namely, there is a winding number defined by

$$W[U] = \frac{1}{8\pi^2} \int dt dk_x dk_y \operatorname{Tr} \left(U^{-1} \partial_t U \left[U^{-1} \partial_{k_x} U, U^{-1} \partial_{k_y} U \right] \right). \quad (9.90)$$

If the Floquet operator is the identity, this implies that the quasi-energies of the Floquet spectrum are zero and, therefore, in the Floquet zone there is a gap to all states in $\epsilon = 0$. The time evolution operator interpolates between the identity for $t = 0$ and the identity for $t = T$. If in the process the gap remains open, and there are edge states, these should remain during the evolution. We expect that this property remains invariant during the evolution, which implies that the information on the edge states must be, somehow, contained in the time evolution operator, $U(\mathbf{k}, t)$. Indeed, it may be shown that the winding number $W[U]$ gives the number of edge states across the gap. The case of a general Floquet operator $U(\mathbf{k}, T) \neq 1$, may also be considered.

9.2.6 *Berry-Floquet phase*

Let us assume that the periodic Hamiltonian, $H(t)$, with frequency ω, has an additional slow variation in time. By "slow" we mean in comparison to the period, $2\pi/\omega$. We then consider $H[t, \mathbf{R}(t)]$, where the parameters $\mathbf{R}(t)$ vary slowly. The wave function

$$\psi(t) = e^{i\gamma} e^{-\frac{i}{\hbar} \int^t \epsilon(t')dt'} \phi[\mathbf{R}(t)], \quad (9.91)$$

where $\phi[\mathbf{R}(t)]$ is a Floquet state (adiabatic) that remains valid for a time interval of several periods, while the parameters $\mathbf{R}(t)$ do not change appreciably:

$$(H[t, \mathbf{R}(t)] - i\hbar \partial_t) \phi[\mathbf{R}(t)] = \epsilon(t)\phi[\mathbf{R}(t)]. \quad (9.92)$$

The Schrödinger equation reads

$$i\hbar \frac{\partial}{\partial t} \psi(t) = e^{-\frac{i}{\hbar} \int^t \epsilon(t')dt'} e^{i\gamma} \left(\epsilon(t)\phi(t) - \hbar\dot{\gamma}\phi(t) + i\hbar\dot{\mathbf{R}} \cdot \frac{\partial}{\partial \mathbf{R}} \phi[\mathbf{R}(t)] \right)$$

$$= H[t, \mathbf{R}(t)] e^{-\frac{i}{\hbar} \int^t \epsilon(t')dt'} e^{i\gamma} \phi[\mathbf{R}(t)] \quad (9.93)$$

$$\Leftrightarrow \dot{\gamma}\phi = i\dot{\mathbf{R}} \cdot \frac{\partial}{\partial \mathbf{R}} \phi. \quad (9.94)$$

We now left-multiply both sides of (9.94) by $\langle\langle\phi[\mathbf{R}(t)]|$,

$$\dot{\gamma} = i\langle\langle\phi[\mathbf{R}(t)]| \frac{\partial}{\partial\mathbf{R}} |\phi[\mathbf{R}(t)]\rangle\rangle \cdot \dot{\mathbf{R}}. \tag{9.95}$$

We here used the space-time multiplication as defined in (9.46). The Berry phase accumulated in a closed cycle is

$$\gamma = i\oint\langle\langle\phi(\mathbf{R})| \frac{\partial}{\partial\mathbf{R}} |\phi(\mathbf{R})\rangle\rangle \cdot \delta\mathbf{R}. \tag{9.96}$$

By making an analogy to the static case studied in section 2.8, one may also define a Hall conductivity for Floquet states.

9.3 Instantaneous periodic pulses

A simple way to change the properties of a system consists in introducing a perturbation that acts over short time periods (short pulses). The Hamiltonian contains a time independent term, H_0, and at certain times, that may be regular or random, the system is under the influence of a perturbation, H_1, that only lasts a short time. To simplify, we may take it as instantaneous. If the pulses are time periodic, the non-perturbed Hamiltonian acts during a time interval, T, then the perturbation acts, and the process is repeated indefinitely. Also we may study a non-periodic situation, taking a sequence of intervals $T_1, T_2, T_3, \cdots$ that may be random in some specified way. Here we will consider exclusively the periodic case.

As before, we define a Floquet time evolution operator that characterizes the evolution over one period. Considering that the term H_1 is of the form of a sum of Dirac delta functions in instants separated by T, we may write

$$U(T) = U(t = T, 0) = e^{-iH_0T}e^{-iH_1}. \tag{9.97}$$

Here H_0 and H_1 are the matrices for terms of the total Hamiltonian, that is supposed non-interacting to simplify. Note that the pulses do not need to be instantaneous. A similar expression is obtained if the perturbation acts during a finite time, T_p, as long as during that interval the perturbation is time independent. In that case, the total period would be $T + T_p$ and the Floquet operator could be written as

$$U(T + T_p) = U(t = T + T_p, 0) = e^{-iH_0T}e^{-iH_1T_p}. \tag{9.98}$$

That would simply lead to a renormalization of the perturbation parameters. We will just consider instantaneous pulses. Instead of the pulses period, we may consider their frequency defined by $\omega = 2\pi/T$.

9.3.1 Eigenvalues of the Floquet operator

The eigenvalues of the Floquet operator may be obtained diagonalizing H_0 and H_1. We introduce two unitary matrices that diagonalize H_0 and H_1,

$$R_0^\dagger H_0 R_0 = (H_0)_d \,,$$
$$R_1^\dagger H_1 R_1 = (H_1)_d \,, \qquad (9.99)$$

which allows us to write

$$U(T) = R_0 R_0^\dagger e^{-iH_0 T} R_0 R_0^\dagger R_1 R_1^\dagger e^{-iH_1} R_1 R_1^\dagger$$
$$= R_0 \left(e^{-i\lambda_0 T}\right) R_0^\dagger R_1 \left(e^{-i\lambda_1}\right) R_1^\dagger \,, \qquad (9.100)$$

where λ_0 and λ_1 are the eigenvalues of H_0 and H_1, respectively. We may relate the Floquet operator with the effective Hamiltonian

$$U(T) = e^{-iH_{eff}T} \,, \qquad (9.101)$$

and $U(T)$ had eigenvalues

$$\lambda_i = e^{-i\epsilon_i T} \,, \qquad (9.102)$$

where ϵ_i are the quasi-energies.

9.3.2 Effective Hamiltonian

In general, the Floquet operator for a problem where the applied perturbation at equal times has amplitude λV, is defined by

$$U(T) = e^{-iH_0 T} e^{-i\lambda V} = e^{-iH_{eff}T} \,. \qquad (9.103)$$

Using the Baker-Campbell-Hausdorff formula,

$$e^A e^B = e^{A+B+\frac{1}{2}[A,B]+\frac{1}{12}[A,[A,B]]+\frac{1}{12}[[A,B],B]+\cdots} \,, \qquad (9.104)$$

we may conclude that the effective Hamiltonian is given by

$$H_{eff} = H_0 + \frac{\lambda}{T}V - i\frac{\lambda}{2}[H_0,V] - \frac{T\lambda}{12}[H_0,[H_0,V]] - \frac{\lambda^2}{12}[[H_0,V],V] + \cdots \qquad (9.105)$$

In the limit when the period is very small (or the frequency $\omega = 2\pi/T$ is large) and additionally the intensity of the perturbation is small $\lambda << 1$, the effective Hamiltonian is simply given by

$$H_{eff} \sim H_0 + \frac{\lambda}{T}V \,, \qquad (9.106)$$

and the problem has a description equivalent to a static problem with an effective Hamiltonian with a coupling constant renormalized by the frequency of the periodic pulses.

We may, alternatively, consider a time symmetric protocol, *i.e.*, $\hat{H}(t) = \hat{H}(T - t)$. Then the evolution operator may be written as

$$\hat{U}(T) = e^{i\lambda\hat{V}/2} e^{-i\hat{H}_0 T} e^{i\lambda\hat{V}/2} . \tag{9.107}$$

We may take as example the Kitaev chain and illustrate the change of topology as a result of the periodic perturbation. To write the Floquet Hamiltonian we use the identity

$$\exp\hat{Y}\exp\hat{X}\exp\hat{Y} = \exp\{\hat{X} + 2\hat{Y} - \frac{1}{6}[[\hat{X},\hat{Y}],\hat{Y}] + \frac{1}{6}[\hat{X},[\hat{X},\hat{Y}]] + \cdots\} \tag{9.108}$$

and we obtain the effective Hamiltonian to second order,

$$\hat{H}_{\text{eff}} = \hat{H}_0 + \frac{\lambda}{T}\hat{V} - \frac{T\lambda}{12}[\hat{H}_0,[\hat{H}_0,\hat{V}]] +$$
$$+ \frac{\lambda^2}{24}[[\hat{H}_0,\hat{V}],\hat{V}] + \cdots . \tag{9.109}$$

Using (i) usual commutation relations $\{\hat{c}_i, \hat{c}_j\} = 0$ e $\{\hat{c}_i, \hat{c}_j^\dagger\} = \delta_{ij}$, where $\{\hat{A}, \hat{B}\} = \hat{A}\hat{B} + \hat{B}\hat{A}$ is an anti-commutator and δ_{ij} a Kronecker delta, and (ii) explicit forms for the static and pulse terms, $\hat{H}_0$ in the form of the Kitaev Hamiltonian (with parameters in general non-spatially uniform) and $\hat{H}_1$ of the form of a chemical potential (in general non-uniform, that is, of the form $\hat{H}_1 = \sum_j \mu_j n_j$, where n_j is the local density), we need to calculate four commutators:

$$[\hat{c}_a^\dagger \hat{c}_b, \hat{c}_d^\dagger \hat{c}_e] = \delta_{bd}\,\hat{c}_a^\dagger\hat{c}_e - \delta_{ae}\,\hat{c}_d^\dagger\hat{c}_b,$$
$$[\hat{c}_a^\dagger \hat{c}_b^\dagger, \hat{c}_d^\dagger \hat{c}_e] = -\delta_{be}\,\hat{c}_a^\dagger\hat{c}_d^\dagger - \delta_{ae}\,\hat{c}_d^\dagger\hat{c}_b^\dagger,$$
$$[\hat{c}_a \hat{c}_b, \hat{c}_d^\dagger \hat{c}_e] = \delta_{bd}\,\hat{c}_a\hat{c}_e + \delta_{ad}\,\hat{c}_e\hat{c}_b,$$
$$[\hat{c}_a^\dagger \hat{c}_b^\dagger, \hat{c}_d^\dagger \hat{c}_e] = \delta_{bd}\,\hat{c}_a^\dagger\hat{c}_e - \delta_{be}\left(\hat{c}_a^\dagger\hat{c}_d - \delta_{ad}\right) -$$
$$- \delta_{ad}\,\hat{c}_b^\dagger\hat{c}_e + \delta_{ae}\left(\hat{c}_b^\dagger\hat{c}_d - \delta_{bd}\right). \tag{9.110}$$

The second order effective Hamiltonian is, therefore, obtained

$$\hat{H}_{\text{eff}} = \sum_{i=1}^{L}\left[\left(\tilde{J}_i\,\hat{c}_i^\dagger\hat{c}_{i+1} + \tilde{J}_i'\,\hat{c}_i^\dagger\hat{c}_{i+2} + \tilde{\Delta}_i\,\hat{c}_{i+1}^\dagger\hat{c}_i^\dagger + \right.\right.$$
$$\left.\left. + \tilde{\Delta}_i'\,\hat{c}_{i+2}^\dagger\hat{c}_i^\dagger + \text{H.c.}\right) - \tilde{\mu}_i\,\hat{c}_i^\dagger\hat{c}_i\right], \tag{9.111}$$

where $\tilde{J}_i$ $(\tilde{J}_i')$, $\tilde{\Delta}_i$ $(\tilde{\Delta}_i')$ and $\tilde{\mu}_i$ are renormalized hopping terms between nearest (second) neighbors, first (second) neighbor pairing terms and local potential, respectively. They are given by

$$\tilde{J}_i = -J_i \Big\{ 1 +$$
$$+ \lambda(V_{i+1} - V_i)\big[\lambda(V_{i+1} - V_i)/2 + T(\mu_{i+1} - \mu_i)\big]/12 \Big\},$$
$$\tilde{J}'_i = T\lambda\big[J_i J_{i+1}(V_{i+2} - 2V_{i+1} + V_i) +$$
$$- \Delta_i \Delta^*_{i+1}(V_{i+2} + 2V_{i+1} + V_i)\big]/12,$$
$$\tilde{\Delta}_i = -\Delta_i \Big\{ 1 +$$
$$+ \lambda(V_{i+1} + V_i)\big[\lambda(V_{i+1} + V_i)/2 + T(\mu_{i+1} + \mu_i)\big]/12 \Big\},$$
$$\tilde{\Delta}'_i = T\lambda(V_{i+2} + V_i)(J_i \Delta_{i+1} + \Delta_i J^*_{i+1})/12$$
$$\tilde{\mu}_i = \mu_i + \lambda/T \, V_i - \lambda T\Big[|J_{i-1}|^2(V_i - V_{i-1}) +$$
$$- |J_i|^2(V_{i+1} - V_i) + |\Delta_{i-1}|^2(V_i + V_{i-1}) +$$
$$+ |\Delta_i|^2(V_{i+1} + V_i)\Big]/6, \tag{9.112}$$

where a more general case is considered, where all the terms are, in general, complex and have spatial dependence.

This expression may be simplified considering an Hamiltonian with homogeneous parameters, under the action of pulses that are the result of adding to the Hamiltonian a local potential with amplitude V, that we take as being $V = 1$. This term affects the electronic density in a homogeneous way, when the pulses are applied. In this case, the expansion to second order results in the Floquet Hamiltonian

$$\hat{H}_{\text{eff,h}} = -\sum_{i=1}^{L} \Big[(J_1 \, \hat{c}^\dagger_i \hat{c}_{i+1} + J_2 \, \hat{c}^\dagger_i \hat{c}_{i+2} + \Delta_1 \, \hat{c}^\dagger_{i+1} \hat{c}^\dagger_i +$$
$$+ \Delta_2 \, \hat{c}^\dagger_{i+2} \hat{c}^\dagger_i + \text{H.c.}) + \tilde{\mu} \, \hat{c}^\dagger_i \hat{c}_i \Big], \tag{9.113}$$

where the renormalized constants are $J_1 = J$, $J_2 = -T\lambda|\Delta|^2/3$, $\Delta_1 = \Delta\big[1 + \lambda(\lambda + 2T\mu)/6\big]$, $\Delta_2 = -T\lambda\Delta\text{Re}(J)/3$, $\tilde{\mu} = \mu + \lambda/T - 2\lambda T|\Delta|^2/3$.

In momentum space we may write

$$\hat{H}_{\text{eff,h}} = 1/2 \sum_k (\hat{c}^\dagger_k, \hat{c}_{-k})\mathcal{H}_k \begin{pmatrix} \hat{c}_k \\ \hat{c}^\dagger_{-k} \end{pmatrix}, \tag{9.114}$$

where

$$\mathcal{H}_k = -\big[\tilde{\mu}/2 + J_1 \cos(k) + J_2 \cos(2k)\big]\tau_z + \big[\Delta_1 \sin(k) + \Delta_2 \sin(2k)\big]\tau_y, \tag{9.115}$$

with τ_α Pauli matrices in Nambu space, The diagonalization leads to the dispersion

$$E_k^2 = \left[\tilde{\mu}/2 + J_1 \cos(k) + J_2 \cos(2k)\right]^2 + \left[\Delta_1 \sin(k) + \Delta_2 \sin(2k)\right]^2. \quad (9.116)$$

In the high frequency regime (with $\mu = 0$) we recover the phase diagram of the Kitaev chain, with the gap closing at $\lambda/T = 2$. In this regime, the parameters are simply renormalized by T. If we consider smaller values of the frequency, the second neighbor terms become relevant and lead to values of the winding number 0, ± 1, and ± 2. The winding numbers correspond to the number of times the vector $\mathbf{h}(k) = (0, h_y(k), h_z(k))$ circles around the point $(0, 0)$. Circling zero, once, or twice, leads to winding numbers $W = 0, 1$ and 2, respectively. Note that, in the static case, the winding number took the values $W = 0, 1$.

Further reading:

On the Loschmidt echo:

- T. Gorin, T. Prosen, T. H. Seligmanm and M. Znidaric, *Physics Reports* **435**, 33 (2006).
- V. Mukherjee, S. Sharmam and A. Dutta, *Physical Review* B **86**, 020301(R) (2012).
- F. Andraschko, and J. Sirker, *Physical Review* B **89**, 125120 (2014).

On the work done on a system while energy exchange occurs:

- E. Mascarenhas, H. Bragança, R. Dorner, M. França Santos, V. Vedral, K. Modi, and J. Goold, *Physical Review* E **89**, 062103 (2014).

On the robustness of edge states and topological invariants after a transformation of the Hamiltonian:

- A. Rajak, and A. Dutta, *Physical Review* E **89**, 042125 (2014).
- P. D. Sacramento, *Physical Review* E **90**, 032138 (2014).
- L. D'Alessio, and M. Rigol, *Nature Communications* **6**, 8336 (2015).
- M. S. Foster, V. Gurarie, M. Dzero, and E. A. Yuzbashyan, *Physical Review Letters* **113**, 076403 (2014).
- P. D. Sacramento, *Physical Review* E **93**, 062117 (2016).

On periodic perturbations, Floquet systems, and the creation of topological edge states:

- T. Oka, and H. Aoki, *Physical Review* B **79**, 081406 (2009).
- T. Kitagawa, E. Berg, M. Rudner, and E. Demler, *Physical Review* B **82**, 235114 (2010).
- Netanel H. Lindner, Gil Refael, and Victor Galitski, *Nature Physics* **7**, 490 (2011).
- Liang Jiang, Takuya Kitagawa, Jason Alicea, A. R. Akhmerov, David Pekker, Gil Refael, J. Ignacio Cirac, Eugene Demler, Mikhail D. Lukin, and Peter Zoller, *Physical Review Letters* **106**, 220402 (2011).
- Mark S. Rudner, Netanel H. Lindner, Erez Berg, and Michael Levin, *Physical Review* X **3**, 031005 (2013).
- M. Bukov, L. D'Alessio, and A. Polkovnikov, *Advances in Physics* **64**, 139 (2015).
- Frederik Nathan, and Mark S. Rudner, *New Journal Physics* **17**, 125014 (2015).
- Q.-J. Tong, J.-H. An, J. Gong, H.-G. Luo, and C. H. Oh, *Physical Review* B **87**, 201109 (2013).
- M. Benito, A. Gómez-León, V. M. Bastidas, T. Brandes, and G. Platero, *Physical Review* B **90**, 205127 (2014).

Examples of instantaneous perturbations:

- M. Thakurathi, A. A. Patel, D. Sen, and A. Dutta, *Physical Review* B **88**, 155133 (2013).
- P. Qin, C. Yin, and S. Chen, *Physical Review* B **90**, 054303 (2014).
- Tilen Cadez, Rubem Mondaini, and Pedro D. Sacramento, *Physical Review* B **99**, 014301 (2019).

Appendix A

Physical realization of Kitaev model

Superconductors with triplet pairing are not abundant in nature and, in particular, a one-dimensional superconductor with p wave pairing does not exist in nature. However, this type of superconductor may be obtained combining materials that are abundant. An example consists on a semiconductor wire with strong spin-orbit coupling in contact with a traditional superconductor with s wave symmetry, in the presence of an external magnetic field.

Let us consider first the semiconductor with spin-orbit of the Rashba type. In momentum space, and placing the chemical potential outside the gap, the spin-orbit interaction lifts the spin degeneracy except at the momenta zero and π and the two spin components become separated. Applying to the system an intense magnetic field, the two spin components separate completely in energy and a gap appears between the two components. We may, now, adjust the chemical potential in a way that it is located in this gap. Introducing now the superconducting coupling due to the proximity effect with a two-dimensional or three-dimensional superconductor, pairing between the electrons located near the Fermi surface takes place. Due to the magnetic field, these electrons are preferentially of a given spin component: the system has effectively polarized spins and the pairing occurs predominantly between electrons with the same spin, as required in Kitaev model.

Let us consider, then, an Hamiltonian of the form

$$\hat{H} = \int dx \begin{pmatrix} \psi_{x,\uparrow} \\ \psi_{x,\downarrow} \\ \psi_{x,\uparrow}^{\dagger} \\ \psi_{x,\downarrow}^{\dagger} \end{pmatrix}^{\dagger} H \begin{pmatrix} \psi_{x,\uparrow} \\ \psi_{x,\downarrow} \\ \psi_{x,\uparrow}^{\dagger} \\ \psi_{x,\downarrow}^{\dagger} \end{pmatrix}, \tag{A.1}$$

where

$$H = \begin{pmatrix} T - g_L B_z & \alpha\partial_x & 0 & \Delta_s \\ -\alpha\partial_x & T + g_L B_z & -\Delta_s & 0 \\ 0 & -\Delta_s & -T + g_L B_z & -\alpha\partial_x \\ \Delta_s & 0 & \alpha\partial_x & -T - g_L B_z \end{pmatrix}. \tag{A.2}$$

Here $T = -1/(2m)\partial_x^2 - \mu$, α is the spin-orbit interaction, g_L is the Landé factor, B_z is the magnetic field along the z axis and Δ_s is the superconducting pairing, resulting from the proximity effect. μ denotes the chemical potential and m the mass of the electrons. We may now decompose the Hamiltonian matrix in two components $H = H_0 + V$, defined as

$$H_0(x) = \begin{pmatrix} T - g_L B_z & 0 & 0 & 0 \\ 0 & T + g_L B_z & 0 & 0 \\ 0 & 0 & -T + g_L B_z & 0 \\ 0 & 0 & 0 & -T - g_L B_z \end{pmatrix} \tag{A.3}$$

and

$$V(x) = \begin{pmatrix} 0 & \alpha\partial_x & 0 & \Delta_s \\ -\alpha\partial_x & 0 & -\Delta_s & 0 \\ 0 & -\Delta_s & 0 & -\alpha\partial_x \\ \Delta_s & 0 & \alpha\partial_x & 0 \end{pmatrix}. \tag{A.4}$$

Consider, now, that the dominant term is the one of the magnetic field and that the spin-orbit and pairing terms are comparatively small. It is possible to perform a canonical transformation that eliminates, to a given order, the terms of smaller amplitude. The Hamiltonian is transformed in the following form

$$\tilde{H} = e^S H e^{-S} = H + [S, H] + \cdots \tag{A.5}$$

Choosing $[H_0, S] = V$ we obtain to first order,

$$\tilde{H} = H_0 + \frac{1}{2}[S, V] + \cdots \tag{A.6}$$

Determining the operator S, we may calculate the effective Hamiltonian. Let us look for a solution of the type

$$S = \begin{pmatrix} 0 & a & 0 & b \\ e & 0 & c & 0 \\ 0 & f & 0 & d \\ g & 0 & h & 0 \end{pmatrix}. \tag{A.7}$$

In first order, we obtain

$$a = d = e = h = -\frac{\alpha}{2g_L B_z}\partial_x \tag{A.8}$$

and the other coefficients vanish.

Calculating the commutator of the operator S with the perturbation V, we obtain

$$\tilde{H}(x) = \begin{pmatrix} T - g_L B_z & 0 & -\frac{\alpha \Delta_s}{2 g_L B_z} & 0 \\ 0 & T + g_L B_z & 0 & \frac{\alpha \Delta_s}{2 g_L B_z} \\ -\frac{\alpha \Delta_s}{2 g_L B_z} & 0 & -T + g_L B_z & 0 \\ 0 & \frac{\alpha \Delta_s}{2 g_L B_z} & 0 & -T - g_L B_z \end{pmatrix}, \qquad (A.9)$$

where we have neglected the non-diagonal terms proportional to the second spatial derivatives. The two spin components separate and, depending on the sign of the applied magnetic field, only one component survives. Considering, for example, $B_z > 0$, we obtain a simplified Hamiltonian similar to the Kitaev model

$$\tilde{H}(x) = \begin{pmatrix} T - g_L B_z & -\frac{\alpha \Delta_s}{2 g_L B_z} \\ -\frac{\alpha \Delta_s}{2 g_L B_z} & -T + g_L B_z \end{pmatrix}. \qquad (A.10)$$

Further reading:

- F. Pientka, L. I. Glazman and von F. Oppen, *Physical Review* B **88**, 155420 (2013).
- S. Nakosai, Y. Tanaka and N. Nagaosa, *Physical Review* B **88**, 180503 (2013).
- P. M. R. Brydon, S. D., Sarma , H. - Y. Hui and J. D. Sau, *Physical Review* B **91**, 064505 (2015).
- H. - Y. Hui, P. M. R. Brydon, J. D. Sau, S. Tewari and S. D. Sarma, *Scientific Reports* **5**, 8880 (2015).

Appendix B

Fermi surface topology

At zero temperature, the highest energy level occupied by fermions defines the Fermi energy. This energy separates the occupied from the unoccupied states. This concept is particularly transparent if the fermionic system is non-interactive. Introducing interactions between the fermions, the concept of the Fermi surface holds, in the so-called Fermi liquids. This result suggests that there may be some sort of topological protection that guarantees the stability of the Fermi surface and its robustness.

There are different classes of fermionic systems. Semiconductors or conventional superconductors are systems where the energy spectrum presents a gap and the chemical potential (Fermi energy at zero temperature) is located in the gap. Other classes have spectra with no gap at the Fermi level. Defining the codimension as the difference between the momenta space diension and the dimension of the energy zeros at the chemical potential, the classes may be characterized by their codimension: Fermi surfaces with codimension 1 (metals, liquid helium He^3 in the normal state), Fermi lines with codimension 2 (superconductors with d wave symmetry), Fermi points with codimension 1 (Weyl points as in helium in the superfluid phase A) and the possibility of states of Fermi condensates with codimension 0 (in some strongly interacting systems).

Frequently it is possible to have transitions between the various classes or a transition between different topological numbers, changing some parameters upon which the Hamiltonian depends. This leads to quantum phase transitions at zero temperature and the band structure at the Fermi level is changed. The transitions may be associated with some spontaneous symmetry breaking, such as in a transition to a magnetic or superconducting state, or may be the result of a change in the parameters that may be treated perturbatively (but eventually having high values), such as in a metal-insulator transition.

Indeed, there is a property that provides a topological invariant. We may identify it considering the Green function in momentum and energy space. We will see that there is an invariant that signals the existence of the Fermi energy.

The Green function may be defined as

$$G(\lambda, t) = -i\langle 0|Tc_\lambda(t)c_\lambda^\dagger(0)|0\rangle, \tag{B.1}$$

where $\lambda = (\boldsymbol{p}, \sigma)$ are, for example, momentum and spin degrees of freedom, t is time and T is the time-ordering operator defined by

$$G(\lambda, t) = -i\theta(t)\langle 0|c_\lambda(t)c_\lambda^\dagger(0)|0\rangle + i\theta(-t)\langle 0|c_\lambda^\dagger(0)c_\lambda(t)|0\rangle. \tag{B.2}$$

The operator c destroys an electron. The operators are written in the Heisenberg representation

$$c_\lambda(t) = e^{iHt}c_\lambda e^{-iHt}, \tag{B.3}$$

where H is the Hamiltonian assumed, to simplify, as time independent. The Green function gives the probability amplitude to have in the groundstate an electron at time 0 and time t, with parameters λ, both for positive and negative times. There are different types of Green functions but are not relevant for this discussion, and we will omit here their definitions.

If we have an electronic band occupied by a single electron, the Green function is given by

$$G(\lambda, t) = -i\theta(t)e^{-i\epsilon_\lambda t}. \tag{B.4}$$

Here, ϵ_λ is the energy of an electron with quantum numbers λ. If we perform a Fourier transform to energy space,

$$G(\lambda, E) = \int_{-\infty}^{\infty} dt e^{iEt} G(\lambda, t), \tag{B.5}$$

we find

$$G(\lambda, E) = \frac{1}{E - \epsilon_\lambda + i\delta}. \tag{B.6}$$

To guarantee convergence, we have added a positive infinitesimal δ.

In the case of a non-interactive degenerate electron system, with states occupied until the Fermi energy, similar calculations lead to a Green function

$$G(\boldsymbol{k}, E) = \frac{1}{E - \xi_\mathbf{k} + i\delta_\mathbf{k}}. \tag{B.7}$$

Here the energies are measured with respect to the chemical potential (or Fermi energy at zero temperature) and

$$\xi_{\mathbf{k}} = \epsilon_{\mathbf{k}} - \mu \, . \tag{B.8}$$

Note that, now, the convergence factor is different, depending if the energy is above or below the Fermi level, and $\delta_{\mathbf{k}} = \mathrm{sgn}(\xi_{\mathbf{k}})\delta$.

The Green function is a quantity that is useful in many calculations, For example, consider an operator of a particle such as $\boldsymbol{J} = \int d^3 x \boldsymbol{J}(\boldsymbol{x})$. We may define a local operator in second quantization as

$$\boldsymbol{J}(\boldsymbol{x}) = \sum_{\alpha, \beta} \psi_\beta^\dagger(\boldsymbol{x}) \boldsymbol{J}_{\beta\alpha}(\boldsymbol{x}) \psi_\alpha(\boldsymbol{x}) \, . \tag{B.9}$$

The average of the operator in the groundstate may, then, be expressed in terms of the Green function as

$$\langle 0 | \boldsymbol{J}(\boldsymbol{x}) | 0 \rangle = -i \lim_{t' \to t^+} \lim_{x' \to x} Tr \left[J(\boldsymbol{x}) G(\boldsymbol{x}, t; \boldsymbol{x}, t+) \right] \, . \tag{B.10}$$

Also, for example, the density operator may be written as

$$\langle n(x) \rangle = -i Tr [G(x, t; x, t^+)] \, , \tag{B.11}$$

where t^+ means a larger time but infinitesimally close to t.

Let us consider, now, the Green function defined for imaginary frequencies (or energies) such that $E = i\omega$. Then the inverse of the Green function may be written as

$$G^{-1}(\mathbf{k}, \omega) = i\omega - \xi_{\mathbf{k}} + \mu \, . \tag{B.12}$$

Let us focus exclusively on the case of the Fermi surface and, to simplify, let us consider a two-dimensional space. In this case, we have

$$G^{-1} = i\omega - \frac{p_x^2 + p_y^2}{2m} + \mu \, , \tag{B.13}$$

considering a parabolic dispersion. We have, then, a three-dimensional space defined by the momenta p_x, p_y and a third dimension that is ω. The Fermi surface is, in this case, a line of zero energy, $\omega = 0$.

Let us fix, for now, $p_y = 0$ and consider a circulation in the plane p_x, ω. We may choose polar coordinates $p_x = \rho \cos \varphi, \omega = \rho \sin \varphi$ around the Fermi surface (line in this case with codimension 1), and linearize the energy spectrum $p_x^2/(2m) - \mu = v_F p_x$. Then, we write

$$G^{-1} = i\rho \sin \varphi - v_F \rho \cos \varphi \, . \tag{B.14}$$

Writing $G = |G| e^{i\Phi}$, we see that when the angle φ varies by 2π until 0, the angle $-\Phi$ changes in the same way. Therefore, there is a circulation (vorticity) of the Green function around the Fermi line. Carrying out the circulation in a way that does not include the Fermi surface (that is, not circulating around the zero energy), the value of the circulation is, then, zero. Therefore, there is an invariant that counts the number of circulations around the Fermi line or, equivalently, counts the number of multiples of 2π of the change of the angle.

This topological invariant may be defined in a way equivalent that is more easily generalizable to other situations. An invariant, N_1, may be defined as

$$N_1 = Tr \int_C \frac{dl}{2\pi i} G(\mu, p) \partial_l G^{-1}(\mu, p), \qquad (B.15)$$

where the trace is over any internal degrees of freedom of the Green function and C is a closed path. If this wraps the Fermi surface, it may give origin to a non-zero vorticity. It is easy to verify that N_1 reproduces the vorticity of the trajectory around the Fermi surface.

Consider again the example of the three-dimensional space (p_x, p_y, ω). If we decrease the value of the chemical potential, the radius of the Fermi surface decreases until it converges in a point when the chemical potential is located at the lowest energy point of the electronic band. Decreasing even further the chemical potential, there is no Fermi surface any more and the system has a gap at the chemical potential. The nature of the Fermi surface changed. This is an example of a *Lifshitz transition* and results in a transition from a metallic state (gapless) to an insulator (with a gap in the spectrum) with no symmetry change. In this case a topological transition took place, since the topological invariant N_1 is now zero, due to the disappearance of the Fermi surface.

Let us consider, now, a transition from a normal conductor to a conventional superconductor with s wave pairing. This transition has a different nature, since there is a symmetry breaking.

The Green function has now a matrix structure due to the appearance of the Cooper pairs. It may be written as

$$G^{-1} = \begin{pmatrix} i\omega - \frac{p^2}{2m} + \mu & \Delta \\ \Delta^* & i\omega + \frac{p^2}{2m} - \mu \end{pmatrix}. \qquad (B.16)$$

The energy spectrum is given by

$$E^2 = \left(\frac{p^2}{2m} - \mu \right)^2 + |\Delta|^2. \qquad (B.17)$$

This spectrum has a gap and there is no Fermi surface, since the chemical potential is located in the gap, as a result of the electron pairing. Let $\Delta = 0$.

In this case, the inverse Green function decouples in two sectors: one for the particles and one for the holes. It is easy to see that $N_1 = 1$ for the particles and $N_1 = -1$ for the holes. Summing the two contributions, the total topological number vanishes $N_1 = 0$. In the phase with gap the topological invariant also vanishes, therefore, the transition to the superconducting phase preserves the invariant.

Further reading:

- G. E. Volovik, *The Universe in a Helium Droplet*, Clarendon Press, Oxford (2003).
- G. E. Volovik, *Quantum Phase Transitions from Topology in Momentum Space*, Lect. Notes Phys. **718**, 31-73 (2007).

Index

S matrix, 182–186

$\mathbb{Z}_2$
  group, 6, 8, 17
  index, 8, 54, 55, 98, 132, 134
  superconductor, 98
  topological phase, 105

π modes, 242

abelian, 4, 5
Adler-Bell-Jackiw, 126
Affleck-Kennedy-Lieb-Tasaki, 143
Altland-Zirnbauer, 62
Andreev, 91
anomalous velocity, 17
antiunitary, 19, 21, 54–56, 185

Baker-Campbell-Hausdorff, 245
Balian-Werthammer, 72
Bardeen-Cooper-Schrieffer (BCS), 71, 194
Bell states, 190
Berry
  connection, 12, 13, 25, 27, 100, 131, 180
  curvature, 12, 16, 30, 31, 33, 106, 123, 130, 132, 230, 238
  phase, 11, 14, 15, 31, 132, 146, 148, 159, 163, 244
Bethe ansatz, 142, 143
Bogoliubov-de Gennes, 65, 67, 70, 98, 99, 101, 195

Bott (periodicity theorem), 9, 26
bulk-boundary, 34, 35, 106

Chern, 8, 29, 31, 34, 35, 51, 99, 100, 106, 114, 125, 178–181, 208, 229, 231, 238, 239, 241
chiral
  anomaly, 126, 127
  charge, 122, 123, 126, 134
  transformation, 158
chirality, 35, 36, 103, 122, 127, 178, 181
Clifford, 43, 76
codimension, 121, 255, 257
conductance, 37

dimensional reduction, 42, 107
dimerization, 83, 142, 145, 149
Dirac
  cone, 30, 34, 48, 52, 96, 129, 177, 180, 181, 235
  Hamiltonian, 121, 128, 179
  mass, 20, 30, 35, 48
  ring, 131
Drumhead states, 133
duality, 142, 157, 162

edge states, 25, 26, 34, 35, 43, 44, 50, 85, 95, 125, 126, 133, 145, 146, 156, 159, 167, 168, 177–179, 197, 205, 227–229, 239, 241, 243

entanglement, 189, 191, 193, 194
entanglement
    spectrum, 192, 197, 198, 205

Faraday effect, 167, 172
Fermi
    -Dirac, 70
    arcs, 124
    surface, 103, 121, 130, 230, 239,
        251, 255–258
fidelity, 200, 201, 207, 222
fidelity
    classical, 201
    operator, 205, 206, 208
    partial states, 201, 203, 204
    spectrum, 205–209
    susceptibility, 210, 211
Floquet
    Hamiltonian, 246, 247
    operator, 241–245
    states, 233–235, 238, 239, 243
    system, 232, 236, 237, 242
    zone, 235, 238, 239, 241

graphene, 20, 40, 49
Grassmannian, 7
Green function, 256–258
gyrotropic material, 174

Haldane, 48, 49, 142, 143, 167, 181
Hall, 34, 37, 39, 40, 105, 186, 197,
    213, 244
hedgehog, 150
Heisenberg model, 148
helicity, 40
Hellman-Feynman, 33, 225
homotopy
    class, 2, 5
    group, 2, 4, 5, 17
honeycomb lattice, 40, 58

impurities, 39, 105, 107–110, 112,
    116, 128, 194–196, 203, 204
insulator
    anomalous Hall, 47, 51
    chiral topological, 54

higher order, 61
spin-Hall, 54
topological, 8, 53, 57

Jackiw-Rebbi, 35
Jordan-Wigner, 139, 141, 156, 163,
    222
Josephson, 89, 93

Kane-Mele, 53, 55, 134
Kitaev, 25, 74, 75, 80, 81, 83, 86, 141,
    156–159, 209, 211, 227, 246, 248,
    251, 253
Kramers pair, 55, 105, 107
Kubo, 34

Landau, 52, 127, 128, 189
Lieb-Shultz-Mattis, 142
Lifshitz transition, 258
Loschmidt echo, 222, 224

Möbius, 6
magnetoresistance, 128
Magnus (expansion), 236
Majorana, 71, 72, 75, 82–84, 89, 91,
    93, 95, 103, 110, 139, 140, 159,
    211–213, 215–217, 227–229, 241
Meissner, 214
monopole, 28, 31, 100
multiband, 198

Nambu, 72–74
Nielsen-Ninomiya, 48, 122
nodal line, 20, 131–133
nodal ring, 130, 134
non-abelian, 4, 96, 161, 211
non-reciprocal, 167, 171

pairing
    $p + ip$, 93, 96, 98, 100, 241
    singlet, 73, 101, 200, 252, 258
    superconducting, 68, 73, 198, 241
    triplet, 73, 101, 102, 107, 112, 200,
        213, 251
parity
    fermionic, 72, 78–80, 216

symmetry, 19–21, 32, 102
particle-hole, 66, 72, 99, 107, 242, 259
Pfaffian, 56, 110
photonic crystal, 177
polarization, 13, 14, 172
pseudo-spin, 22, 47, 61, 69, 70, 99,
  101, 122, 134

quantum work, 225, 226

Raghu, 167, 177, 181
Rashba, 110, 251
reduced density matrix, 191, 193,
  195, 196, 198

Sato and Fujimoto, 103, 106, 112,
  116, 198, 207, 228, 229
Schmidt decomposition, 191, 192
Schrödinger, 11, 14, 41, 183, 234, 236,
  243
semimetal, 121, 130, 136
Shannon entropy, 191, 192
Shockley, 23, 82–85, 162
singular transformation, 162
skyrmion, 151
spin-orbit, 54, 56, 74, 103–107, 110,
  241, 251, 252
Su-Schrieffer-Heeger, 83, 85, 86
survival probability, 221, 227, 229
symmetry
  charge conjugation, 20, 67, 74
  chiral, 6, 8, 20, 44, 61, 133
  class, 61, 81, 98, 140
  particle-hole, 20, 66, 73, 98

reflection, 22, 91
space inversion, 19, 31, 122, 127
time reversal, 18, 54, 55, 69, 73, 90,
  92, 128, 167, 178, 186
translation, 142, 145

topological charge, 132
topological insulator, 240
topology
  higher order, 61
  strong, 57
  weak, 57
twisted boundary conditions, 114, 163

valence bond, 145
von Neumann entropy, 191–194, 205,
  207
vortex, 96, 132, 152–154, 156, 215,
  216

Wannier (states), 13
Weyl
  point, 20, 121–124, 126, 128
  representation, 128
  ring, 131
winding number, 3, 24, 40, 61, 72, 75,
  81, 107, 133, 135, 143, 158, 159,
  243, 248

XY model, 141, 152, 154, 156

Zak phase, 14, 25, 160, 161
Zeeman, 91, 103, 128, 199

* 9 7 8 9 8 1 1 2 3 7 2 1 8 *